235
Advances in Polymer Science

Editorial Board:
A. Abe · A.-C. Albertsson · K. Dušek · W.H. de Jeu
H.-H. Kausch · S. Kobayashi · K.-S. Lee · L. Leibler
T.E. Long · I. Manners · M. Möller · E.M. Terentjev
M. Vicent · B. Voit · G. Wegner · U. Wiesner

Advances in Polymer Science

Recently Published and Forthcoming Volumes

Silicon Polymers
Volume Editor: Muzafarov, A.M
Vol. 235, 2011

Chemical Design of Responsive Microgels
Volume Editors: Pich, A., Richtering, W.
Vol. 234, 2011

Hybrid Latex Particles
Volume Editors: van Herk, A.M., Landfester, K.
Vol. 233, 2011

Biopolymers
Volume Editors: Abe, A., Dušek, K., Kobayashi, S.
Vol. 232, 2010

Polymer Materials
Volume Editors: Lee, K.-S., Kobayashi, S.
Vol. 231, 2010

Polymer Characterization
Volume Editors: Dušek, K., Joanny, J.-F.
Vol. 230, 2010

Modern Techniques for Nano- and Microreactors/-reactions
Volume Editor: Caruso, F.
Vol. 229, 2010

Complex Macromolecular Systems II
Volume Editors: Müller, A.H.E., Schmidt, H.-W.
Vol. 228, 2010

Complex Macromolecular Systems I
Volume Editors: Müller, A.H.E., Schmidt, H.-W.
Vol. 227, 2010

Shape-Memory Polymers
Volume Editor: Lendlein, A.
Vol. 226, 2010

Polymer Libraries
Volume Editors: Meier, M.A.R., Webster, D.C.
Vol. 225, 2010

Polymer Membranes/Biomembranes
Volume Editors: Meier, W.P., Knoll, W.
Vol. 224, 2010

Organic Electronics
Volume Editors: Meller, G., Grasser, T.
Vol. 223, 2010

Inclusion Polymers
Volume Editor: Wenz, G.
Vol. 222, 2009

Advanced Computer Simulation Approaches for Soft Matter Sciences III
Volume Editors: Holm, C., Kremer, K.
Vol. 221, 2009

Self-Assembled Nanomaterials II
Nanotubes
Volume Editor: Shimizu, T.
Vol. 220, 2008

Self-Assembled Nanomaterials I
Nanofibers
Volume Editor: Shimizu, T.
Vol. 219, 2008

Interfacial Processes and Molecular Aggregation of Surfactants
Volume Editor: Narayanan, R.
Vol. 218, 2008

New Frontiers in Polymer Synthesis
Volume Editor: Kobayashi, S.
Vol. 217, 2008

Polymers for Fuel Cells II
Volume Editor: Scherer, G.G.
Vol. 216, 2008

Polymers for Fuel Cells I
Volume Editor: Scherer, G.G.
Vol. 215, 2008

Photoresponsive Polymers II
Volume Editors: Marder, S.R., Lee, K.-S.
Vol. 214, 2008

Silicon Polymers

Volume Editor: Aziz M. Muzafarov

With contributions by

A. Bockholt · M.A. Brook · A. Feigl · E.Sh. Finkelshtein
F. Ganachaud · J.B. Grande · M.L. Gringolts · Md.A. Hoque
Y. Kakihana · Y. Kawakami · S. Kirchmeyer · A. Miyazato
S.A. Ponomarenko · B. Rieger · S. Tateyama
N.V. Ushakov · J. Weis

Editor
Aziz M. Muzafarov
N.S. Enikolopov Institute of Synthetic Polymer Materials
Russian Academy of Sciences
Profsoyuznaya ul. 70
117393 Moscow
Russia
aziz@ispm.ru

ISSN 0065-3195 e-ISSN 1436-5030
ISBN 978-3-642-16047-9 e-ISBN 978-3-642-16048-6
DOI 10.1007/978-3-642-16048-6
Springer Heidelberg Dordrecht London New York

Library of Congress Control Number: 2010936199

© Springer-Verlag Berlin Heidelberg 2011
This work is subject to copyright. All rights are reserved, whether the whole or part of the material is concerned, specifically the rights of translation, reprinting, reuse of illustrations, recitation, broadcasting, reproduction on microfilm or in any other way, and storage in data banks. Duplication of this publication or parts thereof is permitted only under the provisions of the German Copyright Law of September 9, 1965, in its current version, and permission for use must always be obtained from Springer. Violations are liable to prosecution under the German Copyright Law.
The use of general descriptive names, registered names, trademarks, etc. in this publication does not imply, even in the absence of a specific statement, that such names are exempt from the relevant protective laws and regulations and therefore free for general use.

Cover design: WMXDesign GmbH, Heidelberg

Printed on acid-free paper

Springer is part of Springer Science+Business Media (www.springer.com)

Volume Editor

Aziz M. Muzafarov

N.S. Enikolopov Institute of Synthetic Polymer Materials
Russian Academy of Sciences
Profsoyuznaya ul. 70
117393 Moscow
Russia
aziz@ispm.ru

Editorial Board

Prof. Akihiro Abe

Professor Emeritus
Tokyo Institute of Technology
6-27-12 Hiyoshi-Honcho, Kohoku-ku
Yokohama 223-0062, Japan
aabe34@xc4.so-net.ne.jp

Prof. A.-C. Albertsson

Department of Polymer Technology
The Royal Institute of Technology
10044 Stockholm, Sweden
aila@polymer.kth.se

Prof. Karel Dušek

Institute of Macromolecular Chemistry
Czech Academy of Sciences
of the Czech Republic
Heyrovský Sq. 2
16206 Prague 6, Czech Republic
dusek@imc.cas.cz

Prof. Dr. Wim H. de Jeu

Polymer Science and Engineering
University of Massachusetts
120 Governors Drive
Amherst MA 01003, USA
dejeu@mail.pse.umass.edu

Prof. Hans-Henning Kausch

Ecole Polytechnique Fédérale de Lausanne
Science de Base
Station 6
1015 Lausanne, Switzerland
kausch.cully@bluewin.ch

Prof. Shiro Kobayashi

R & D Center for Bio-based Materials
Kyoto Institute of Technology
Matsugasaki, Sakyo-ku
Kyoto 606-8585, Japan
kobayash@kit.ac.jp

Prof. Kwang-Sup Lee

Department of Advanced Materials
Hannam University
561-6 Jeonmin-Dong
Yuseong-Gu 305-811
Daejeon, South Korea
kslee@hnu.kr

Prof. L. Leibler

Matière Molle et Chimie
Ecole Supérieure de Physique
et Chimie Industrielles (ESPCI)
10 rue Vauquelin
75231 Paris Cedex 05, France
ludwik.leibler@espci.fr

Prof. Timothy E. Long

Department of Chemistry
and Research Institute
Virginia Tech
2110 Hahn Hall (0344)
Blacksburg, VA 24061, USA
telong@vt.edu

Prof. Ian Manners

School of Chemistry
University of Bristol
Cantock's Close
BS8 1TS Bristol, UK
ian.manners@bristol.ac.uk

Prof. Martin Möller

Deutsches Wollforschungsinstitut
an der RWTH Aachen e.V.
Pauwelsstraße 8
52056 Aachen, Germany
moeller@dwi.rwth-aachen.de

Prof. E.M. Terentjev

Cavendish Laboratory
Madingley Road
Cambridge CB 3 OHE, UK
emt1000@cam.ac.uk

Maria Jesus Vicent, PhD

Centro de Investigacion Principe Felipe
Medicinal Chemistry Unit
Polymer Therapeutics Laboratory
Av. Autopista del Saler, 16
46012 Valencia, Spain
mjvicent@cipf.es

Prof. Brigitte Voit

Institut für Polymerforschung Dresden
Hohe Straße 6
01069 Dresden, Germany
voit@ipfdd.de

Prof. Gerhard Wegner

Max-Planck-Institut
für Polymerforschung
Ackermannweg 10
55128 Mainz, Germany
wegner@mpip-mainz.mpg.de

Prof. Ulrich Wiesner

Materials Science & Engineering
Cornell University
329 Bard Hall
Ithaca, NY 14853, USA
ubw1@cornell.edu

Advances in Polymer Sciences Also Available Electronically

Advances in Polymer Sciences is included in Springer's eBook package *Chemistry and Materials Science*. If a library does not opt for the whole package, the book series may be bought on a subscription basis. Also, all back volumes are available electronically.

For all customers who have a standing order to the print version of *Advances in Polymer Sciences*, we offer the electronic version via SpringerLink free of charge.

If you do not have access, you can still view the table of contents of each volume and the abstract of each article by going to the SpringerLink homepage, clicking on "Browse by Online Libraries", then "Chemical Sciences", and finally choose *Advances in Polymer Science*.

You will find information about the

- Editorial Board
- Aims and Scope
- Instructions for Authors
- Sample Contribution

at springer.com using the search function by typing in *Advances in Polymer Sciences*.

Color figures are published in full color in the electronic version on SpringerLink.

Aims and Scope

The series *Advances in Polymer Science* presents critical reviews of the present and future trends in polymer and biopolymer science including chemistry, physical chemistry, physics and material science. It is addressed to all scientists at universities and in industry who wish to keep abreast of advances in the topics covered.

Review articles for the topical volumes are invited by the volume editors. As a rule, single contributions are also specially commissioned. The editors and publishers will, however, always be pleased to receive suggestions and supplementary information. Papers are accepted for *Advances in Polymer Science* in English.

In references *Advances in Polymer Sciences* is abbreviated as *Adv Polym Sci* and is cited as a journal.

Special volumes are edited by well known guest editors who invite reputed authors for the review articles in their volumes.

Impact Factor in 2009: 4.600; Section "Polymer Science": Rank 4 of 73

Preface

The chemistry of organosilicon polymers is in the uprising stage of its development; the prospects of further growth and improvement opening before it are endless. Almost every review or study devoted to organosilicon polymers contained this phrase for several tens of years to date, and in every case, it reflected the real state of affairs without fail. It is just as true for today when the chemistry of organosilicon polymers enters yet another stage of its development.

Among the main tendencies characterizing this new stage, one can name a higher level of directed control of organosilicon polymers' structure, as well as implementing elements of selectivity and preliminary self-organization. The new synthetic approaches are based on modern experimental techniques and new methods of properties investigation of the created polymers and polymer-based materials. The unique qualities of organosilicon polymers ensure their being in high demand in almost every aspect of human activity and serve as a powerful driving force for further development of their synthesis. Areas of implementation, as well as the specific characteristics of particular materials achieved by previous generations of organosilicon scientists, are in constant need of being perfected. Further expansion and improvement of polymers possessing useful practical properties are the natural need of this science field, providing a steady connection between the science labs and the real world.

One of the separate branches that have reached a qualitatively different stage of their development are sol-gel technologies, which had come a long way from the "black box" method to the understanding of the chemistry of the process and the order of the major part of the stages that happen under various conditions. Even more successful were the scientists who created the so-called liquid silicon – a polysilane polymer that, under the influence of UV, transforms into polycrystalline silicon, which allows a fundamental change in the production of integrated circuits.

The rundown of recent achievements would be inconclusive without the mentioning of the unique process of forming of siloxane bond via the interaction of hydrosilane with alkoxysilane groups with the release of corresponding alkane. It is equally impossible not to take notice of the remarkable progress of the hybrid polymers, created by polymerization of silyl derivatives of ferrocene. It is obvious that even the most superficial recitation of actively developing fields clearly demonstrates the uprise of polymerization approaches which provide the most effective control of the forming polymers. As for condensation processes, apart from the aforementioned unique reaction of catalytic condensation of hydro- and alkoxysilanes, one must take note of the evident progress in the synthesis of dendrimers and hyperbranched polymers that, due to the specifics of their chemical nature, are the most fast-developing molecular systems in this particular area of polymer chemistry.

This introduction does not include all the spectacular achievements of the last decades, just as not all of the aforementioned fields of study are present in this book. Silicon-containing dendrimers and hyperbranched polymers are well represented in a separate volume which has been taken notice of among polymer as well as organosilicon scientific circles. We have also neglected the thriving area of unsaturated organosilicon polymers in hopes of becoming readers of a separate specialized paper one day. By making – and not making – certain provisos, we acknowledge the fact that the selection of materials for this volume has been subjective, and the volume itself is a caption of a fast moving object that does not quite allow for a complete comprehension of this object, but enables one to feel its movement and the main vectors of development.

The first chapter is devoted to the advances in polysilanes. The recent remarkable progress in this field only serves to emphasize the actuality and inexhaustible nature of silicon chemistry. This chapter may present a perfect illustration of how the traditional and would-be thoroughly explored systems can regain our attention due to the development of new technological methods.

In the second chapter, silicon is used as an element that does not take part in complex conjugated structures built of aromatic subunits. Instead, due to the high reactivity of its functional groups and stability of silicon–carbon bonds, silicon serves as a skeleton holding those laced structures; it becomes the element via which they interact with the surface and themselves.

The third chapter reviews the dynamic of the design of polymer structures created using polymerization methods. The material gathered in this section convinces us yet again that this is but a beginning for this exuberantly developing area of silicon–carbon polymers.

The fourth chapter of this book reminds us that good things come in small packages. It is devoted to probably the most promising method of siloxane polymers synthesis, particularly with regard to unlimited capability of designing macromolecules of complex architecture. This is even more remarkable, because up to date this field lacked high selectivity of the reactions.

The concluding fifth chapter touches upon the issues and advances in constructing of the most thermodynamically stable polysiloxanes built of cage-like subunits. The chemistry of siloxane cages has a long history, but their polymer biography is in the very beginning.

To my regret, it was not possible for this book to include the earlier planned chapters on the progress of the basics of silicon chemistry, namely the methods of production of organosilicon monomers, particularly chloride-free and bioinspired methods of silicon polymers synthesis. I hope, however, that in view of the attention that the Springer publishing house pays to this branch of chemistry, those materials will appear in the journal *Silicon* if not in another collected volume.

I am convinced that this book will be of interest not only to those working with silicones, but also to the whole polymer community and material science specialists, as it contains a lot of new and fascinating data. I would be particularly pleased with the attention of young researchers who are bound to find the spirit of constant renewal in this old but highly dynamic area of chemistry contagious.

In conclusion, I would like to thank all the authors who accepted the invitation to participate in this volume and Springer publishing house for providing an opportunity to work on this book, as well as for their patience and cooperation.

Moscow, Russia   *Aziz M. Muzafarov*

Contents

**Modern Synthetic and Application Aspects of Polysilanes:
An Underestimated Class of Materials?** 1
A. Feigl, A. Bockholt, J. Weis, and B. Rieger

**Conjugated Organosilicon Materials for Organic Electronics
and Photonics** 33
Sergei A. Ponomarenko and Stephan Kirchmeyer

Polycarbosilanes Based on Silicon-Carbon Cyclic Monomers 111
E.Sh. Finkelshtein, N.V. Ushakov, and M.L. Gringolts

**New Synthetic Strategies for Structured Silicones Using
$B(C_6F_5)_3$** 161
Michael A. Brook, John B. Grande, and François Ganachaud

**Polyhedral Oligomeric Silsesquioxanes with Controlled
Structure: Formation and Application in New Si-Based
Polymer Systems** 185
Yusuke Kawakami, Yuriko Kakihana, Akio Miyazato,
Seiji Tateyama, and Md. Asadul Hoque

Index 229

Modern Synthetic and Application Aspects of Polysilanes: An Underestimated Class of Materials?

A. Feigl, A. Bockholt, J. Weis, and B. Rieger

Abstract Polysilanes can be synthesized by various methods still lacking satisfactory product control in respect to quality and purity. Nonetheless polysilanes are promising in a number of applications like e.g. organic electronic devices. This review summarizes recent progress in the synthesis and possible applications of polysilanes.

Keywords Dehydocoupling · Masked disilenes · OLEDs · OFEDs · Polysilanes · Polysilane synthesis · Wurtz-Coupling

Contents

1	Introduction	2
2	Synthesis	2
	2.1 Inorganic Polysilanes: Polyhydrosilanes Si_nH_{2n}, Si_nH_{2n+2}	3
	2.2 Organic Polysilanes: Poly(organosilane)s Si_nR_{2n}, Si_nR_{2n+2}	4
3	Structure Determination of Polysilanes	19
4	Physical Properties	19
5	Applications	21
	5.1 Applications Based on the Reactivity of Polysilanes	21
	5.2 Applications Based on the Photophysical and Electronic Properties of Polysilanes	22
6	Conclusions	28
	References	28

B. Rieger (✉), A. Feigl, A. Bockholt, and J. Weis
Institut für Siliciumchemie, WACKER-Lehrstuhl für Makromolekulare Chemie, Technische Universität München, Lichtenbergstrasse 4, 85748 Garching b. München
e-mail: rieger@tum.de

1 Introduction

Even though polysilanes as a type of polymer with a backbone consisting exclusively of silicon have been a known class of substance since the early 1920s [1, 2], the scientific efforts to find a route towards structurally controlled, high molecular weight polysilanes have been of limited success.

In contrast to the well known chemistry of forming Si–O, Si–N or Si–C bonds, the formation of Si–Si bonds to yield polymeric species still needs further development.

All common methods of synthesis are either unable to yield high molecular weights while guaranteeing narrow or monomodal mass distributions, or they are intolerant towards some functional groups. This limits the spectrum of available polysilanes.

Additionally, silane monomers containing Si–Cl or Si–H bonds are moisture and air sensitive or even pyrophoric while oligomeric precursors containing Si–Si-bonds can be slightly sensitive. The preparation of polysilanes is challenging and difficult to transfer to an industrial scale and application.

This might be the reason why polysilanes are still not as commonly used as, for example, silicones, though their electro- and photochemical properties resulting from a σ-conjugation along the Si–Si-chain nominate them as promising candidates for technical applications [3].

The last comprehensive review of polysilanes was given 20 years ago by Miller and Michl [3]. Since then, considerable work has been carried out to find a 'perfect' method for their synthesis. Special interest has also been directed to the characterization of polysilanes to explain their special physical properties.

The purpose of this review is to summarize the more recent progress in the synthesis of polysilanes, to give an updated overview about applications and to point out that polysilanes hold great potential once their synthesis can be improved. Many methods describing the synthesis of polysilanes are similar to the routes towards oligosilanes. To keep the scope of this review, the latter will not be discussed here.

2 Synthesis

As polysilanes can bear different substituents R along the Si-backbone, they can be distinguished from inorganic polysilanes, with R being only H, and organic polysilanes, R being a functional group and/or H (Fig. 1).

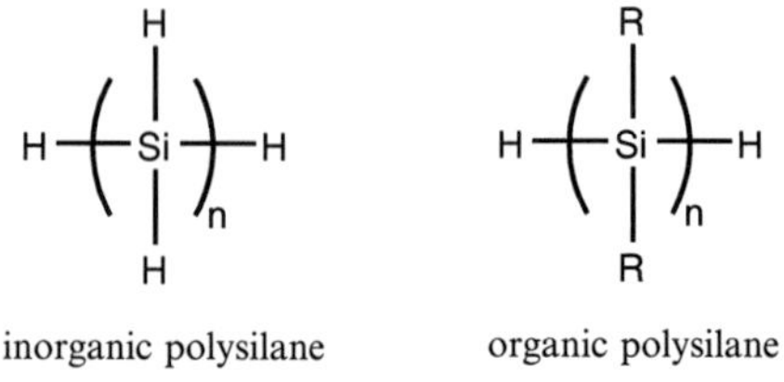

Fig. 1 Inorganic and organic polysilane

inorganic polysilane

organic polysilane

The synthetic routes towards inorganic and organic polysilanes differ and will thus be described in separate sections.

2.1 Inorganic Polysilanes: Polyhydrosilanes Si_nH_{2n}, Si_nH_{2n+2}

After the fundamental work of Stock between 1916 and 1926 [4], in which he established silane chemistry and characterized mono- to octasilanes, Fehér and co-workers tried to find a reliable route to synthesize silanes and polyhydrosilanes in the early 1970s.

As the yields were very low, Fehér and co-workers set up a technical apparatus to produce oligomeric raw silanes on a larger scale from magnesia silicide [5] (Fig. 2).

Decomposing Mg_2Si with phosphoric acid at elevated temperatures yields a mixture of silanes $((SiH_2)_n; n = 1 - 15)$ which can be trapped at $-70°C$. To handle the pyrophoric and explosive products, techniques using special valves and storage vessels had to be developed [6]. Thus in 1971, Fehér and co-workers were able to synthesize $Si_{15}H_{32}$, the polyhydrosilane with the longest silicon chain known until then. Analysis of the obtained silane fractions was accomplished by GC [5, 7].

Publications on this type of approach towards silanes, and polysilanes in particular, ceased after 1985 [8]. This can be explained by the great practical effort that had to be applied to obtain the desired products.

In 2008, a new synthetic method for the production of polysilanes was published. $SiCl_4$ and H_2 are reacted in a plasma process to yield perhalogenated polysilanes, which can then be transformed by hydrogenation with $LiAlH_4$, as proposed by Auner et al., to yield a high molecular weight polyhydrosilane (HPS) (Fig. 3).

The reported plasma process is highly sophisticated. The plasma is continually stimulated, pulsed, directed through electromagnetic fields and a microwave resonance chamber to ensure a homogenous mixture and thus a more uniform range of products.

The obtained mixture still contains low molecular weight silanes such as Si_2Cl_6 and Si_3Cl_8, oligosilanes containing more than three silicon atoms and high molecular weight polysilanes. The products were distilled and identified by cryoscopy energy dispersive X-ray spectroscopy (EDX) and nuclear magnetic resonance spectroscopy (NMR) for further use in the hydrogenation to HPS [9].

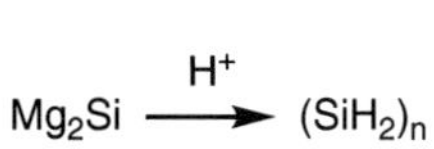

Fig. 2 Decomposition of magnesia silicide

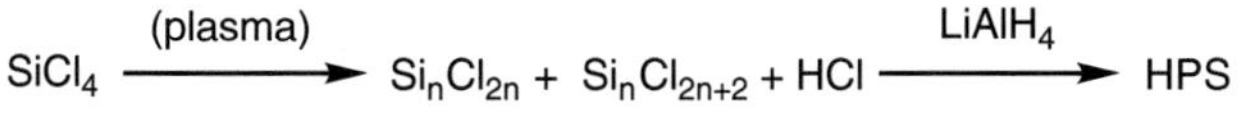

$$Mg_2Si \xrightarrow{H^+} (SiH_2)_n$$

$$SiCl_4 \xrightarrow{(plasma)} Si_nCl_{2n} + Si_nCl_{2n+2} + HCl \xrightarrow{LiAlH_4} HPS$$

Fig. 3 Plasma process for the synthesis of HPS (high molecular weight polysilane)

The advantage of this method over the decomposition of Mg_2Si with acids is that by removal of low molecular weight products before hydrogenation, production of pyrophoric SiH_4 can be avoided.

Nonetheless, this method is not quite suited for easy laboratory application as it utilizes complex plasma techniques.

2.2 Organic Polysilanes: Poly(organosilane)s Si_nR_{2n}, Si_nR_{2n+2}

Unlike the limited synthetic routes towards hydrogenated polysilanes, a relatively broad spectrum of methods is available for alkyl and aryl substituted polysilanes. The first compound containing a silicon–silicon bond and bearing organic substituents was hexaethyldisilane described by Friedel and Ladenburg in 1869 who reacted Si_2I_6 with $ZnEt_2$ [10].

2.2.1 Wurtz-Type Coupling of Chlorosilanes

The so-called Wurtz-type reductive dehalogenative coupling reaction was first observed by Kipping [1, 11] and is the most common method for the synthesis of polysilanes. A dichlorodiorganosilane is reacted with a slight stoichiometric excess of dispersed sodium in an inert, high-boiling solvent by refluxing to yield the desired polysilane (Fig. 4).

Until now, the solvent most commonly used is toluene [3]. Due to the harsh reducing conditions, functionalities that can be introduced via the monomer are limited. Generally, aryl, alkyl, silyl [3] and other intrinsically stable groups like ferrocenyl [12] or fluoroalkyl groups [13] can be used. The only way to introduce more sensitive functionalities is to use protecting groups which can withstand the reaction conditions and be removed after the polymerization [14–17].

Depending on the substituents on the silicon, the polymerization can achieve conversions up to 90% [18, 19]. Additionally, there is no control over the product molecular weight, which varies from below 1,000 to several million grams per mole, having polydispersities from 1.5 up to 10 for the separated fractions [19].

The polysilanes synthesized usually show a trimodal mass distribution containing a low molecular weight cyclic and a medium and a high molecular weight fraction of linear polymer. The oligomeric fraction can easily be separated from the other two by solvent extraction, whereas the fractionation of the medium and high molecular weight products requires more effort.

These features have motivated considerable research on the improvement of the Wurtz-type reductive coupling by tuning the reaction conditions, especially by variation of temperature and solvents [20–23].

Fig. 4 Wurtz-type coupling of chlorosilanes

$$R_2SiCl_2 \xrightarrow{Na^0} (SiR_2)_n + NaCl$$

In general, the reaction temperature has two major influences on the polymers: on the one hand it changes the yields and on the other it changes the length of the polymer and thus the weight distribution. Lowering the temperature results in lower yields, increased molecular weight and a narrower weight distribution [18, 19].

As the low temperature synthesis yields a more homogenous polymer with respect to polydispersity and molecular weight, toluene can be exchanged by lower boiling solvents such as, e.g. THF. Solvent effects, due to polarity and stabilization of the active species in the polymerization, are marginal compared to the temperature effect. The results of a Wurtz-type polysilane synthesis conducted at ambient temperature are given in Table 1, illustrating the low yields and PDI [19].

Holder and co-workers have investigated the influence of chiral solvents on the polymerization of dichloromethylphenylsilane. Chiral solvents are expensive and generally not available in quantities that are sufficient for other than small to medium scale polymerizations. As an exception, optically active limonene as an inexpensive and relatively inert solvent was found to be able to stabilize the growing polymer chains during the reaction, thus delivering higher molecular weights than an optically inactive solvent. At a reaction temperature of 90°C, the molecular weight (and PDI) were even twice as high as in an optically inactive solvent while broadening the PDI as can be seen in Table 2 [24].

Table 1 Examples of final yields and molecular weight parameters of the isolated polysilanes obtained from the Wurtz-type reductive couplings of dichlorodi-n-hexylsilane (DCDHS), dichloro-n-hexylmethylsilane (DCHMS), dichloromethyl-n-propylsilane (DCMPrS), dichloromethyl-n-octylsilane (DCMOS) and dichloromethylphenylsilane (DCMPS) in THF at 22°C published by Holder and Jones [19]

Dichlorosilane	Yield/%	M_n	M_w	PDI
DCDHS	53	17,800	42,800	2.5
DCHMS	52	8,700	17,700	2.0
DCMPrS	40	13,700	33,900	2.5
DCMOS	58	11,440	49,300	4.3
DCMPS	64	25,600	61,500	2.4

Table 2 Average yields, molecular weights and polydispersity indices from the synthesis of polymethylphenylsilane in optically active and racemic limonene at various temperatures found by Holder and co-workers [24]

Temp./°C	Limonene solvent	Yield/%	M_n	M_w	M_w/M_n
70	Active	16	1,610	24,100	15.5
70	Racemic	17	1,600	22,000	14.3
80	Active	25	1,400	50,000	37.0
80	Racemic	25	1,100	40,800	37.0
90	Active	24	1,700	80,000	50.2
90	Racemic	17	1,700	39,200	22.9

Apart from changing the reaction temperature and the variation of solvents, it is also possible to influence the coupling reaction by adding sodium-ion-sequestering species such as 15-crown-5 (acting as a phase transfer catalyst for sodium) [18] or diglyme to improve the polymer yields [21, 25]. Additionally, activation by ultrasonic treatment can facilitate low-temperature polymerization and even gives access to polysilanes with monomodal weight distributions [26, 27].

It seems that the 'ideal' Wurtz-type coupling reaction of chlorosilanes combines all recent scientific results, using an optical active solvent at low temperatures and additives that increase the yield of desired polymer. To what extent the combination of these promising synthetic methods is able to deliver a high molecular weight monodisperse polysilane in high yields remains to be determined.

Consideration of the Reaction Mechanisms

During the last few years, considerable work has been performed to determine a (well defined) reaction mechanism, mainly to find an explanation for the polymodal weight distributions. In a recent review by Jones and Holder progress was reported [19]. The reaction appears to be a condensation reaction involving a chain mechanism [23, 28], suggesting various active species such as silyl radicals, silyl anions, silyl anion radicals and disilenes [29]. It has been discussed whether the inhomogenities in molecular weights result from the many different chain carriers. However, trapping experiments could not corroborate this hypothesis [20, 28, 30].

A proposed mechanism for the Wurtz-type reductive coupling reaction is depicted in Fig. 5. After initiation via a silyl anion radical to a silyl radical, a four stage propagation step occurs to form the polymeric species.[1]

2.2.2 Masked Disilenes

The lack of sufficient control of molecular weight distributions, molecular weight and polymer structure of polysilanes via a Wurtz-type coupling reaction led Sakurai and co-workers to develop an alternative method for polysilane synthesis in 1989 [31].

Because disilenes are not stable enough to be isolated unless they have very bulky substituents, a route to polysilanes analogous to the olefin polymerization (Fig. 6) is not possible [32].

To overcome this stability problem, the high reactivity of disilenes can temporarily be masked by adding a suitable auxiliary group. Roark and Peddle found that an adduct can be formed from dichlorodisilane and a biphenyl anion radical yielding the masked disilene (Fig. 7) [33].

---

[1] For a more detailed and comprehensive description of the mechanism see also the review of Jones and Holder [18].

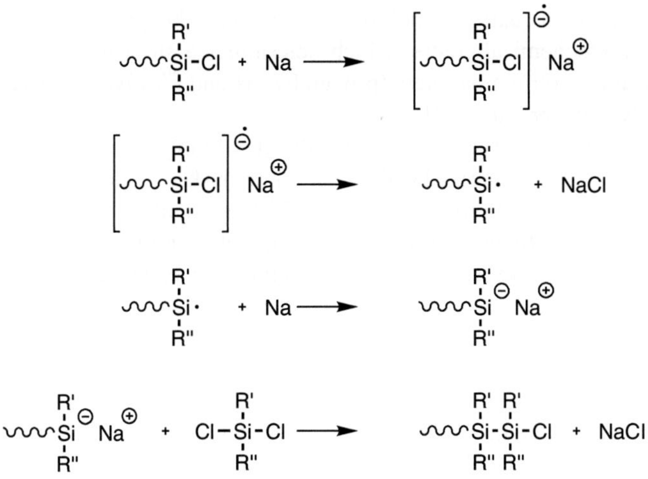

Fig. 5 Proposed chain mechanism for the Wurtz-type reductive coupling of dichlorodiorganosilanes

Fig. 6 Olefin polymerization compared to a hypothetical silylene polymerization

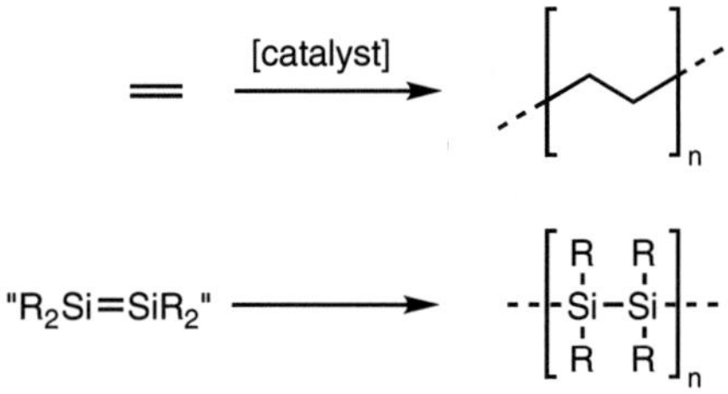

Fig. 7 Masking of a disilene [33]

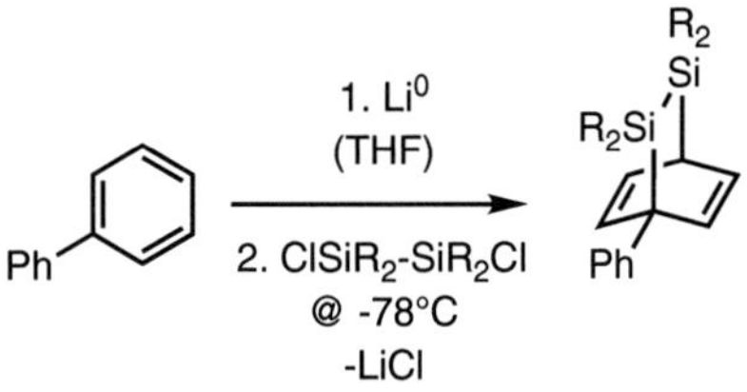

Fig. 8 Anionic polymerization of masked disilenes [31]

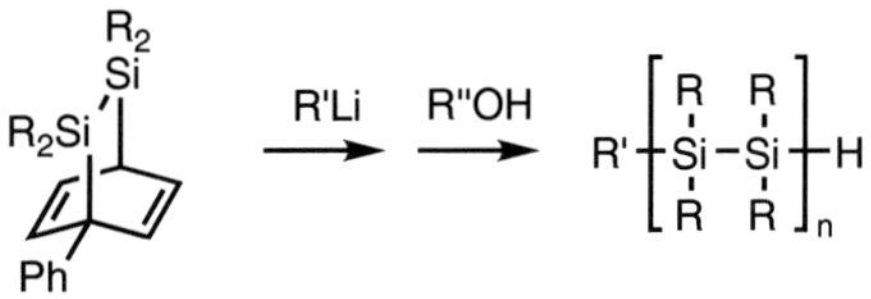

These monomers can be polymerized with a catalytic amount of anionic initiator, such as organolithium compounds or alkali metal alkoxides. Termination is accomplished by adding an alcohol (Fig. 8) [31].

Other molecules such as anthracene, naphthalene or benzene have also been used as masking agents [33], but the biphenyl system has proved to be best with respect to stability and product quality for polymerizations. The phenyl group in the 1-position provides a regiochemical centre, which determines a strict head-to-tail polymerization for asymmetric monomers (proven by ^{1}H and ^{29}Si NMR), giving a highly ordered polysilane product [34].

This anionic polymerization is evidently of a living type as the unterminated polymer can be co-polymerized with either other masked disilenes bearing different functional groups or methylmethacrylate (MMA) to yield a block co-polymer [35]. Furthermore, it was found that the relationship between molecular weight and the degree of monomer conversion is linear, which is a necessary condition for a living polymerization [36].

Varying the initiator can increase the yields and accelerate the reaction. Thus, nearly quantitative conversion can be reached in a few minutes, for example by additionally using potassium-cryptand [2.2.2] [36].

The molecular weights achieved range from 5,000 [37] to 27,000 g mol^{-1} with almost constant narrow PDIs of around 1.5 determined by SEC [38].

As the reaction conditions for the preparation of the masked disilenes are as harsh as those of the Wurtz-type reductive coupling reaction, the introduction of functional groups is not trivial. An elegant way to introduce at least some diversity is via the transformation of amine-functionalized polysilanes [39] to the chloro substituted derivatives. These can be converted to alkyl and aryl groups using the corresponding Grignard compounds (Fig. 9) [35].

This (RRSi–SiRR′)$_n$ substitution pattern is characteristic of polysilanes synthesized by the anionic living polymerization of masked disilenes. For synthetic reasons, the substituents only vary on one of the two silicon atoms in the masked monomer.

Starting from readily available chlorodisilane compound (a) and one equivalent of a reactive compound, e.g. a Grignard (X = MgCl, MgBr) or lithium organic

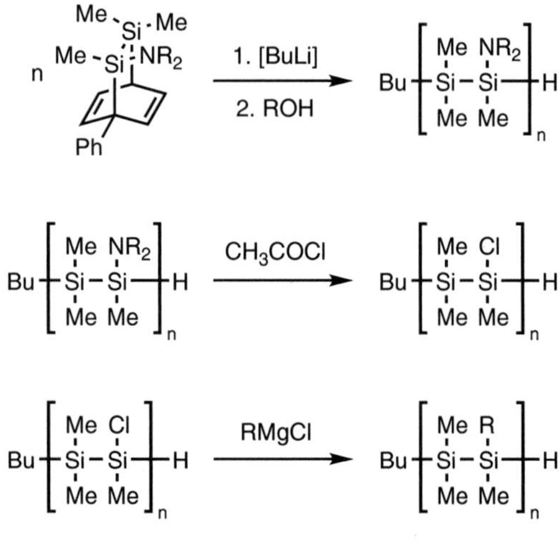

Fig. 9 Transformation of an amine-substituted polysilane

reactant (X = Li), the corresponding precursor for functionalized masked disilenes (b) can be generated bearing the typical pattern of substitution (Fig. 10).

Using two equivalents would lead to the necessity of undesired separation steps for purification and thus is avoided.

Table 3 shows examples of synthesized polysilanes with different substituents R^1, R^2 (Fig. 11) and molecular weights obtained [35, 38].

As the anionic living polymerization of masked disilenes can be controlled to occur in a strict head to tail fashion, Sakurai and co-workers were able to synthesize the first regiochemically regular 'polysilastyrene' (Fig. 12) via the amine functionalization route mentioned above [38].

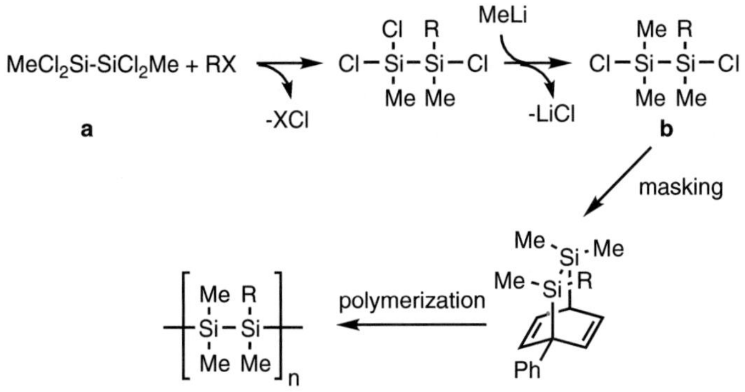

Fig. 10 Polymer synthesis starting from a dichlorodisilane

Table 3 Examples of functionalized polysilanes prepared from masked disilenes

(R^1, R^2)	M_w
n-Bu, Me	110,000
n-Pr, n-Pr	Insoluble
n-Hex, n-Hex	61,000
i-Bu, Me	110,000
Pr$_2$N, Me	Insoluble
Et$_2$N, Me	Insoluble
Bu$_2$N, Me	27,000
n-Hex$_2$N, Me	Insoluble
Ph, Me	15,000

Fig. 11 Substitution pattern of polysilanes prepared from masked disilenes

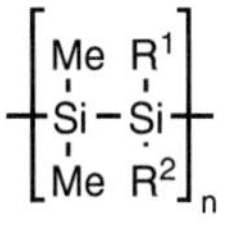

Fig. 12 Comparison of poly(styrene) and "polysilastyrene"

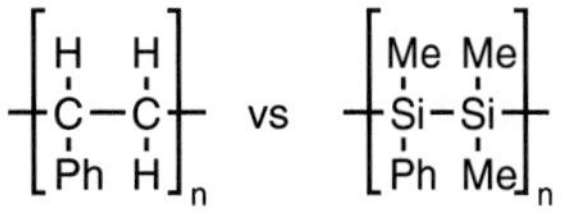

This route towards polysilanes seems to be of high potential concerning PDI and molecular weight. Few recent publications concerning polymerization of masked disilenes can be found containing scientific results with respect to further improvement of synthesis [35, 38, 39].

This method has also been used for the synthesis of defined polysilanes in some applications. For example, Li et al. applied the polymerization of masked disilenes to synthesize an aryl substituted diazene chromophore functionalized polysilane with nonlinear optical properties [40].

Mechanistical Considerations

Investigations on masked disilenes bearing bulky substituents gave access to transient species existing during the reaction, as sterical hindrance inhibits polymerization and allows isolation and description of the intermediates involved. By reaction of a masked di-*iso*-butyldimethyldisilene with one equivalent of methyl lithium followed by quenching with methanol, Sakurai and co-workers were able to investigate the polymerization mechanism [41].

Initiation and propagation both proceed via the attack of an anionic species at the silicon atom at the 3-position of the masked disilene, because the 2-position is sterically hindered. Figure 13 shows a proposed mechanism [35].

Sakurai and Yoshida investigated the mechanism of the anionic living polymerization of masked disilenes and more detailed information can be obtained from their comprehensive review [35].

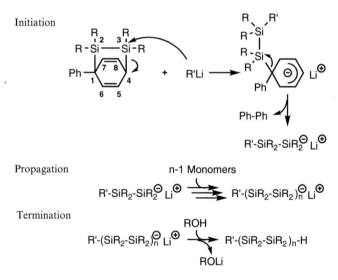

Fig. 13 Proposed mechanism for the anionic polymerization of masked disilenes

2.2.3 Ring Opening Polymerization of Silacycles

Strained silacycles can be used in a ring opening polymerization (ROP) to yield linear polysilanes.

Silacycles are coupling products of the reaction of dichlorodiorganosilanes with alkali metals [42]. This resembles the Wurtz-type coupling reaction of chlorosilanes in educts and reaction conditions and thus tolerates the same narrow range of functional groups. For this reason, it is mainly the methyl and phenyl substituted silanes which have been investigated.

The first to investigate the ROP as a route towards polysilanes bearing organic substituents were Matyjaszewski et al., who found that (PhMeSi)$_4$ readily undergoes polymerization by nucleophilic initiators such as organolithium compounds (Fig. 14).

The molecular weights obtained were up to 100,000 g mol^{-1} with a PDI of ~2. However, a monomodal distribution could not be achieved [43].

This type of polymerization also works for five-membered rings that are not sterically hindered, for example phenyl nonamethyl cyclopentasilane initiated by anionic reagents [44] (Fig. 15).

West and co-workers found that the anionic species generated by the initiator is, to some extent, stabilized by the phenyl groups of the silacycle [45]. However, similar to the polymerization of masked disilenes, a bulky substituent will inhibit polymerization.

Additionally, it is known that cyclic Si$_4$Cl$_8$ can be polymerized in an ROP via radical species to yield an insoluble but moisture sensitive polymer with an estimated degree of polymerization of about 35 and a molecular weight of about 3,500 g mol^{-1} [45]. Due to the insolubility and the high sensitivity to moisture, there have been no further reports on a ring opening type synthesis of perchlorinated polysilanes [42, 45]. Nevertheless, a functionalization by reaction with isopropanol has been accomplished which led to an increased solubility and enabled a SEC-analysis [45].

As mentioned in Sect. 2.1, perchlorinated polysilanes can be used as precursors for HPS. However, the ROP-route using silacycles as monomers, which themselves

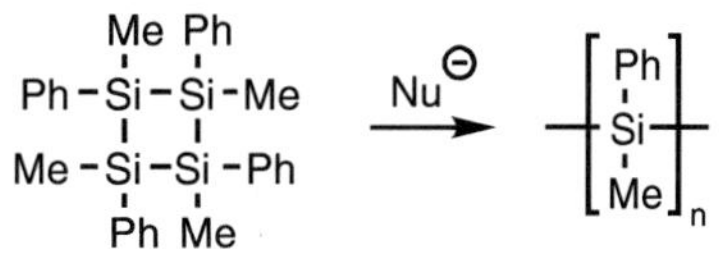

Fig. 14 Ring opening polymerization (ROP) of (SiMePh)$_4$

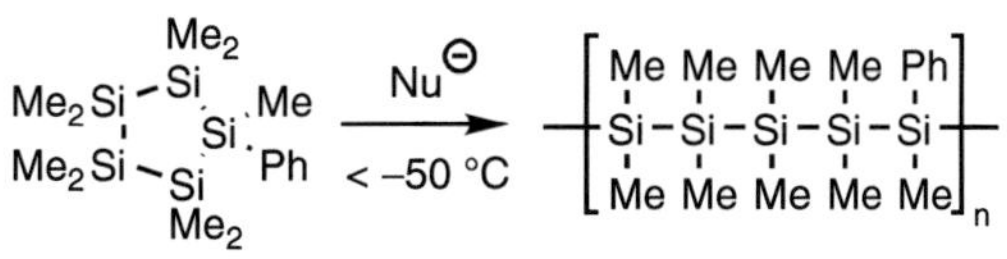

Fig. 15 ROP of phenylnonamethylcyclopentasilane

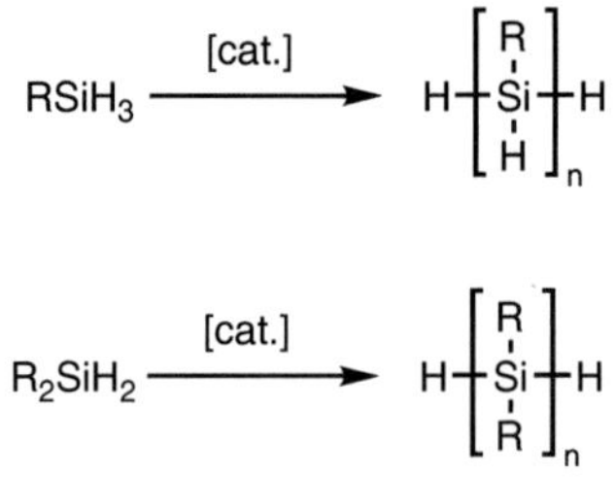

Fig. 16 Dehydrocoupling of primary and secondary silanes to polysilanes (typical catalysts are for example group IV metallocenes)

have first to be synthesized in an additional step, does not seem to be a straightforward approach towards substrates for a hydrogenation.

2.2.4 Dehydrocoupling

Catalytic reactions of silanes to give polysilanes and hydrogen, so-called dehydrocoupling reactions, are a promising route to functionalized, high molecular weight products.

The required monomers for this type of polymerization need to possess at least two Si–H groups. Thus, suitable educts would be primary ($RSiH_3$) and secondary silanes (R_2SiH_2) (Fig. 16).

Nonetheless, the dehydrocoupling of secondary silanes was found to be non-trivial with respect to finding a catalyst that is able to produce more than dimeric or oligomeric species.

Because of apparent differences in reactivity, primary and secondary silanes will be dealt with in separate sections.

Primary Silanes

By far the most common adduct for the preparation of polysilanes via the dehydrocoupling reaction is phenylsilane, $PhSiH_3$, simply because it is readily available. Methylsilane is also available, but due to its gaseous state, the insolubility of polymethylsilane and the resulting preparative difficulties, it is not as attractive a reactant as phenylsilane and has thus not evoked the same scientific interest. Depending on the catalyst used, polyphenylsilanes with varying chain lengths and molecular weight distributions can be obtained. Corey et al. published an M_n of 2,300, M_w of 3,850 and a PDI of 1.7, [46] while Tilley et al. managed to polymerize $PhSiH_3$ with an M_n of 2,310 and an M_w of 12,030 (PDI 5.2) [47].

Other primary silanes, mainly functionalized aryl and alkyl silanes, were successfully polymerized and some examples are listed in Table 4.

Not all publications give information about the calibration standard for the SEC-analysis. The table shows a strong variation of molecular weights and PDIs, even with identical monomers. This is due to different catalysts being applied for the polymerization and will be discussed in the following section.

Table 4 Examples of functionalized polysilanes synthesized by dehydrocoupling of the corresponding hydrosilanes. SEC measurements were calibrated with polystyrene standards. Data given without any information about the calibration standard is marked with an asterisk

Arylsilanes	M_n; PDI
m-MeC$_6$H$_4$-SiH$_3$	900; 2.1 [48]*
p-MeC$_6$H$_4$-SiH$_3$	530; 1.9 [48]*
	1,530; 1.5 [49]
p-MeOC$_6$H$_4$-SiH$_3$	2,330; 1.5 [49]
	800; 3.5 [50]
p-FC$_6$H$_4$-SiH$_3$	3,780; 1.9 [49]
p-Me$_2$NC$_6$H$_4$-SiH$_3$	400; 1.5 [49]
	590; 1.7 [50]
p-(i-Pr)OC$_6$H$_4$-SiH$_3$	2,150; 1.3 [49]
m-CF$_3$C$_6$H$_4$-SiH$_3$	1,680; 1.5 [48]*
p-CF$_3$C$_6$H$_4$-SiH$_3$	630; 2.4 [48]*
	6,400; 1.4 [49]
p-MeSC$_6$H$_4$-SiH$_3$	4,450; 2.2 [50]
3,5-Me$_2$C$_6$H$_3$-SiH$_3$	1,460; 1.5 [49]
3,5-(CF$_3$)$_2$C$_6$H$_3$-SiH$_3$	830; 1.2 [49]
C$_6$F$_5$-SiH$_3$	730; 1.5 [49]
Alkylsilanes	M_n; PDI
MeSiH$_3$	1,200; 5.3 [51]*
n-Octyl-SiH$_3$	3,330; 1.1 [52]
n-Dodecyl-SiH$_3$	2,180; 1.1 [52]
n-Hexyl-SiH$_3$	900–1,400; n.a. [53]*
Et$_3$Si(CH$_2$)$_2$SiH$_3$	1,190; 1.8 [54]*
MePhSiH-CMeHSiH$_3$	2,130; 1.3 [54]*
PhPhSiH-CMeHSiH$_3$	1,730; 1.7 [54]*
(Si$_6$H$_{11}$)SiH$_3$	4,200; 2.5 [55]

Catalysts for Primary Silanes

When using transition metal catalysts, a variation of the ligands usually leads to good control of the polymerization reactions.

The most intensively studied catalysts for the promotion of dehydrogenative Si–Si-bond formation are metallocene derivatives of group IV metals first reported by Harrod and co-workers [53]. Their general structure can be described as Cp$_2$MR$_2$ (Cp = cyclopentadienyl derivative; M = metal centre; R = organic or inorganic substituent).

Derivatization of those 'simple' metallocenes was then investigated by the groups of Tilley [56], who used mixed ligand complexes (Cp/Cp*) and Corey [57], who applied metallocene chlorides that have to be activated by n-BuLi. This has aroused the interest of more groups and considerable data are available concerning such catalysts. For a comprehensive listing of dehydrocoupling catalysts the reader is referred to Corey's review [58].

Table 5 Comparison of M_n and PDI with increasing atomic number of the central metal. SEC data were referred to polystyrene standards

Catalyst	M_n; PDI
Cp_2TiCl_2/n-BuLi	1,300; 1.1 [57]
Cp_2ZrCl_2/n-BuLi	1,860; 1.6 [59]
Cp_2ZrCl_2/n-BuLi/$B(C_6F_5)$	2,670; 1.9 [59]
Cp_2HfCl_2/n-BuLi/$B(C_6F_5)$	3,050; 2.8 [59]

It has been shown that a variation of the central metal, the ligands or the activation method has a strong influence on the polymer properties.

These trends will be briefly discussed in the following section.

The reactivity of the catalysts increases with greater atomic number. An enhanced reactivity yields a polymer with higher molecular weight, but its weight distribution broadens. An example is given in Table 5.

Furthermore, the data from Table 5 show that the activation of the catalyst plays an important role in controlling molecular weight and weight distributions. This will be discussed later in more detail.

Catalysts with mixed ligands also provide higher molecular weights than simple bis-cyclopentadienyl systems, the consequences of this higher reactivity being broader PDIs. Corey and co-workers found that polymerization of $PhSiH_3$ with a catalyst bearing chirally substituted Cp-ligands can increase the molecular weight of the polymer from $M_n = 1,600$ to 2,000 while broadening the PDI from 1.4 to 2.0 [60].

With most catalysts, an undesirably long induction period is observed. This can be overcome with only marginal influence on the desired polymer properties by variation of the substituents on the central metal of the metallocene. The exchange of Me for $Si(SiMe_3)_3$ on $CpCp^*ZrMe_2$ reduces the induction period from 52 to 24 h [61]. The induction period can also be shortened by using chloro substituted metallocenes and the appropriate activation agents [57].

The use of 'modern' metallocenes, such as those used in olefin polymerization, is thought to be ineffective in dehydrocoupling reactions. Corresponding scientific studies investigated *ansa*-metallocenes such as chiral bridged bis-indenyl complexes, but found them either to be as reactive as the 'common' dehydropolymerization catalysts at best or sterically too crowded and thus non-reactive at worst [62, 63].

Secondary Silanes

In contrast to the broad variety of catalysts and synthesized polymers that can be found with primary silanes, there are fewer reports about secondary silanes. Secondary silanes are sterically more hindered than primary silanes and thus the homocoupling is much slower. This is the reason why mostly oligomeric species are obtained in dehydrocoupling reactions of a secondary silane with group IV transition metal catalysts.

Table 6 gives examples of secondary silanes that have been successfully polymerized with a dehydrocoupling catalyst.

The problems of polymerizing secondary silanes seem to be not only finding a silane with a suitable substitution to allow a polymerization but also finding the right catalyst, as in the case for the polymerization of primary silanes.

Other Catalysts for Primary and Secondary Silanes

Although in the majority of reports about dehydrocoupling reactions group IV metallocenes have been used, other metals and ligand systems have been applied to the polymerization of primary as well as secondary silanes. Indeed, the first complex found to be active in dehydrogenative coupling was Wilkinson's catalyst ($(PPh_3)_3RhCl$) [66]. Differentiation can be made between catalysts that are able to produce dimers up to oligomers and those able to produce polymeric species.

Table 7 shows examples of polymers obtained by non-metallocene catalysts.

Independent of the catalyst used, the polysilanes obtained are inhomogeneous and control of structure and molecular weights remains difficult. Most promising

Table 6 Examples of secondary silanes successfully polymerized by dehydrocoupling catalysts. SEC data were referred to polystyrenene standards

Secondary silane	M_n; PDI
PhMeSiH$_2$	1,380; 1.3 [64]
Ph$_2$SiH$_2$ (comonomer ε-caprolactone)	1,510; 1.2 [65]
(Si$_6$Me$_{11}$)SiH$_2$SiH$_2$Me	1,800; 1.4 [55]
1,4-(SiMeH$_2$)$_2$(Si$_6$Me$_{10}$)	900; 2.0 [55]

Table 7 Examples of silanes polymerized by non-metallocene catalysts. SEC data were referred to polystyrene standards, else marked with an asterisk

Catalyst	Silane monomer	M_w
[Mo]	PhMeSiH$_2$	1,800 [64]*
[Rh]	HexSiH$_3$	1,380 [67]
[Ni]+ Lewis acid	PhSiH$_3$	700–4,400 [68]
[Ni]	PhSiH$_3$	370–1,800 [69]*
[Pt]	Et$_2$SiH$_2$	2,100–39,000 [70]
[Pt]	MeHexSiH$_2$	6,800–25,000 [71]
[Nd]	PhSiH$_3$	520–4,800 [72]
[Nd]	HexSiH$_3$	500–900 [72]
[Nd]	MeSiH$_3$	500–7,300 [73]

are approaches towards the polymerization of secondary silanes to give a possibility for α–ω functionalization via hydrosilylation.

A comprehensive list of the used catalyst systems can be found in the review of Corey [58].

Mechanistical Considerations

It is desirable to model the catalysts for optimum product control in the reaction. Therefore, the mechanism of the dehydrocoupling reaction of silanes has been investigated and several models have been proposed.

Based on the information obtained by reacting hydrogen terminated oligomers and analysing the product distributions, different species such as silylenes [74] or metal silylene complexes [58] have been proposed to exist during the polymerization.

The first mechanism to be widely accepted was proposed by Tilley et al., which is called a σ-bond metathesis (Fig. 17) [75]. In this model a metallocene precursor is first transformed to a metallocene hydride that enters the hypothetic catalytic cycle, forming a metal silyl species via an initial σ-bond metathetic step with the simultaneous production of hydrogen. The second σ-bond metathetic step forms the Si–Si bond and is supposed to regenerate the catalytically active metal hydride.

The metathetic step involving the Si–Si bond formation is the reaction rate determining step as its transition state is sterically most crowded in the case of bulky substituents R. This agrees with the observation that primary silanes are usually more reactive than secondary silanes. For the same reason, linear chain growth would be favoured over branching as long as monomer is still present. As the reaction steps are all reversible, the 'M–H'-species should be able to insert into a Si–H-bond of the polysilane leading to a degradation of the already formed polymer. Larger oligomers could then react intra-molecularly to give cyclic species.

Harrod and Dioumaev observed reduced $[Cp_2Zr^{III}-R]_n$ species after the activation of dichlorozirconocene with n-BuLi and proposed an activation mechanism explaining the formation of the anticipated catalytically active complex (Fig. 18) [76].

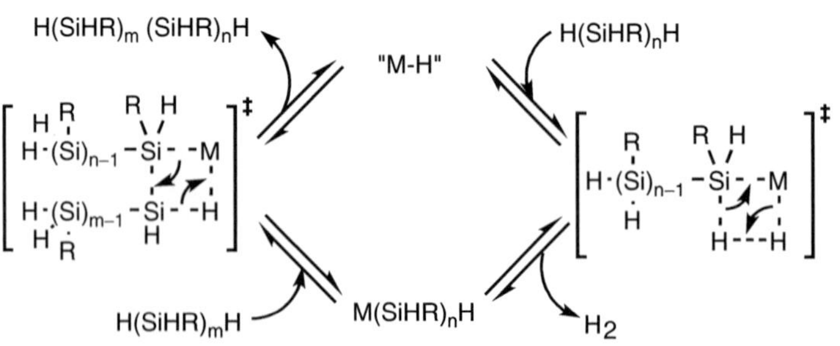

Fig. 17 σ-bond mechanism proposed by *Tilley et al.*[75]

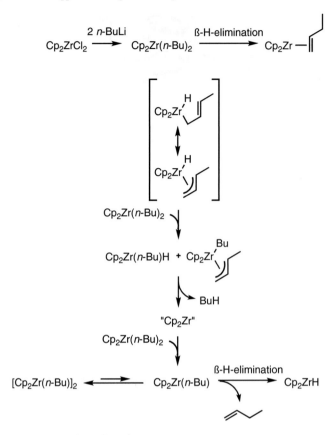

Fig. 18 Decomposition of $Cp_2Zr(n-Bu)_2$

A dibutylzirconocene is formed via reaction of the dichlorometallocene with n-BuLi, which then liberates butane by hydrogen abstraction from the second ligand. After this, the π-complex formed rearranges by insertion of the metal into a C–H-bond. Two ligand exchange reactions with undecomposed dibutylzirconocene, comproportionation and liberation of another butane and finally β–H-elimination yield the active Zr-hyride species.

Furthermore, Harrod proposed a mechanism for 'cation-like' silylzirconocenes, which are formed after activation of the metallocene and addition of the monomer. This mechanism involves one-electron oxidation/reduction steps as well as σ-bond metathesis depicted in Fig. 19 [59].

Zr^{III}-species could simply be by-products [77] and not be involved in the catalytic cycle as they could not be observed in all dehydrocoupling reactions [78]. Furthermore, no evidence for the existence of silyl radicals has been provided.

The many uncertainties show that the mechanism describing dehydrocoupling is still worthy of investigation. Especially the mechanism of activation and thus the genuinely active catalyst species during the reaction has still not been identified.

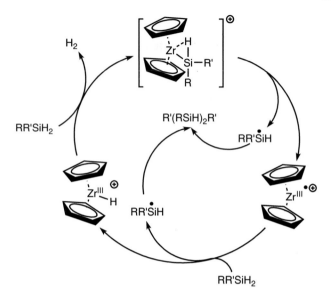

Fig. 19 One-electron redox mechanism proposed by *Harrod*

Cl(SiR$_2$)$_n$Cl + Li(SiR'$_2$)$_m$Li ⟶ Cl(SiR$_2$)$_n$(SiR'$_2$)$_m$Li ⟹ polysilane

Fig. 20 Stepwise condensation of chlorinated and lithiated oligosilanes to polysilanes

For the dehydropolymerization with catalysts other than group IV metallocenes, no mechanistic proposals have been made up to now.

2.2.5 Miscellaneous

In addition to the synthetic methods mentioned above, some other reactions for synthesizing polysilanes have been described and will briefly be discussed in the following paragraph.

Electrochemical reduction of diorganodichlorosilanes with simultaneous ultrasonic treatment is able to yield the corresponding polysilanes by stepwise chain elongation. With this technique, Ishifuna and co-workers were able to obtain polymethylphenylsilane with an M_n of 3,600 and a PDI of 2.3 from a dichlorotrisilane at room temperature. Lowering the temperature to $-10°C$ increased the molecular weight to an M_n of 5,500 and an even narrower PDI of 1.5 was achieved. This type of polymerization has also been shown to produce a polymer with a monomodal weight distribution [79].

Another alternative is the stepwise condensation of acyclic difunctional reagents such as dilithiated oligosilanes with dichloro-oligosilanes (Fig. 20).

The polysilanes obtained by this method were analysed by osmometry and had an M_n of 5,000–10,000. This method proved to be quite ineffective and polymer yields of more than 50% were not achieved [80].

3 Structure Determination of Polysilanes

With a slight dependence on the Si-substituents, the preferred main chain conformation in most polysilanes is an all-anti type arrangement, which is why polysilanes adopt helical structures. Fogarty and co-workers have developed a nomenclature to characterize the conformers, using n-Si$_4$Et$_{10}$ as a model substance [81, 82]. Investigations by circular dichroism spectroscopy showed no preferred helix direction throughout the whole polymer backbone [42]. This corroborates earlier investigations proposing breaks in the helical structure by gauche turns that are responsible for the inhomogeneous molecular weight distributions in the Wurtz-type reductive polymerization of chlorosilanes [83].

Furthermore, attempts have been made to synthesize stereo regular polysilanes. Chiral group IV metallocenes have been successfully applied in the stereoselective olefin polymerization, but their use in the dehydrocoupling reaction has only yielded polysilanes with an atactic structure [60]. Even though the tacticities found indicate stereoirregularity, typical signals for isotactic, syndiotactic and heterotactic chain segments have been observed by ^{29}Si-NMR. Those signals have their origin in separately resonating small polymer chain segments, bearing statistically arranged substituents [42].

As the ring opening polymerization of masked disilenes seems to occur in a strict head to tail fashion, highly ordered polysilanes have been obtained showing two sharp signals in the ^{29}Si-NMR spectrum indicating a completely syndiotactic polymer [34, 35].

Thus, it is possible to prepare structurally controlled polysilanes although the method is limited by the functionalities accessible with the masked disilene route. Surprisingly, to date, this has not been investigated any further.

4 Physical Properties

Most polysilanes are soluble polymers depending on their crystallinity, which can be tuned by the side groups and their functionalities. Their glass transition temperatures may range from −72°C to +120°C depending on the substituents. Generally, it can be stated that polysilanes with long alkyl chains are elastomers and phenyl groups or short alkyl chains will increase crystallinity [42].

Thermal treatment up to 300°C does not decompose most polysilanes whereas heating over 1,000°C leads to the formation of ceramic silicon carbide [42, 84].

The backbone of polysilanes exclusively consists of catenated Si atoms and their physical properties largely depend on the nature of the Si–Si bond and on the substituents present.

Hyperconjugated electrons along the main chain are responsible for semiconducting and photoconducting properties. These are of interest for applications discussed in Sect. 5.

The special bonding situation in polysilanes can be used to explain most of their properties and will be described in the following paragraph.

A red shift in the absorption maxima and a gradual increase in the molar absorption coefficients (ε) can be observed with increasing chain length. This is due to the σ-conjugation along the silicon main chain depending on substituent and conformational effects that can be explained by molecular orbital theory (MO). The LCAO, Sandorfy model C and FE (free electron) methods have been used.

The combination of sp^3 hybrids affords the formation of the molecular orbitals along the silicon main chain [3]; d-orbitals do not contribute to the formation of the bonds. The combination of hybrid atom orbitals leads to the delocalized σ and σ* molecular orbitals responsible for the σ-conjugation. This delocalization lowers the LUMO while raising the HOMO with increasing chain length; the observed red shift in the absorption maxima results (Fig. 21).

Furthermore, substituents significantly stabilize the LUMO. While alkyl substituents with hyperconjugation effects do this without any significant influence on

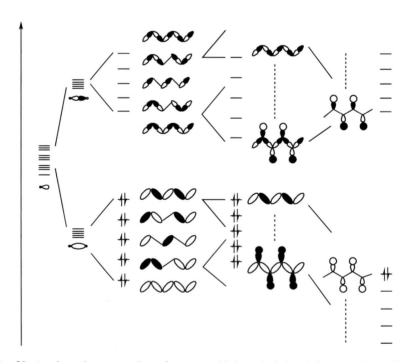

Fig. 21 A schematic presentation of σ-type orbitals and their relative energies in linear oligosilanes[3]

the HOMO, aryl groups destabilize the HOMO due to conjugative effects that are more prominent than those of the hyperconjugation.

Compared to the C–C bond, the Si–Si bond has a more electropositive nature and heterolytic dissociation occurs more easily due to the low energy σ–σ^* excitation [85]. The absorption bands of polysilanes in the near UV range between 290 and 410 nm and they are temperature and structure dependant [86]. Since the σ–σ^* transition is allowed, the absorption coefficients of their optical absorption spectra are large [42]. Although the Si–Si bond has a similar bond strength to the C–C bond [87], cleavage of the Si–Si bond by UV excitation can easily occur in the solid state as well as in solution. Both may lead to decomposition of the polymeric species [3, 88, 89].

Absorption and emission UV spectra of common polysilanes are well described and have been reprinted in various reviews and reports [3, 35, 36, 61, 87].

The band gap in polysilanes is dependent on the electrochemical properties of the substituents and ranges around 4 eV compared to 8 eV for a saturated carbon skeleton [3].

5 Applications

Polysilanes were first synthesized in the early 1920s, but no investigation of their application was made until the late 1970s. The discovery of the two-step transformation of polydimethylsilane to β-SiC fibres by Yajima et al. [90] marked the beginning of interest in their properties and thus the application of polysilanes. After the discovery of the semiconducting behaviour of polymethylphenylsilane by West et al. [91], the cornerstone for the application of polysilanes in electronics was laid.

Because of their special chemical and electro-optical properties, polysilanes have been proposed as materials for many applications. These can be classified into two main categories:

1. Applications based on the reactivity of polysilanes
2. Applications based on the photophysical and electronic properties of polysilanes

Because of the great progress in organic semiconductors and the extensive search for new materials by the electronic industry, the latter of the aforementioned categories has been the more active field of investigation.

5.1 Applications Based on the Reactivity of Polysilanes

Polysilanes react readily with surfaces and molecules containing nucleophilic oxygen. Thus, they can be used as adhesion promoters on glass or oxide materials. The preparation of polymer composites reinforced by glass particles is one example of

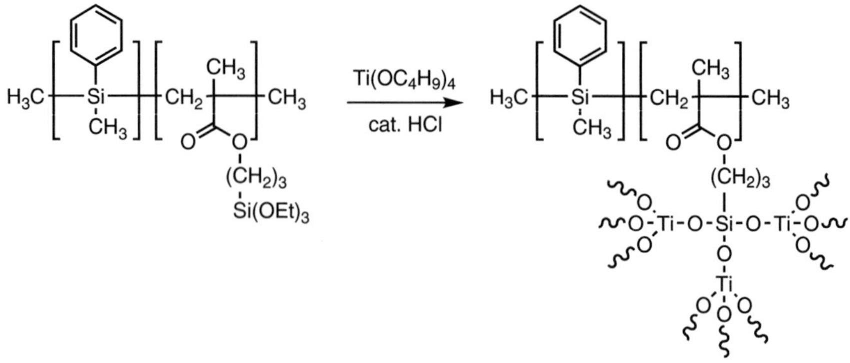

Fig. 22 Formation of polysilane–titania hybrids

this use [92]. Another interesting example is the formation of titania based nanostructured hybrid xerogels or nanoporous films. In a first step, a block co-polymer is formed by the photopolymerization of 3-methacryl-oxypropyltriethoxysilane with polymethylphenylsilane as macro-photoinitiator (see below). The product bears triethoxysilyl groups which are reacted with titanium(IV) butoxide in a second step (Fig. 22). Then it is dried in air to form titania by a sol–gel process. During the formation of the xerogel, the polysilane effects a structuring of the product on the nano scale. The polysilane is bound to titania via Si–O–Ti-bonds formed during the process [93].

The elemental composition of polysilanes suggests a use of the material as a precursor for silicon carbide (SiC). Particularly the preparation of SiC fibres has been investigated [90, 94], but SiC composites with glass and with alumina have also been prepared [95, 96].

In addition, inorganic polyhydrosilanes (Si_nH_{2n+2}, Si_nH_{2n}) can be used as precursors for the preparation of amorphous silicon: the compounds are easily decomposed by laser irradiation or by treatment at elevated temperatures. If laser irradiation is used, the amorphous silicon can even be structured [97]. Polysilanes can be coated from solution or printed by ink jet techniques and then be converted into silicon for the use in transistors or solar cells [98–101].

5.2 Applications Based on the Photophysical and Electronic Properties of Polysilanes

Polysilanes have unique optical, electronic and photophysical properties due to the delocalization of the σ-electrons along the silicon chain (see above). This delocalization was first discovered while studying the absorption spectra of oligo- and polysilanes. Polysilanes show distinctly red shifted absorption bands compared to the homologous series of polyolefins. Furthermore, the absorption bands depend strongly on the chain length, the conformation and the substituents of the polysilane [102, 103]. It was deduced that there has to be a conjugation along the σ-bonds of the silicon chain, which was later confirmed by molecular orbital calculations. The σ-conjugation offered an explanation for the facile cleavage of the silicon chain by near UV radiation and it suggested the possibility to oxidize polysilanes and use the oxidized polymer chain as a charge carrier material. These findings stimulated a lot of work on the application of polysilanes as photoresists and as UV sensitive polymerization starters as well as photoconductors and hole transporting materials in electronic devices, such as organic light emitting diodes (OLEDs) and organic photovoltaics (OPV).

As mentioned above, the cleavage of Si–Si-bonds in polysilanes upon irradiation with UV-light is very efficient and results in the formation of silicon radicals. These radicals react with olefins to initiate radical polymerizations. Especially the polymerization of methylmethacrylate and styrene with a variety of polysilanes as photoinitiators has been studied in detail [104]. The advantage of this kind of initiation is the possibility to prepare polysilane–polyolefin hybrids [105]. All modifications of radical polymerizations, such as the atom transfer radical polymerization (ATRP), are possible with polysilanes as photoinitiators [106].

The facile cleavage of Si–Si bonds by UV radiation is also made use of in the application of polysilanes as photoresists. This application has been well studied [107–110]. The technology covers a wide range from the structuring of metal films to the fabrication of micro lenses.

5.2.1 Polysilanes in Electronic Devices

Polysilanes in OLEDs

Polysilanes can be employed as hole transporting materials or emitters in OLED devices, more specifically in polymer OLEDs. Polymer OLEDs are prepared by spin coating techniques on transparent substrates like indium-tin-oxide (ITO) coated glass serving as anode. The basic design of a polymer OLED is shown in Fig. 23.

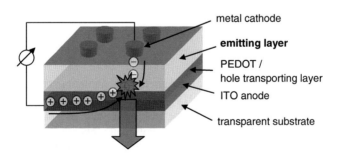

Fig. 23 Polymer OLED

On the transparent substrate and ITO, a hole transporting layer such as poly(ethylene dioxythiophene)– poly(styrene sulfonate) (PEDOT:PSS) is coated, topped by an emitting layer and a metal cathode. The emitting layer contains the p–n junction formed by a mixture of an electron transporting polymer, a hole transporting polymer and an emitting material. Applying a voltage to the electrodes injects electrons and holes into the device which recombine in the emitting layer and emit light. Polysilanes can serve as hole transporting materials as well as emitting materials in OLEDs. More specifically, polysilanes can serve as triplet harvesting materials in OLEDs containing phosphorescent emitting materials such as transition metal complexes. All these different uses are explained in more detail in the following sections.

Polysilanes as Hole Transporting Materials

Kido et al. have reported the application of polymethylphenylsilane as a hole transporting layer in multi-layer OLEDs. In contrast to the OLED shown in Fig. 23, the emitting layer in these early devices was not a mixture of polymers but was prepared by layer-by-layer deposition of the active materials. In the work by Kido, aluminum-tris(8-hydroxyquinoline) (Alq3) was used as both electron transporting material and emitter. It was applied onto the hole transporting polysilane layer by vacuum deposition [111].

Other pioneers in polysilane OLEDs were Suzuki (NTT Basic Research Laboratories, Japan) and Haarer (Physikalisches Institut, Universität Bayreuth) [112]. Based on polysilanes, they and others designed numerous multi-layer OLEDs which contained a variety of fluorescent and phosphorescent dyes as emitters. Some fluorescent dyes used were lanthanoid complexes [113] or derivatives of coumarine, perylene, phenoxazole or 4-dicyanomethylene-2-methyl-6-(p-dimethylaminostyryl)-4H-pyrane (DCM) [114–117]. Benzophenone was used as a phosphorescent dye [118].

The electronically excited states of the emitter can be populated via direct exciton formation at the emitter molecule. Alternatively, the electronic properties of polysilanes facilitate an energy transfer from excited states of the polysilane to the emitter molecule.

Polysilanes as UV Emitters

Another intensively studied area of research with numerous contributions from Japan is the use of polysilanes as UV emitters in OLEDs. In polysilane based UV OLEDs with a layered design, the polysilane acts as semiconducting and emitting layer. The first UV OLED using polymethylphenylsilane was published in 1995 [119] but electroluminescence was only detectable at very low temperatures (77 K). Luminescence was drastically reduced at higher temperatures, the reason for this being a thermally induced conformational change of the silicon chain [120]. The change in conformation induces a mitigation of the delocalization of σ-electrons, thus leading to the formation of local defects and non-radiating states. In order to get UV OLEDs operating at room temperature, conformationally stable all-trans polysilanes and polysilanes with a high glass transition temperature were employed. The first successful room temperature UV OLED was prepared with polydimethylsilane. Since it is insoluble in all organic solvents, it had to be applied by vacuum deposition [121]. Room temperature UV OLEDs, prepared by solution based processes, used defect free, linear high molecular weight ($M_w > 5 \times 10^5$ g mol^{-1}) polymethylphenylsilane [122] and poly(bis(4-butylphenyl)silane [123]. The lifetime of the prepared UV OLEDs is described as short compared to the state of the art polymer OLEDs.

Polysilanes for Triplet Harvesting

Polysilanes are also applicable as matrix materials in phosphorescent OLEDs. Mixtures of polysilanes and triplet emitters are sufficient to effect an energy transfer from polysilane triplet states to emitter triplet states, thus amplifying the luminescence of the device. It has been shown that if polysilanes have electrophosphorescent side chains consisting of triplet emitters, the energy transfer from polysilane to emitter is most effective [124]. Thus the beneficial electronic properties of polysilanes are perfectly combined with the spectroscopic properties of transition metal based triplet emitters. The compounds described are derivatives of polymethylphenylsilanes, (Fig. 24) which are covalently attached to triplet emitters with iridium as metal centre. The polymers were applied in OLEDs with an ITO/active layer/Ca/Ag layer sequence. The active layer contained a fraction of 70% by weight of the

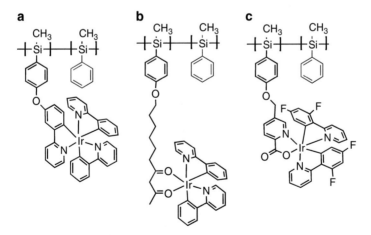

Fig. 24 Polysilanes with covalently attached iridium based triplet emitters. Polymers (**a**) and (**b**) give *green phosphorescent* OLEDs while polymer (**c**) emits *blue light*

polymers (a), (b) or (c) and an electron transporting material on the basis of oxadiazoles. Unfortunately, no performance data and no lifetimes of the devices are specified.

5.2.2 Polysilanes in Photovoltaics

The semiconducting and photoconducting properties of polysilanes also led to investigations of polysilanes as components in organic photovoltaics. The design of polymer solar cells closely resembles OLED devices (Fig. 25). Here a photoactive layer formed by a mixture of an electron transporting polymer and a hole transporting polymer rather than an emitting layer is employed. The polymer mixture undergoes a micro-phase separation after spin coating and thus forms a so-called p–n bulk heterojunction. In this case, absorption of light leads to generation of charge carriers in the photoactive layer. These charge carriers are transported to the electrodes and thus cause a current to flow.

Polysilanes can be part of the active layer or, such as in multi-layered OLEDs, serve as a hole transporting layer. There are a number of publications describing polysilanes as a photoactive layer or as a hole transporting material in organic solar cells but in most of them polysilanes are merely mentioned as a general example of an organic semiconductor. They are, for example, not specified in the published examples of patents and are only included to widen the scope of the claims. Only the publications mentioned below cover the application of polysilanes in organic photovoltaics in more detail.

Deviating from the principal of bulk heterojunction solar cells have been prepared using only polysilanes. However, polysilanes are preferentially employed as mixtures with other organic semiconductors and electron transporting materials such as fullerenes as photoactive layer [125, 126]. If polysilane–fullerene mixtures are used, higher efficiencies are observed than with pure polysilanes. The efficiencies depend directly on the fullerene concentration in the mixture [127, 128]. Instead of fullerenes, anthracenes can be used as electron transporting materials in mixtures with polysilanes. With 4.2%, the highest efficiency of a polysilane solar cell was observed with a polysilane–anthracene co-polymer [129]. It is generally accepted that purification and doping of polysilanes result in higher efficiencies of the corresponding organic solar cells [130, 131] or at least lead to higher conductivities.

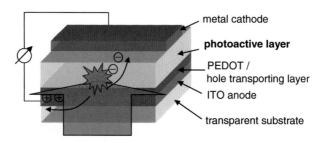

Fig. 25 Organic solar cell

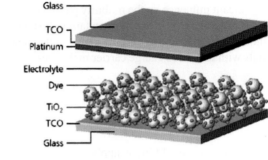

Fig. 26 Dye sensitized solar cell according to Grätzel (TCO: transparent conducting oxide, e.g. indium tin oxide) (Source: Heise online)

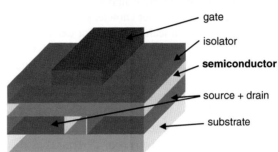

Fig. 27 Layout of an organic field effect transistor (OFET)

Polysilanes have also been employed in dye sensitized solar cells (Grätzel cells, (Fig. 26) [132, 133].

They can serve as a component in the polymer electrolyte, but more obvious is the use as photo sensitizer in mixtures with the dye or as co-polymers with dye. To summarize, the knowledge about the use of polysilanes in organic solar cells is marginal. In particular, studies of the correlation between the band gap of polysilanes and the performance of corresponding solar cells have not been made.

5.2.3 Polysilanes in Organic Field Effect Transistors

Polysilanes have also been proposed for use as semiconductors in organic field effect transistors (OFETs) [134–136], but even less is known about the use of polysilanes in this application.

One possible design of an OFET is shown in Fig. 27.

A field effect transistor (FET) controls the flow of electrons or holes from source to drain by affecting the conductivity of the semiconducting material by varying a voltage applied across the gate and source. Any *p*- or *n*-type organic semiconductor can be used in OFETs; the *p*-type polysilanes can be employed pure or in mixtures with other organic polymers as the semiconducting layer [137].

The minor attention paid to polysilanes as semiconductor in OFETs may be caused by the low charge carrier mobility in polysilanes, typical values being about 10^{-4} cm^2 V^{-1} s^{-1} [138]. This is three orders of magnitude lower than in other organic semiconductors. The substituents attached to the silicon backbone have no significant influence on the mobility parameter, so a dramatic increase of the mobility is not to be expected with standard polysilanes. However, new concepts such as those mentioned above for polysilane–anthracene co-polymers may result in materials with enhanced charge carrier mobilities more suitable for OFETs.

6 Conclusions

The multitude of different applications proposed for polysilanes is astonishing. However, to the best of our knowledge, commercialization has not yet been achieved in any of the shown technologies. For example, organic electronic devices are still based on organic semiconductors such as polythiophenes.

Important reasons for not applying polysilanes are probably the difficulties in controlling their synthesis with regard to molecular weight, PDI and impurities as well as the high costs of these elaborate methods and the purification processes involved. Additionally, there is little known about polysilane degradation and, thus, life time shortening in semiconductor devices. This in turn may be due to the limited supply of structurally diverse polysilanes accessible by the routes known.

If a reliable low cost synthetic process for the production of polysilanes can be developed a commercialization of this promising class of materials might be possible. This is especially true if this new process allows a free choice of substituents and thus yields more stable polysilanes.

References

1. Kipping FS (1921) J Chem Soc 119:830
2. Kipping FS (1923) J Chem Soc 125:2291
3. Miller DR, Michl J (1989) Chem Rev 89:1359
4. Stock A (1926) Z Elektrochem 32:341
5. Fehér F, Schinkitz D, Schaaf J (1971) Z Anorg Allg Chem 383:303
6. Fehér F, Schinkitz D, Wronka G (1971) Z Anorg Allg Chem 384:226
7. F Fehér, D Schinkitz, H Strack (1971) Z Anorg Allg Chem 358:202
8. Fehér F, Baier H, Enders B, Krancher M, Laakmann J, Ocklenburg FJ, Skrodski D (1985) Z Anorg Allg Chem 530:191
9. Auner GA, Bauch C, Lippold G, Deltschew R (2008) DE 102006034061 A1 CAN 115: 115263
10. Friedel C, Ladenburg A (1869) C R Hebd Seances Acad Sci 68:920
11. Kipping FS (1924) J Chem Soc 125:2291
12. Ziegler JM, Rozell JM, Pannell KH (1987) Macromolecules 6:399
13. Fujino M, Hisaki T, Fujiki M, Matsumoto N (1992) Macromolecules 1079
14. Ziegler JM (1986) Polym Prepr 27:109
15. Ziegler JM, Harrah LA (1987) Macromolecules 20:601
16. Horguchi R, Onishi Y, Hayase S (1988) Macromolecules 21:304
17. Miller DR, Sooriyakumaran R (1988) Macromolecules 21:3120
18. Jones RG, Holder SJ (2000) Synthesis of polysilanes by the wurtz reductive-coupling reaction. In: Silicon-containing polymers. Kluwer, Dordrecht

19. Jones RG, Holder SJ (2006) Polym Int 55:711
20. Ziegler JM (1987) Polym Prepr 27:109
21. Cragg RH, Jones RG, Swain AC, Webb SJ (1990) J Chem Soc Chem Commun 1147
22. Miller MD, Ginsberg EJ, Thompson D (1993) Polym J 25:807
23. Jones RG, Budnik U, Holder SJ, Wong WKC (1996) Macromolecules 29:8036
24. Holder SJ, Achilleos M, Jones RG (2006) J Am Chem Soc 128:12418
25. Gauthier S, Worsford DJ (1989) Macromolecules 22:2213
26. Kim HK, Matyjaszewski K (1988) J Am Chem Soc 110:3321
27. Matyjaszewski K, Chen YL, Kim HK (1988) Inorganic and organometallic polymers. In: ACS Symp Series. ACS, Washington DC
28. Worsfold DJ (1988) Inorganic and organometallic polymers. In: ACS Symp Series. ACS, Washington DC
29. Gauthier S, Worsfold DJ (1990) Mechanistic studies in polysilane polymerization. In: Silicon-based polymer science: a comprehensive resource. In: ACS Series. ACS, Washington DC
30. Matyjaszewski K (1987) Polym Prepr 28:224
31. Sakamoto K, Obata K, Hirata H, Nakajima M, Sakurai H (1989) J Am Chem Soc 111:7641
32. Raabe G, Michl J (1989) The chemistry of organic silicon compounds, Part 2 Wiley, Chichester
33. Roark DN, Peddle GJD (1972) J Am Chem Soc 94:5837
34. Sakamoto K, Yoshida M, Sakurai H (1990) Macromolecules 23:4494
35. Sakurai H, Yoshida M (2000) Synthesis of polysilanes by new procedures. Part 1: Ring opening polymerisations and the polymerisation of masked disilenes. In: Silicon-containing polymers. Kluwer, Dorndrecht
36. Sakurai H, Sakamoto K, Funada Y, Yoshida M (1993) Polym Prepr 34:218
37. Sanji T, Kawabata K, Sakurai H (2000) J Organomet Chem 611:32
38. Sanji T, Isozaki S, Yoshida M, Sakamoto K, Sakurai H (2003) J Organomet Chem 685:65
39. Sakurai H, Honbori R, Sanji T (2005) Organometallics 24:4119
40. Li J, Li Z, Tang HD, Zeng HY, Qin JG (2003) J Organomet Chem 685:258
41. Sakamoto K, Yoshida M, Sakurai H (1984) Polymer 35:4990
42. Chandrasekhar V (2005) Polysilanes and other silicon containing Polymers. In: Inorganic and organometallic polymers. Springer, Berlin Heidelberg New York
43. Cypryrk M, Gupta Y, Matyjaszewski K (1991) J Am Chem Soc 113:1046
44. Suzuki M, Kotani J, Gyobu S, Kaneko T, Saegusa T (1994) Macromolecules 27:2360
45. Koe JR, Powell DR, Buffy JJ, Hayase S, West R (1998) Angew Chem 37:1441
46. Grimmond BJ, Corey JY (2002) Inorg Chim Acta 330:89
47. Imori T, Tilley TD (1994) Polyhedron 13:2231
48. Banovetz JP, Suzuki H, Waymouth RM (1993) Organometallics 12:4700
49. Hashimoto H, Obara S, Kira M (2000) Chem Lett 188
50. Obora Y, Tanaka M (2000) J Organomet Chem 595:1
51. Mu Y, Harrod JF, John F (1991) Inorg Organomet Oligomers Polym, Proc IUPAC Symp Macromol 33:23
52. Rosenberg L, Kobus DN (2003) J Organomet Chem 685:107
53. Aitken C, Harrod JF, Samuel E (1985) J Organomet Chem 279:C11
54. Shankar R, Saxena A, Brar AS (2001) J Organomet Chem 628:262
55. Hengge E, Gspaltl P, Pinter E (1996) J Organomet Chem 521:145
56. Woo HG, Tilley TD (1989) J Am Chem Soc 111:8043
57. Corey JY, Zhu XH (1992) J Organomet Chem 439:1
58. Corey JY (2004) Adv Organomet Chem 51:1
59. Dioumaev VK, Harrod JF (1996) J Organomet Chem 521:133
60. Grimmond BJ, Corey JY (2000) Organometallics 19:3776
61. Gray GM, Corey JY (2000) Synthesis of polysilanes by new procedures. Part 2: Catalytic dehydropolymerisation of hydrosilanes. In: Silicon-containing polymers. Kluwer, Dordrecht
62. Shaltout RM, Corey JY (1995) Tetrahedron 51:4309
63. Shaltout RM, Corey JY (1995) Main Group Chem 1:115
64. Minato M, Matsumoto T, Ichikawa M, Ito T (2003) Chem Commun 2968

65. Kim B, Woo HG, Kim W, Li H (2006) J Chem Technol Biotechnol 81:782
66. Ojima I, Inaba S, Kogure T, Nagai Y (1973) J Organomet Chem 55:C7
67. Berris C, Diefenbach SP (1992) U.S. 5003100 CAN 115:115263
68. Fontaine FG, Kadkhodazadeh T, Zargarian D (1998) Chem Commun 1253
69. Fontaine FG, Zargarian D (2002) Organometallics 21:401
70. Tanaka M, Bannu P (1998) JP 10 67859 CAN 128:244528
71. Chauhan BPS, Shimizu T, Tanaka M (1997) Chem Lett 785
72. Sakakura T, Lautenschlager HJ, Nakajima M, Tanaka M (1991) Chem Lett 913
73. Kobayashi T, Sakakura T, Hayashi T, Yamura M, Tanaka M (1992) Chem Lett 1157
74. Aitken CT, Harrod JF, Samuel E (1986) J Am Chem Soc 108:4059
75. Tilley TD (1993) Acc Chem Res 26:22
76. Dioumaev VK, Harrod JF (1997) Organometallics 16:1452
77. Dioumaev VK, Harrod JF (1996) J Organomet Chem 521:133
78. Lunzer F, Marschner C, Landgraf S (1998) J Organomet Chem 568:253
79. Ishifune M, Kashimura S, Kogai Y, Fukuhara Y, Kato T, Bu HB, Yamashita N, Murai Y, Murase H, Nishida R (2000) J Organomet Chem 611:26
80. Wesson JP, Williams TC (1981) J Polym Sci A Polym Chem 19:65
81. Fogarty HA, Ottosson CH, Michl J (2000) J Mol Struct 556:105
82. Fogarty HA, Ottosson CH, Michl J (2000) J Mol Struct Theochem 506:243
83. Jones RG, Wong WKC, Holder SJ (1998) Organometallics 17:59
84. Yajima S, Hayashi J, Omori M (1975) Chem Lett 931
85. Michl J (1990) Acc Chem Res 23:127
86. Michl J, West R (2000) Structure and spectroscopy of polysilanes. In: Silicon containing polymers. Kluwer, Dordrecht
87. West R (1982) Organopolysilanes. In: Comprehensive organometallic chemistry. Pergamon, Oxford
88. Miller DR (1989) Angew Chem Int Ed Engl 28:1733
89. West R (1986) J Organomet Chem 300:327
90. Yajima S, Hasegawa Y, Hayashi J, Iimura M (1978) J Mater Sci 13:2569
91. West R (1981) J Am Chem Soc 103:7352
92. Orefice RL, Arnold JJ, Miller TM, Zamora MP, Brennan AB (1997) Polym Prepr 38:157
93. Matsuura K, Miura S, Naito H, Inoue H, Matsukawa K (2003) J Organomet Chem 230:6851
94. West R, Ikuo N, Xing-Hua Z (1984) Polym Prepr 25:4
95. Langguth K (1995) Ceram Int 21:237
96. Langguth K, Bockhle S, Müller E, Röwer G (1995) J Mater Sci 30:5973
97. Okamoto K, Shinohara M, Yamanishi T, Miyazaki S, Hirose M (1994) Appl Surf Sci 79:57
98. Kotaro Y, Tazawa S, Kitsuno Y, Kawasaki K (1992) JP 6191821A2 CAN 121:219559
99. Fukujama K, Kitsuno Y, Sakawaki A, Takadera T, Kotaro Y (1998) JP 2000031066A2 CAN 132:95820
100. Yamamoto H, Takeuchi Y, Matsuki Y, Kato H, Hayakawa H, Endo M (2002) JP 2004186320A2 CAN 141:91765
101. Aoki T (2004) JP 2005219981A2 CAN 143:204477
102. Harrah LH, Ziegler JM (1985) J Polym Sci Polym Lett Ed. 23:209
103. Trefonas P, Damewood JR, West R, Miller RD (1985) Organometallics 4:1318
104. Peinado C, Alonso A, Catalina F, Schnabel W (2000) Macromol Chem Phys 201:1156
105. Matsuura K, Matsukawa K, Kawabata R, Higashi N, Niwa M, Inoue H (2002) Polymer 43:1549
106. Pyun J, Matyjaszewski K (2001) Chem Mater 12:3436
107. Hamada Y, Tabei E, Mori S, Yamamoto Y, Noguchi N, Aramata M, Fukushima M (1998) Synth Metal 97:273
108. Sakurai Y, Okuda S, Nagayama N, Yokoyama M (2001) J Mater Chem 11:1077
109. Sakurai Y, Okuda S, Nishiguchi H, Nagayama N, Yokoyama M (2003) J Mater Chem 13:1862
110. Hashimoto K, Nomura N (1991) JP 03139650 A CAN 116:245278
111. Kido J, Nagai K, Okamoto Y, Skotheim T (1991) Appl Phys Lett 59:2760
112. Suzuki H, Meyer H, Simmerer J, Yang J, Haarer D (1993) Adv Mater 5:743

113. Kido J, Nagai K, Okamoto K (1993) J Alloys Compd 192:30
114. Suzuki H, Meyer H, Hoshino S, Haarer D (1995) J Appl Phys Lett 78:2648
115. Suzuki H, Hoshino S (1996) J. Appl Phys Lett 79:8816
116. Kamata N, Ishii R, Tonsyo S, Terunuma D (2002) Appl Phys Lett 81:4350
117. Seoul C, Park J, Lee J (2003) Polym Prepr (Am Chem Soc, Div Polym Chem) 44:435
118. Hoshino S, Suzuki H (1996) Appl Phys Lett 69:224
119. Fujii A, Yoshimoto K, Yoshido M, Ohomori Y, Moshino K (1995) Jpn J Appl Phys 34:L1365
120. Ebihara K, Koshihara S, Miyazawa T, Kira M (1996) Jpn J Appl Phys 35:L1278
121. Hattori R, Sugano T, Fujiki T (1996) Jpn J Appl Phys 35:L1509
122. Xu Y, Fujino T, Naito H, Oka K, Dohmaru T (1998) Chem Lett 299
123. Yuan C, Hoshino S, Toyoda S, Suzuki H, Fujiki M, Matsumoto N (1997) Appl Phys Lett 71:3326
124. Tokito S, Shirane K, Kamachi M (2003) WO 2003/092334 CAN 139:351099
125. Kanai M, Tanaka H, Sako S (1991) JP 03181184 A CAN 116:162057
126. Kanai M, Tanaka H, Sakou H (1991) DE 4039519 A1 CAN 116:95256
127. Lee J, Seoul C, Park J, Youk JH (2004) Synth Metal 145:11
128. Rybak A, Jung J, Ciesielski W, Ulanski J (2006) Mater Sci 24:527
129. Haga Y, Harada Y (2001) Jpn J Appl Phys 1 40:855
130. Fukushima M, Aramata M, Mori S (1998) JP 3275736 B2 CAN 129:41835
131. Yamaguchi F, Ueda M, Fujisawa K (1999) JP 11012362 A CAN 130:154129
132. Furukawa M, Kaimoto T, Matsuo S (2004) JP 2004063238 CAN 140:184763
133. Ohshita J, Kangai S, Yoshida H, Kunai A, Kajiwara S, Ooyama Y, Harima Y (2007) J Organomet Chem 692:801
134. Nakayama T, Naito K (1998) JP 10319450 A CAN 130:87978
135. Nishizawa H, Uchikoga S, Shuichi N, Yoshihiko H, Hayase S (1993) EP 528662 A1 CAN 119:215411
136. Ogier SD, Veres J, Zeidan M (2007) WO 2007082584 A1 CAN 147:224628
137. (1992) JP 5275695 A2 CAN 120:313472
138. Okumoto H, Yatabe T, Richter A, Peng J, Shimomura M, Kaito A, Minami N (2003) Adv Mater 15:716

Conjugated Organosilicon Materials for Organic Electronics and Photonics

Sergei A. Ponomarenko and Stephan Kirchmeyer

Abstract In this chapter different types of conjugated organosilicon materials possessing luminescent and/or semiconducting properties will be described. Such macromolecules have various topologies and molecular structures: linear, branched and hyperbranched oligomers, polymers, and dendrimers. Specific synthetic approaches to access these structures will be discussed. Special attention is devoted to the role of silicon in these structures and its influence on their optical and electrical properties, leading to their potential application in the emerging areas of organic and hybrid electronics.

Keywords Anthradithiophene · Dendrimer · Electroluminescence · Oligothiophene · Organic field-effect transistor (OFET) · Organic light-emitting diode (OLED) · Organic solar cells · Pentacene · Photoluminescence · Poly(1 · 4-phenylene vinylene) · Silafluorene · Silole

Contents

1	Introduction	36
2	Linear Conjugated Organosilicon Oligomers	37
	2.1 Silicon-Containing Thiophene Oligomers	38
	2.2 Organosilicon Oligoacene Derivatives	43
	2.3 Silole-Based Oligomers	50
	2.4 Silicon Analogs of Oligo(p-Phenylenevinylene)s	57
3	Branched Conjugated Organosilicon Oligomers	59

S.A. Ponomarenko
Enikolopov Institute of Synthetic Polymeric Materials of Russian Academy of Sciences (ISPM RAS), Profsoyuznaya st. 70, Moscow 117393, Russia
e-mail: ponomarenko@ispm.ru

S. Kirchmeyer (✉)
H.C. Starck Clevios GmbH, Chempark Leverkusen, Building B 202, Leverkusen 51368, Germany
e-mail: stephan.kirchmeyer@hcstarck.com

4	Conjugated Organosilicon Dendrimers	64
5	Hyperbranched Conjugated Organosilicon Polymers	70
6	Linear Conjugated Organosilicon Polymers	74
	6.1 Polymers with Silicon Atoms in the Side Chains	74
	6.2 Silanylene-Containing Polymers	83
	6.3 Silol-Containing Polymers	88
7	Conclusions and Outlook	98
References		99

Abbreviations

[C70]PCBM	([6,6]-Phenyl C71-butyric acid methyl ester)
2T	$2,2'$-Bithiophene
3AC	Anthracene
3D	Three-dimensional
3T	$2,2':5',2''$-Terthiophene
4AC	Tetracene
4T	$2,2':5',2'':5'',2'''$-Quaterthiophene
5AC	Pentacene
5T	$2,2':5',2'':5'',2''':5''',2''''$-Quinquethiophene
6T	$2,2':5',2:5'',2''':5''',2'''':5'''',2'''''$-Sexithiophene
7T	$2,2':5',2'':5'',2''':5''',2'''':5'''',2''''':5''''',2''''''$-Septithiophene
η_{EL}	External electroluminescence quantum efficiency
Φ_F	Luminescence quantum yield
Ac	Acetyl
ADT	Anthradithiophene
AFM	Atomic force microscopy
AIE	Aggregation induced emission
Alq$_3$	Tris(8-quinolinolato) aluminum(III) complex
BS	Dibenzosilole
Bu	n-Butyl
tBu	$tert$-Butyl
n-BuLi	n-Butyl lithium
t-BuLi	$tert$-Butyl lithium
CEE	Cooling-enhanced emission
CIE	International Commission on Illumination
CV	Cyclic voltammogram
Cz	Carbazolyl
D–A complex	Donor–acceptor complex
Dec	n-Decyl
DFT	Density functional theory
DMS	Dimethylsilyl
DMSO	Dimethyl sulfoxide
DSC	Differential scanning calorimetry
EDOT	3,4-Ethylenedioxythiophene

EL	Electroluminescence
Et	Ethyl
ET	Electron transport
eV	Electron volt
Fe(acac)$_3$	Iron(III) acetylacetonate
FET	Field-effect transistor
FF	Fill factor
HB	Hyperbranched
Hex	*n*-Hexyl
HOMO	Highest occupied molecular orbital
HTL	Hole-transporting layer
IP	Ionization potential
I_{sc}	Short circuit current
ITO	Indium tin oxide
LDA	Lithium di(iso-propyl)amide
LEC	Light-emitting electrochemical cell
LOPV	Ladder oligo(*p*-phenylenevinylene)
LUMO	Lowest unoccupied molecular orbital
MALDI–TOF	Matrix assisted laser desorption ionization–time-of-flight mass spectrometry
Me	Methyl
MEH-PPV	Poly[2-methoxy-5-(2′-ethyl-hexyloxy)-1,4-phenylene vinylene]
M_n	Number-averaged molecular weight
M_w	Weight-averaged molecular weight
NiCl$_2$(dppe)	1,2-Bis(diphenylphosphino)ethane nickel(II) chloride
NIR	Near infrared
NPB	*N,N*′-Bis(1-naphthyl)-*N,N*′-diphenylbenzidine
NPD	4,4′-Bis[*N*-1-naphthyl-*N*-phenylamino]-biphenyl
Oct	*n*-Octyl
OEt	Ethoxy
OFET	Organic field-effect transistor
OLED	Organic light-emitting diode
OligoT	Oligothiophene
OMe	Methoxy
OPV	Organic photovoltaics
OTFT	Organic thin film transistor
P3HT	Poly(3-hexylthiophene)
PBD	2-(4-Biphenylyl)-5-(4-*tert*-butylphenyl)-1,3,4-oxadiazole
PCBM	[6,6]-Phenyl C61-butyric acid methyl ester
PCE	Power conversion efficiency
PEDOT	Poly(3,4-ethylenedioxythiophene)
PF	Polyfluorene
Ph	Phenyl
PL	Photoluminescence
PMMA	Poly(methyl methacrylate)

PPV	Poly(1,4-phenylene vinylene)
ppy	2-Phenylpyridine
PS	Polystyrene
PSS	Poly(styrene sulfonate)
PTV	Polythiophenevinylene
PVK	Poly(*N*-vinyl carbazole)
Py	Pyridyl
SAM	Self-assembled monolayer
SAMFET	Self-assembled monolayer field-effect transistor
SBAr	Silicon-bridged biaryl
SCE	Saturated calomel electrode
SiF	Silafluorene
T	Thienyl
TES	Triethylsilyl
TGA	Thermal gravimetric analysis
TIPS	Triisopropylsilyl
TMS	Trimethylsilyl
TPD	*N*,*N'*-Diphenyl-*N*,*N'*-di(*m*-tolyl)biphenyl-4,4'-diamine
TPS	Triphenylsilyl
TPSppy	2-(4'-(Triphenylsilyl)biphenyl-3-yl)pyridine
TS	Dithienosilole
TVS	Trivinylsilyl
Und	*n*-Undecyl
UV-vis	Ultraviolet-visible
V_{oc}	Open-circuit voltage

1 Introduction

Organic electronics has been a fast growing field of science and technology since the beginning of the twenty-first century [1, 2]. It is designed for cost efficient and flexible lightweight large area devices, the basic units of which are organic field-effect transistors (OFETs), also known as organic thin film transistors (OTFTs) [3], organic light-emitting diodes (OLEDs) [4], and organic photovoltaic cells (OPVs) or solar cells [5–7]. They can also be combined with sensing elements [8] lasers, etc. In general, these devices are not intended to outperform contemporary inorganic (silicon) electronics. They will have lower performance due to material limitations, e.g., reduced charge carrier mobility in OTFT will limit the ability to process high frequencies. However, organic electronics will have its own market niche based on its flexibility, low weight, and, eventually, low cost as a result of substitution of expensive lithography, wet processing, and other technologies used in conventional silicon electronics by cheap roll-to-roll, ink-jet, gravure, or other printing techniques.

A major difference between silicon and organic electronics concerns the electronic structure of the semiconducting materials employed. Silicon as semiconductor is doped with elements like boron or phosphorus, which determines the type of charge carrier by introduction of excess electrons or holes. In organic intrinsic (undoped) semiconductors HOMO–LUMO energy levels and their relative positions to the corresponding energy levels of the electrodes – ionization potentials (IPs) determine the type of the main charge carriers. As a consequence, the chemical structure of the organic semiconductor strongly influences whether an organic semiconductor is p-type (hole conducting) or n-type (electron conducting).

Conjugated organosilicon materials with semiconducting properties resemble a broad class of "organic" rather than silicon semiconductors. As typical for intrinsic organic semiconductors, the introduction of silicon atoms into the conjugated organic structure changes the HOMO–LUMO energy levels and influences their optical and semiconducting properties. Prerequisite is a direct covalent linkage between the silicon atom and the organic conjugated core. Nevertheless, in addition to conjugated structural parts, most organic semiconductors contain nonconjugated groups, which influence solubility, aggregation, crystallization, and film-forming properties. Therefore, another option is to attach organosilicon fragments to units which do not interact electronically with the chromophore but improve self-organization and morphology of the semiconductor during processing. Such an approach might be especially important for solution processing techniques. Direct conjugated linkage of silicon to chromophores might also impact the morphology, especially in the case of (hyper)branched or dendritic molecules, where silicon acts as the branching centers, but in comparison to electronic effects morphology effects are usually less dominant.

In the following, different types of conjugated organosilicon materials will be discussed which differ in topology and molecular structure: linear, branched, and hyperbranched (HB) polymers, oligomers, and dendrimers. This will comprise all materials containing conjugated organosilicon or organic units as well as silicon atoms or organosilicon fragments in the same molecular structure without direct electronic interaction.

2 Linear Conjugated Organosilicon Oligomers

Among linear conjugated organosilicon oligomers, two classes of molecules can be distinguished, which were widely investigated and show promising semiconducting and luminescent properties: (1) silicon-containing thiophene oligomers and (2) organosilicon oligoacene derivatives. It should be noted that oligothiophenes and oligoacenes (especially pentacene) themselves are among the best organic semiconductors [9, 10]. Modification with silicon will add specific features to their properties which will be discussed in the following. Apart from that, the introduction of silicon into aromatic structures creates a new building block which can be used to construct conjugated oligomers and polymers: a silacyclopentadiene also called "silole." Oligomers based on silole itself as well as its most important derivatives, such as dibenzosilole (BS) and dithienosilole (TS) and more recently developed silicon analogs of oligo(p-phenylenevinylenes), also open opportunities for new electronic properties.

2.1 Silicon-Containing Thiophene Oligomers

The synthesis of silicon-containing thiophene oligomers was comprehensively reviewed in 1997 [11]. In the scope of this chapter we will consider recent and most important examples of these materials and their application in organic electronics and photonics.

In contrast to benzene, thiophene itself shows no luminescence, but it's oligomers starting from 2,2′-bithiophene (2T) are luminescent and may find applications in organic photonics and electronics. Among α,α'-oligothiophenes the luminescence quantum yield (Φ_F) increases with increasing conjugation length and is accompanied by a significant red shift of the luminescence maximum. The addition of silylene or disilylene units to the α,α'-position will significantly increase Φ_F of linear 2T- and 3T-derivatives, but decreases Φ_F for 6T-derivatives as compared to pristine oligothiophenes. Hadziioannou and coworkers have reported a series of trimethylsilyl- and pentamethyldisilanyl-oligothiophenes (Fig. 1) [12]. The fluorescence quantum yield Φ_F for these molecules reached 23% for Me$_3$Si-T2-Si$_2$Me$_5$ compared to 1–2% measured for 2T; in contrast, the Φ_F for 6T (32%) exceeded the value for Me$_3$Si-T6-Oct$_2$SiMe$_3$ (25%) (Fig. 2). Bearing in mind that the fluorescence maxima of oligothiophenes strongly depend on their conjugation length, substitution of the oligothiophenes with organosilicon units may help to tune its spectral characteristics and efficiency.

α-Trimethylsilyl groups can be easily cleaved from oligothiophene units either chemically to the corresponding oligomers [13] or electrochemically to yield polymers [14, 15]. Silyl substituents with longer alkyl groups seem to be more

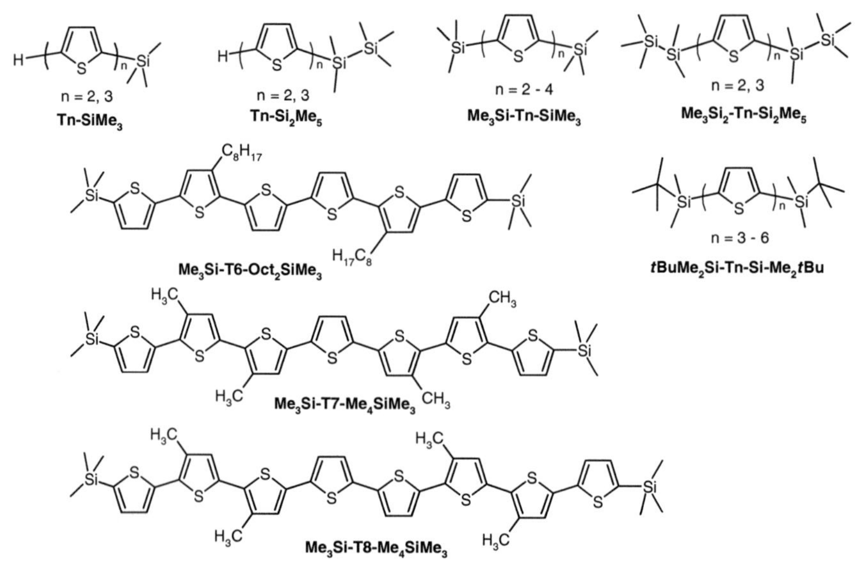

Fig. 1 Silylated oligothiophenes

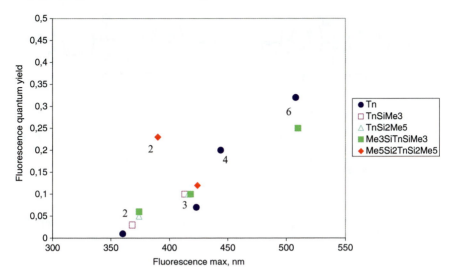

Fig. 2 Fluorescence quantum yield Φ_F vs fluorescence wavelength maximum for a series of oligothiophenes and their organosilicon derivatives (based on data from [12]). Number of conjugated thiophene units is marked on the chart near the corresponding data set

stable: a series of α,α′-bis(dimethyl-*tert*-butylsilyl) oligothiophenes tBuMe$_2$Si-Tn-Si-Me$_2t$Bu ($n = 3$–6) was reported by Barbarella et al. (Fig. 1) [16]. The quater- and sexithienylsilanes were prepared by Fe(acac)$_3$ (iron(III) acetylacetonate) mediated oxidative coupling of corresponding lithium derivatives, while quinquethienylsilane was obtained by the Stille reaction. All oligomers were highly soluble in most organic solvents which allowed their easy synthesis and purification. Vacuum-evaporated thin films of the oligomers with $n = 4$–6 displayed field-effect transistor activity, with charge mobilities increasing with the substrate deposition temperatures. The best OFET performance was achieved from the quinquethiophenesilane tBuMe$_2$Si-T5-Si-Me$_2t$Bu, which was characterized to have a mobility up to 2×10^{-4} cm^2 V^{-1} s^{-1} and an on/off ratio greater than 10^3 combined with good device stability in air for several months. The lower mobility of these oligomers compared to α,α′-dialkyloligothiophenes (i.e., 0.5 cm^2 V^{-1} s^{-1} for Dec-6T-Dec [17] or 1.1 cm^2 V^{-1} s^{-1} for Et-6T-Et) [18] can be explained by a significant steric hindrance caused by bulky triisopropylsilyl groups. This is evidenced by unusual triclinic crystallization of these compounds, in which the conjugated backbone shows strong deviation from coplanarity [19].

Many examples reported in the literature evidence an improved solubility of organosilicon modified oligothiophenes when compared to their unmodified derivatives. It was found that α-trimethylsilyl substituents will increase the solubility of bi-, ter-, and quaterthiophenes. More bulky α-dimethyl-*tert*-butylsilyl substituents improve the solubility up to a chromophore length of sexithiophenes. In order to synthesize soluble oligothiophenes with longer chromophores (i.e., septi- and octithiophenes, see Fig. 1), additional modifications will be necessary, e.g., by adding

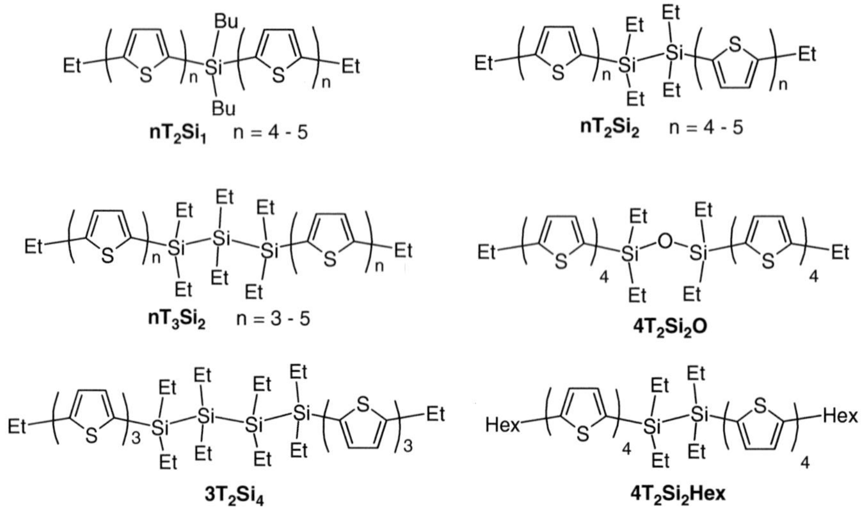

Fig. 3 Oligothiophenesilane dimers

several additional 3-methylsubstituents to the thiophene rings. However, adjacent trimethylsilyl groups to 3-methylthien-2,5-diyl units will quickly cleave from the core unit [13].

Ohshita et al. have reported on a series of oligothiophenesilane dimers bridged by mono-, bi-, or trisilanylene units with the intention to trace the influence of σ–π conjugation between the oligothiophene units and the silanylene bridge on the semiconducting properties of these materials (Fig. 3) [20, 21]. The FET mobilities of vapor-deposited films in top contact OFETs were found to be enhanced with the oligothiophene chain length and reached 5.1×10^{-2} cm^2 V^{-1} s^{-1} for 5T$_2$Si$_3$. This tendency follows the trend of unmodified oligothiophenes, which behave similarly [22]. In summary the influence of the silylene units is not well pronounced. On the one hand, in a series of quinquethiophenesilane dimers with different silylene bridges (5T$_2$Si$_x$) no clear influence of the Si-chain length on the charge carrier mobility was found. On the other hand, in a series of quaterthiophenes the charge carrier mobility of films was increased in the row 4T$_2$Si$_3$ < 4T$_2$Si$_2$O < 4T$_2$Si$_2$. This indicates a σ–π conjugation between the oligothiophene units and the silanylene bridge of lesser importance than it was expected, and a major influence of other factors, such as film morphology and molecular alignment in the solid film. Quaterthiophenesilane dimers were sufficiently soluble in organic solvents in order to make solution-processed OFETs by spin-coating. While 4T$_2$Si$_3$Hex did not yield any FET mobility, the mobilities of wet coated films of 4T$_2$Si$_3$ were an order of magnitude higher than for vapor evaporated films ($\mu = 4.1 \times 10^{-3}$ and 2.9×10^{-4} cm^2 V^{-1} s^{-1}, respectively). Quinquethiophenesilanes were hardly soluble in organic solvents, making it impossible to process films by spin-coating.

Another type of linear silicon-containing thiophene oligomers are monochlorosilyl derivatives of dialkyloligothiophenes Cl-Si-Spacer-OligoT-End (Fig. 4) [23, 24].

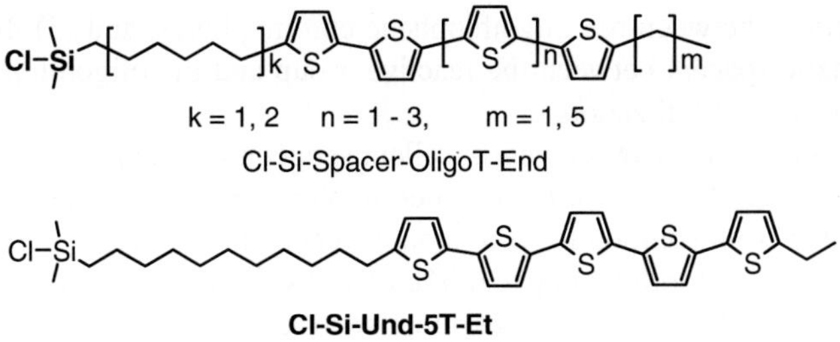

Fig. 4 Monochlorosilyl derivatives of dialkyloligothiophenes: general structure and the most promising material

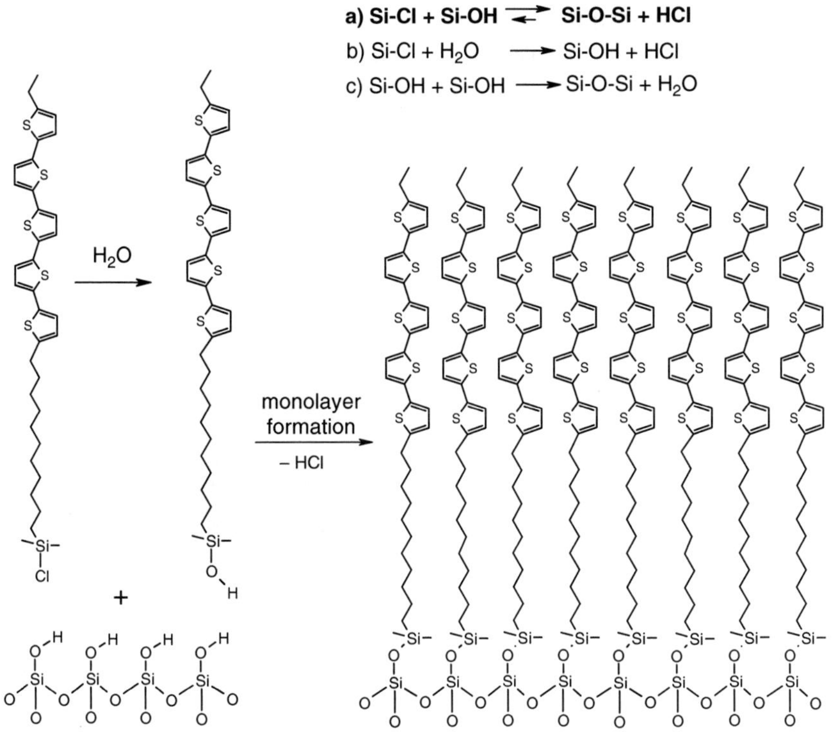

Scheme 1 Schematic representation of a SAM formation by Cl-Si-Und-5T-Et on SiO$_2$

Each of these structures contains the reactive monochlorosilyl group Si–Cl, attached to a semiconducting oligothiophene unit OligoT via flexible aliphatic spacers. Such molecular structures allow crystalline self-assembling monolayers (SAMs) to form on dielectric hydroxylated silicon dioxide [25] or even on oxidated polymer surfaces [26] by self-assembly from solution. A schematic representation of the SAM formation is shown in Scheme 1. Obviously the following factors play crucial roles: (1) the reversibility of the reaction of monochlorosilane with silanole, (2) a strong π–π interactions between the oligothiophene chromophores, and (3) the presence of the aliphatic spacers between the reactive group and the oligothiophene cores, which facilitates crystallization.

The oligothiophene SAMs reveal excellent semiconducting properties similar to those of bulk oligothiophenes. Even under ambient conditions SAM semiconductors assembled from solution to form monolayers and yield field-effect transistors (SAMFETs) with a mobility of up to $0.04\,\mathrm{cm^2\,V^{-1}\,s^{-1}}$ and on/off ratio up to 1×10^8 for 40 μm channel length devices [25]. In a series of the molecules with the structure Cl-Si-Spacer-OligoT-End the mobility increased by a factor of 10 from 4T to 5T oligothiophene units and by a factor of 2–3 with increasing spacer length [24]. It should be noted that the first oligothiophene SAMFETs prepared from nonsilicon-containing bifunctional ter- or quaterthiophenes on organosilicon modified silica or alumina surfaces worked for submicron channel length transistors only, and only in a few cases showed reasonable mobility: $0.0035\,\mathrm{cm^2\,V^{-1}\,s^{-1}}$ for quaterthiophene and $8 \times 10^{-4}\,\mathrm{cm^2\,V^{-1}\,s^{-1}}$ for terthiophene, with the on/off ratio up to 1,800 [27]. However, the more recent approach using Cl-Si-Und-5T-Et assembled on silica was proven to be highly efficient: it was possible to make fully functional SAMFET-based functional 15-bit Code Generators containing over 300 SAMFETs (Fig. 5) with all SAMFETs working simultaneously and with equal (or at least very close) electrical characteristics [28].

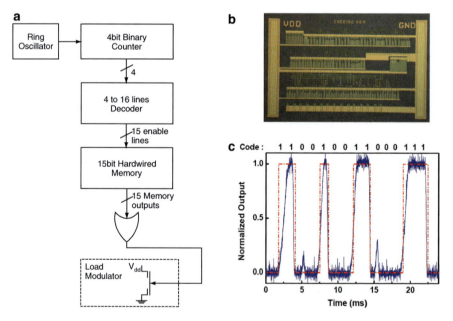

Fig. 5 SAMFET-based functional 15-bit code generator: block diagram (**a**), optical microphotograph (**b**), and output characteristics (**c**). The bit rate was about $1\,\mathrm{kBit\,s^{-1}}$ at a supply voltage of −40 V. The outputted code is indicated at the top and by the *red line* [28]

In all types of linear Si-containing thiophene oligomers, a strong influence of silicon atoms on electronic and optical properties of the oligothiophenes was found, especially for oligomers with shorter chromophores such as bi- and terthiophenes. The introduction of silicon substituents has a more pronounced influence on their solubility and thin film morphology, independent of the presence of electronic coupling between the Si atom and oligothiophene core. Unique properties of organosilicon SAM oligothiophenes pave the way to bottom up organic electronics.

2.2 Organosilicon Oligoacene Derivatives

Various oligoacene derivatives have been modified with organosilicon units by Anthony and other groups [29]. Unlike oligothiophenes with the most reactive protons at the α positions, oligoacenes have reactive sites at the center benzene units, which can be easily modified with various organosilicon groups. In order to release steric interactions between the bulky silane groups and the chromophoric oligoacenes, the silane groups are usually attached via acelylenic extension units.

Anthracene and its derivatives are known for their good luminescence properties. Anthracene can be substituted at 9,10-positions to yield the triisopropylsilylethynyl derivative TIPS-3AC (Fig. 6). This compound crystallizes in films, which can be used as emitter in simple OLED devices showing an intense blue emission with a maximum luminance of $1,000\,\text{cd}\,\text{m}^{-2}$, an efficiency of $1.7\,\text{cd}\,\text{A}^{-1}$ at a luminance of $100\,\text{cd}\,\text{m}^{-2}$ and a bias voltage of 7.8 V [30]. Attachment of two methoxy groups to the chromophore of TIPS-3AC leads to the crystalline compound TIPS-3AC-OMe$_2$. The decreased π–π interaction increases the stacking distance from 3.7 to 5.7 Å, which surprisingly does not significantly alter the OLED performance ($1.4\,\text{cd}\,\text{A}^{-1}$ at a brightness of $100\,\text{cd}\,\text{m}^{-2}$ and bias voltage of 7.0 V). When the triisopropylsilylethynyl substituents are shifted to 1,4-positions in the anthracene

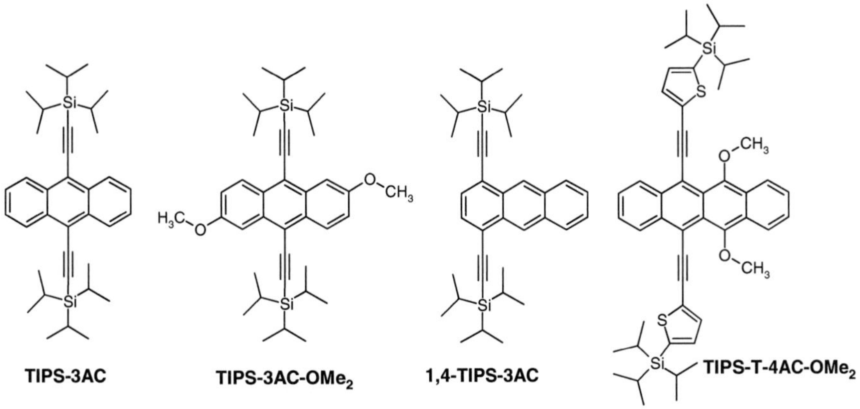

Fig. 6 Triisopropylsilyl-modified oligoacenes for OLEDs

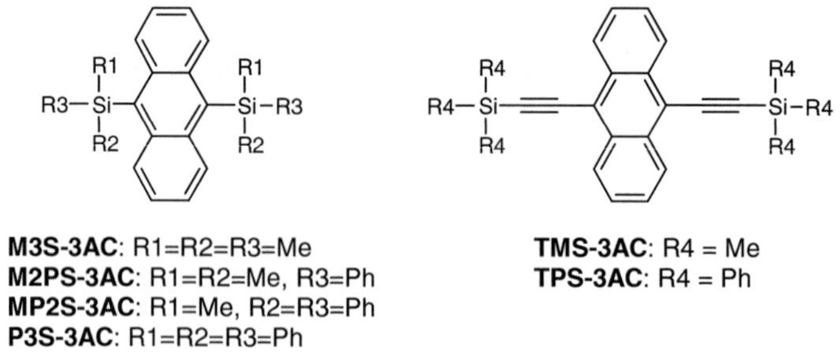

M3S-3AC: R1=R2=R3=Me
M2PS-3AC: R1=R2=Me, R3=Ph
MP2S-3AC: R1=Me, R2=R3=Ph
P3S-3AC: R1=R2=R3=Ph

TMS-3AC: R4 = Me
TPS-3AC: R4 = Ph

Fig. 7 Silylsubstituted anthracene derivatives with bulky phenyl groups

core unit (compound 1,4-TIPS-3AC in Fig. 6) the film is hindered from crystallization, and OLEDs only exhibit a weak green emission (0.4 cd A^{-1} at a brightness of 100 cd m^{-2}). Tetracene derivatives with more bulky organosilicon substituents in the 5,6-positions and electron-donating methoxy groups in the 11,12-positions (compound TIPS-T-4AC-OMe$_2$ in Fig. 6) emit red light in OLEDs [31]. These examples demonstrate how structural design and fine tuning can significantly change phase behavior and electronic properties of the materials that consequently influence the device performance.

Karatsu et al. have reported on photo- and electrooptical investigation of a series of silylsubstituted anthracene derivatives with bulky phenyl groups (Fig. 7) [32]. They showed efficient blue fluorescence with the quantum yield of 77–92% (compared with 36% for the parent anthracene). With an increasing number of phenyl radicals the Stokes shift for phenylsilyl compounds significantly increases from 891 cm^{-1} for M3S-3AC to 1,582 cm^{-1} for P3S-3AC, but slightly decreases for phenylsilylethynyl compounds from 208 cm^{-1} for TMS-3AC to 202 cm^{-1} for TPS-3AC. Multilayered EL devices were prepared using these compounds as a dopant (up to 5%) in a 4,4′ − N,N′-dicarbazolyl-biphenyl (CBP) host which emitted a pure blue color, with the best characteristics for P3S-3AC (CIE coordinates 0.145, 0.155)[1].

Pentacene is a benchmark as semiconductor for thin-film OFETs, showing a mobility in the good quality devices up to 5 cm^2 V^{-1} s^{-1} [33]. However, its drawbacks are insolubility and low oxidation and thermal stability, which may be improved by incorporation of appropriate organosilicon substituents. Pentacenes substituted at the 6,13-positions are easily accessible from pentacenequinone by synthetic methods known since the 1940s [34]. Anthony et al. prepared a series of trialkylsilylethynyl pentacene derivatives from pentacenequinone and corresponding Grignard reagents (Scheme 2) [35, 36]. All derivatives proved to be soluble in common organic solvents, which allow preparation of OFETs by solution processing. The most promising semiconducting properties revealed 6,13-bis(triisopropyl-

---

[1] Coordinated within the CIE 1931 color space chromaticity diagram. CIE – the International Commission on Illumination (abbreviation comes from French "Commission internationale de l'éclairage").

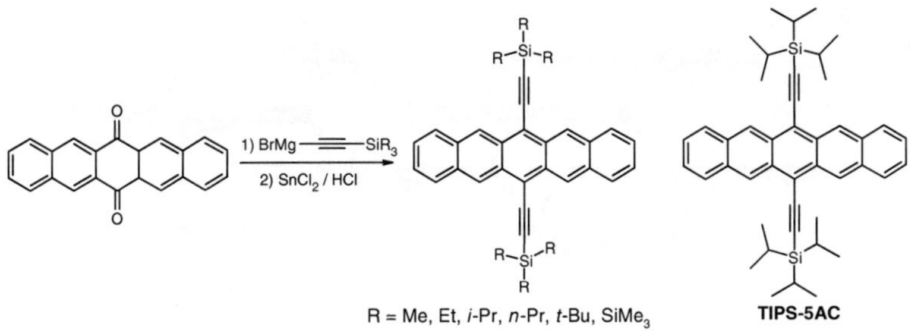

Scheme 2 Synthetic route to organosilicon 6,13-disubstituted pentacene and chemical formula of the most promising derivative for OFETs – TIPS-5AC

silylethynyl) pentacene (TIPS-5AC), although its mobility significantly depended on the preparation method [37]. In the case of thermally evaporated films, the highest hole mobility measured in OFETs was $0.4\,\text{cm}^2\,\text{V}^{-1}\,\text{s}^{-1}$. Solution-deposited TIPS-5AC yields films of significantly higher quality since slow evaporation of the solvent allows the material to self-assemble into large π-stacked arrays. Fast evaporation during spin-casting leads to a lower hole mobility of $0.2\,\text{cm}^2\,\text{V}^{-1}\,\text{s}^{-1}$ with on/off current ratios of 10^6. During drop-casting the solvent is allowed to evaporate slowly and hole mobilities greater than $1\,\text{cm}^2\,\text{V}^{-1}\,\text{s}^{-1}$ and on/off current ratios of greater than 10^7 have been achieved [38]. Electronic properties and electron transfer characteristics of TIPS-substituted oligoacenes have been studied in detail, both experimentally and theoretically [39, 40].

The unique combination of outstanding electrical performance and good solubility of TIPS-5AC was rationalized by favorable 2D π-stacking in the "bricklayer" crystal lattice of this material (Fig. 8a), which is different both from 1D π-stacking in the "slipped-stack" arrangement of some other organosilicon 6,13-disubstituted pentacenes (i.e., triethylsilylethynyl pentacene TES-5AC, Fig. 8b) and from the herringbone structure of pentacene (Fig. 8c). It is well-known that the device performance clearly depends on the crystal packing of the employed semiconductor. Moreover, apparently TIPS-substituents have just the right size for the efficient "bricklayer" packing, since smaller ethyl or larger n-propyl attachment groups lead to 1D "slipped-stack" arrangements, while the largest trimethylsilyl radicals lead to a herringbone structure.

Attempts have been made to deposit TIPS-pentacene from solution as the functional layer in a pentacene/C60 bilayer photovoltaic device. Careful optimization of deposition conditions, optimal concentration of mobile ion dopants, thermal postfabrication annealing, and the addition of an exciton-blocking layer yielded a device with a moderate white-light PCE of 0.52% [41]. Since TIPS-pentacene derivatives rapidly undergo a Diels–Alder reaction with fullerene, the assembly of potentially more efficient bulk-heterojunction photovoltaic devices from TIPS-pentacene and fullerene derivatives were not possible [42]. The energy levels of the TIPS-pentacene-PCBM adduct (PCBM is [6,6]-phenyl C61-butyric acid methyl ester) ineffectively supports the photoinduced charge transfer.

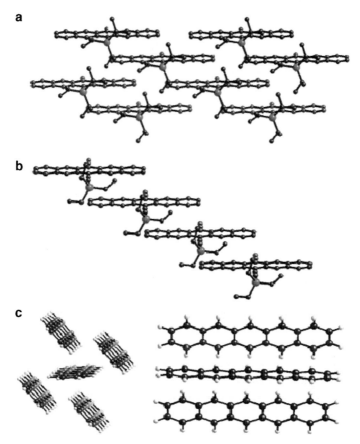

Fig. 8 Solid state ordering of TIPS-5AC (**a**), TES-5AC (**b**), and pentacene (**c**) – reproduced with permission of the American Chemical Society from [29]

Introduction of fluorine substituents into the conjugated core is known as a powerful method to change the polarity of the main charge carriers in organic semiconductors. Some fluorine derivatives of pentacene and TIPS-pentacene have also have been reported (Fig. 9). Solely perfluorinated pentacene 5AC-F14 exhibited n-type behavior with an electron mobility up to $0.22\,cm^2\,V^{-1}\,s^{-1}$ [43, 44]. Contrary to expectation, the partially fluorinated TIPS-pentacene derivatives TIPS-5AC-F4 and TIPS-5AC-F8 showed only p-type mobility, $0.014\,cm^2\,V^{-1}\,s^{-1}$ and $0.045\,cm^2\,V^{-1}\,s^{-1}$, respectively [45]. Devices were prepared by vacuum sublimation and compared to TIPS-5AC (mobility of $0.001\,cm^2\,V^{-1}\,s^{-1}$, deposited under the same conditions). An increasing mobility with an increasing degree of fluorination of TIPS-pentacene was explained by a decreasing π–π interlayer spacing in the crystal packing due to aryl–fluoroaryl interactions (from 3.43 Å for TIPS-5AC to 3.36 Å for TIPS-5AC-F4 and 3.28 Å for TIPS-5AC-F8), while the overall packing arrangement was almost the same for all these compounds.

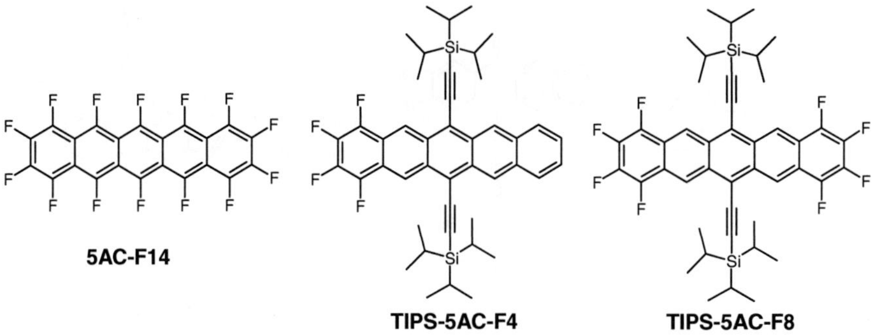

5AC-F14

TIPS-5AC-F4

TIPS-5AC-F8

Fig. 9 Fluorosubstituted pentacene and TIPS-pentacene

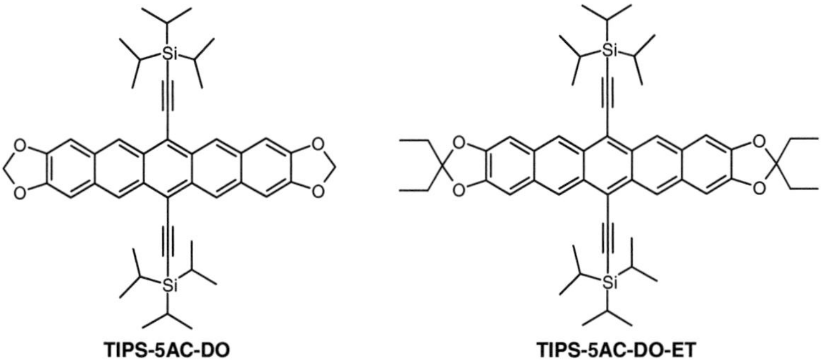

TIPS-5AC-DO

TIPS-5AC-DO-ET

Fig. 10 Dioxolane derivatives of TIPS-pentacene

Some dioxolane derivatives of TIPS-pentacene were successfully applied in OLEDs (Fig. 10). TIPS-5AC-DO used as a guest emitter in the tris(8-quinolinolato) aluminum(III) complex (Alq$_3$) and 4,4'-bis[N-1-naphthyl-N-phenylamino]-biphenyl (NPD) host matrices (common hole-transport materials used in OLEDs) showed efficient energy transfer and bright red photoemission at very low concentrations (less than 0.5 mol%) [46]. However, a higher dopant concentration led to aggregate emissions that motivated the synthesis of the more bulky derivative TIPS-5AC-DO-ET (functionalized with ethyl groups at the dioxolane ring). This compound has an increased chromophore stacking distance in the crystal (5.5 Å vs 3.4 Å for TIPS-5AC-DO) and allowed a concentration up to 2% of dopant in both Alq$_3$ and NPD matrices without formation of aggregates. OLEDs prepared from such composites showed bright red emission with an external electroluminescence quantum yield of 3.3% [47], close to the theoretical maximum and also very close to the highest value reported (3.6%) [48] for a small-molecule red-emissive OLED.

Heterocyclic analogs of pentacene–anthradithiophenes (ADT) are another type of promising semiconductors (with a mobility up to 0.09 cm^2 V^{-1} s^{-1} for vacuum sublimed film of parent ADT and up to 0.15 cm^2 V^{-1} s^{-1} for its dialkyl-substituted analogs), despite the fact that they can only be prepared as mixture of

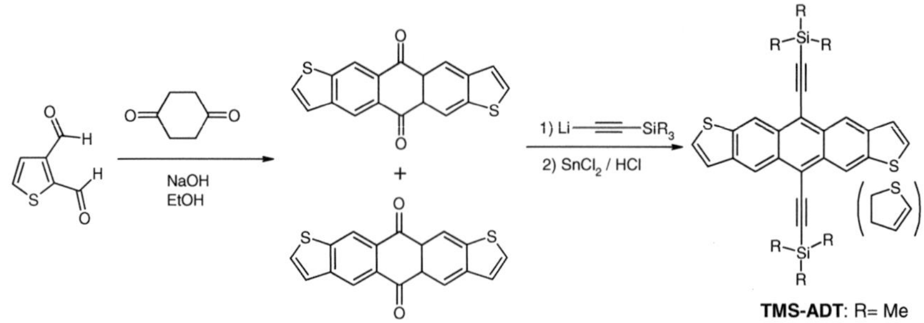

Scheme 3 Synthetic route to silylethynyl-functionalized anthradithiophenes

TMS-ADT: R= Me
TES-ADT: R= Et
TIPS-ADT: R= i-Pr

syn- and *anti-*isomers which cannot be separated [49]. The same problem arises for silylethynyl-functionalized ADTs, a series of which was prepared starting from thiophene-2,3-dicarboxaldehyde and 1,4-cyclohexanedione (Scheme 3) [50]. These compounds showed remarkable solubility when compared to parent ADT, but their semiconducting properties are highly dependent on the substituent groups attached to the central silicon atom. Solution-deposited films of TMS-ADT did not show any field-effect mobility due to lack of π–π-stacking, TIPS-ADT exhibited a relatively low mobility ($<10^{-4}$ cm^2 V^{-1} s^{-1}), while the mobility up to 0.19 cm^2 V^{-1} s^{-1} was reported for TES-ADT [51]. The mobility improved when TES-ADT was deposited by drop-casting, which yielded devices with mobilities as high as 1.0 cm^2 V^{-1} s^{-1} and on/off current ratios of 10^7. This dramatic improvement of semiconducting properties was associated with 2D π-stacking of TES-ADT, similar to TIPS-pentacene.

2,8-Diethyl-5,11-bis(triethylsilylethynyl)anthradithiophene (TES-ADT-ET) was synthesized by a similar synthetic route and successfully used in bulk-heterojunction solar cells in mixture with PCBM [52]. Solvent vapor annealing of these blends leads to the formation of spherulites, which consist of a network of ADT crystallites dispersed in an amorphous matrix primarily of fullerene (Fig. 11). It was shown that the coverage of a device with spherulites directly correlates with its performance. Devices with 82% spherulite coverage reach a PCE of 1%.

Significant improvements in stability and crystallinity were achieved by partial fluorination of silylethynyl-functionalized ADTs (Fig. 12) [53]. These fluorinated materials still behave as p-type semiconductors, but compared to the nonfluorinated derivatives the fluorine introduces a dramatic increase in thermal stability and photostability. TES-ADT-F2 forms highly crystalline films even from spin-cast solutions, leading to devices with maximum hole mobility greater than 1.0 cm^2 V^{-1} s^{-1}. TIPS-ADT-F2 forms large, high-quality crystals that could even serve as a substrate for transistor fabrication. For this compound, a mobility up to 0.1 cm^2 V^{-1} s^{-1} was measured on the free-standing crystals. Recently single-crystal field-effect transistors prepared on the surface of TES-ADT-F2 exhibited outstanding electronic properties: a mobility as high as 6 cm^2 V^{-1} s^{-1}, large current on/off ratios ($I_{on}/I_{off} = 1 \times 10^8$),

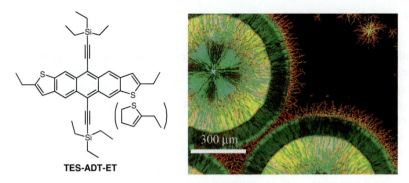

Fig. 11 2,8-Diethyl-5,11-bis(triethylsilylethynyl)anthradithiophene (TES-ADT-ET) and optical image of its blend with PCBM used in efficient photovoltaic devices – reproduced with permission of the American Chemical Society from [52]

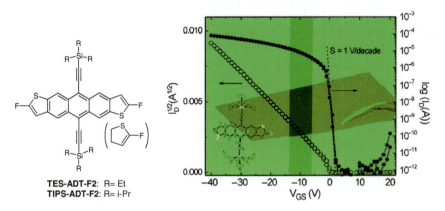

Fig. 12 Partially fluorinated silylethynyl-functionalized anthradithiophenes (*left*). Single crystal of TES-ADT-F2 and its electrical characteristics (*right*) – reproduced with permission of the American Chemical Society from [53]

a small subthreshold slopes ($S = 1\,\text{V}\,\text{dec}^{-1}$), and extremely small hysteresis in the current-voltage characteristics [54]. Optical, fluorescent, and (photo)conductive properties of this and other silylethynyl-functionalized pentacene and ADT derivatives have recently been studied in detail [55]. Soluble silylethynyl-functionalized higher acenes (hexacene and heptacene) [56] and acenedithiophenes with six and seven fused rings [57] have also been reported.

Thus, linear soluble organosilicon oligoacenes have demonstrated a high technological potential for organic electronics. Their application as semiconductors have been demonstrated in OTFTs, OPVs, and in OLEDs. Both the chromophore and the attachments groups influence electronic and morphological properties for the most part in a way that will not be easily predicted and will require subsequent empirical fine tuning of structures and deposition methods to obtain materials with optimal performance.

2.3 Silole-Based Oligomers

Silole (silacyclopentadiene) and its derivatives contain a unique electronic structure leading to excellent photophysical properties [58]. Siloles have a low LUMO energy level combined with a relatively high HOMO level in comparison to other heterocyclic rings commonly used for creation of π-conjugated oligomers and polymers: pyrrole, furan, thiophene, and pyridine (Fig. 13a). The σ*–π* conjugation in the silole ring, arising from interactions of the σ* orbital of the silylene moiety with the π* orbital of the butadiene significantly lowers its LUMO energy (Fig. 13b). As a result 2,5-difunctionalized silole-containing oligomers and polymers exhibit a low band gap when used as structural unit in the conjugated backbone. Indeed, 2,5-disubstituted disilol 2,5-2SEP has an absorption maximum at 417 nm and 2,5-disubstituted quatersilole 2,5–4SEP has a λ_{max} at 443 nm (Fig. 14) [59]. These values are significantly red-shifted compared to a λ_{max} at 301, 351, and 385 nm for bi-, ter-, and quaterthiophenes [12]. In contrast, 1,1-substituted ter- and quatersiloles 1,1–3SMP and 1,1–4SMP have a λ_{max} at 280–290 nm only due to lack of electronic overlap between silicon and the conjugated rings [60].

2,5- and 1,1-Difunctionalized siloles can be prepared by intramolecular reductive cyclization of diethynylsilanes. Starting from these functionalized siloles a number of oligo(2,5-silole)s and oligo(1,1-silole)s have been successfully synthesized [59–61]. However, more often 2,5-difunctional silole monomeric units are combined with other heterocycles, i.e., thiophene, pyrrole, pyridine, etc. A series of thiophene-silole co-oligomers and copolymers have been prepared by Tamao et al. [62]. All oligomers showed a bathochromic (red) shift in the absorption maxima when compared to oligothiophenes, which tends to shift to longer wavelengths with an increasing number of silole units incorporated. In contrast, the conductivity

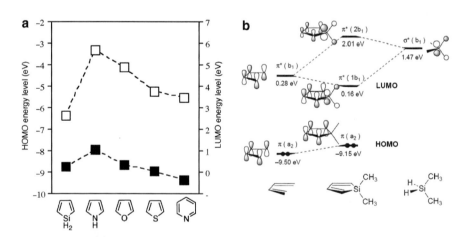

Fig. 13 Relative HOMO and LUMO levels for silole and other heterocycles, based on HF/6-31G* calculations (**a**) and orbital correlation diagram for 1,1-dimethylsilole, based on the PM3 calculations (**b**). Reproduced by permission of The Royal Society of Chemistry [58]

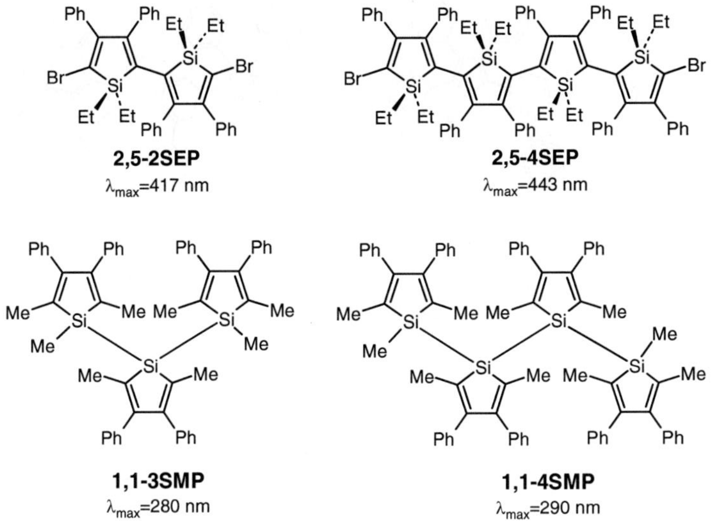

Fig. 14 Examples of oligo(2,5-silole)s and oligo(1,1-silole)s

of the resulting oligomers tends to increase with the number of thiophene units, reaching a maximum of 2.4 S cm^{-1} after doping with iodine. X-ray structural analysis revealed a high coplanarity of the thiophene and silole rings in the co-oligomers. This is not the case for co-oligomers of silole with the more electron rich pyrrole, in which a highly twisted conformation of the three rings of 2,5-dipyrrolylsilole was found [63].

Silole-containing oligomers have been used as electron transporting layers and as emitters to build efficient OLEDs (Fig. 15) [64, 65]. In OLEDs the performance of 2,5-di(2-pyridyl)-1,1-dimethyl-3,4-diphenylsilacyclopentadiene (PySPy) as electron transport (ET) material was found to be enhanced over Alq$_3$, which is one of the best performing electron transporting materials so far [64]. 2,5-Bis-(2′,2″-bipyridin-6-yl)-1,1-dimethyl-3,4-diphenylsilacyclopentadiene (2PyS2Py), used as ET layer, exhibited an electron mobility of 2×10^{-4} cm^2 V^{-1} s^{-1} at a field strength of 0.64 MV cm^{-1} measured by the time-of-flight technique. Incorporation of phenylene and thiophene moieties allowed tuning of the emission color in OLEDs: silole derivatives PSP, SiTSTSi, and TTSTT emitted greenish blue, yellowish-green, and reddish orange light, respectively. 2,5-Di-(3-biphenyl)-1,1-dimethyl-3,4-diphenylsilacyclopentadiene (2PS2P) exhibits a blue fluorescence ($\lambda_{max} = 476$ nm) with a high solid state photoluminescence quantum yield of 85% [66]. In multilayer OLED devices 2PS2P showed emission at 495 nm due to an exciplex formation with N,N'-diphenyl-N,N'-(2-naphthyl)-(1,1′-phenyl)-4,4′-diamine used in the hole-transporting layer (HTL). Interestingly, all these compounds show very weak luminescence in dilute solutions ($\Phi_F = 10^{-2}$–10^{-4}) [67].

1,2-Bis(1-methyl-2,3,4,5-tetraphenyl silacyclopentadienyl)ethane (2PSP) emits a blue–green fluorescence with an absolute quantum yield of 97% as vapor-deposited film [65]. Devices using 2PSP show a very low operating voltage, an

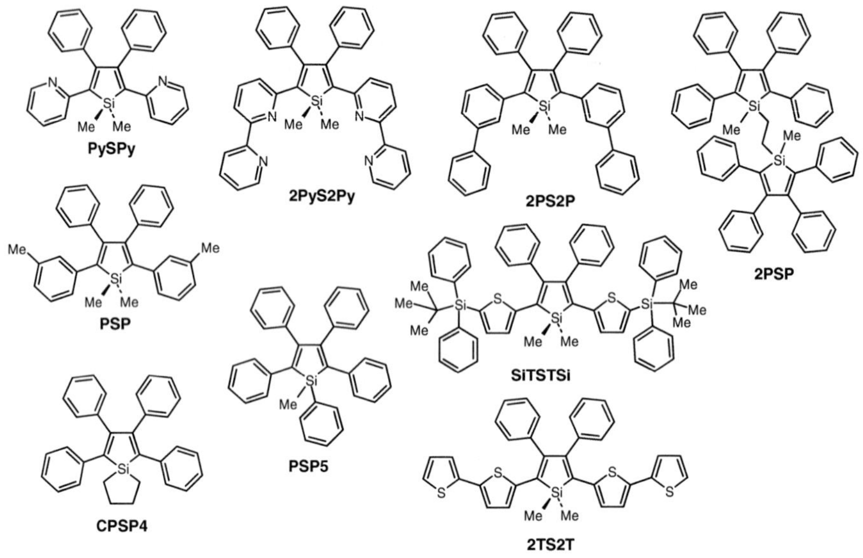

Fig. 15 Silole-containing oligomers used for OLEDs

external electroluminescence quantum efficiency η_{EL} of 4.8%, and an impressive luminous power efficiency of 9 lm W^{-1} at a brightness of 100 cd m^{-2}. Unfortunately, these siloles easily crystallize due to their low glass transition temperatures (T_g). This strong tendency for crystallization contributes to device degradation when these materials are incorporated in OLED structures [68].

1-Methyl-1,2,3,4,5-pentaphenylsilole [69] (PSP5) was reported as the first compound with a huge aggregation induced emission (AIE) – a rare phenomenon, where efficiency of photoluminescence increases by two orders of magnitude upon aggregation [70]. This phenomenon was explained by restricted intramolecular rotations of the phenyl rings in the nanoaggregates (with a large contribution to the nonradiative transition process found in solutions) [71] and observed for a series of similar phenylsiloles [72, 73] as well as for other bulky molecules [74]. PSP5 and other 1,1-disubstituted siloles can be synthesized in a convenient one-pot reaction by lithiation of dithenylacetylene followed by treatment with corresponding dichlorosilane [75]. Multilayer OLED devices, prepared with PSP5 as emitter, showed a blue light emission at 496 nm with a low turn-on voltage (3.4 V), high emission efficiencies (9,234 cd m^{-2}, 12.6 lm W^{-1}, and 12 cd A^{-1}), and high external quantum yield (8%). Further optimization allowed raising the power efficiency of the devices to a record for blue emission value of 20 lm W^{-1} [70]. Among a series of spiro-silacycloalkyl tetraphenylsiloles prepared by Son et al., 1,1'-silacyclopentyl-2,3,4,5-tetraphenylsilole (CPSP4) has shown the most promising properties in OLEDs. A three layer device, comprising N,N'-bis(1-naphthyl)-N,N'-diphenylbenzidine (NPB) as the hole-transport layer, CPSP4 as the emitting layer, and Alq$_3$ as the ET layer, displayed a brightness of 11,000 cd^{-2} at 11 V with a current efficiency of 2.71 cd A^{-1} [73].

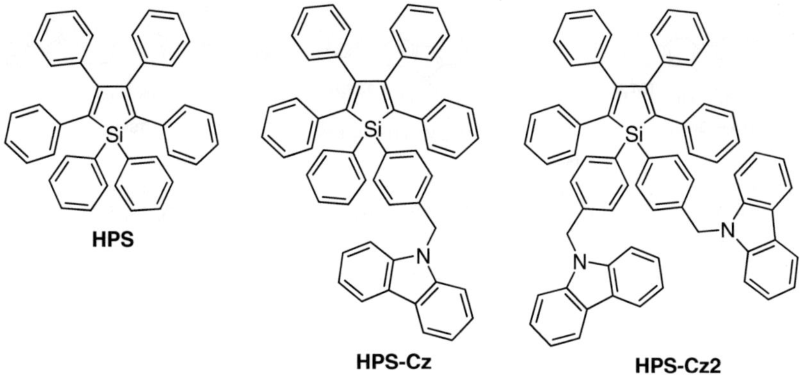

Fig. 16 Hexaphenylsilole (HPS) and its carbazole derivatives

Fig. 17 Important silole-based building blocks are dibenzosilole (BS) and dithienosilole (TS)

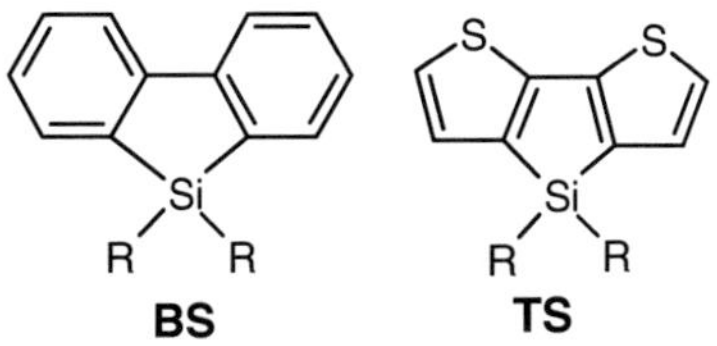

Attaching one or two carbazolyl (Cz) groups to a hexaphenylsilole (HPS) [72] leads to thermally and morphologically stable carbazolylsiloles, HPS-Cz and HPS-Cz2, respectively (Fig. 16), which were tested on OLEDs and OPV cells [76]. These molecules, like their HPS parent, are also AIE-active. Obviously due to the formation of donor–acceptor (D–A) complexes for the asymmetrically substituted HPS-Cz the photoluminescent quantum yield of aggregated HPS-Cz (56%) was 2.4-fold higher than that of HPS-Cz2 (23%). As a result, OLEDs prepared from HPS-Cz were found to be more efficient with a turn-on voltage of 5 V, and emit light with maximal current efficiency of $2.6\,\text{cd}\,\text{A}^{-1}$ at 8.5 V. HPS-Cz2 revealed a higher efficiency in photovoltaic cells. The best devices exhibited a short-circuit current density of $96.5\,\text{mA}\,\text{cm}^{-2}$, an open-circuit voltage of 1.7 V, and a fill factor of 0.21. Although the structure of the cell was far from being optimized, it showed an external photovoltaic efficiency as high as 2.19% (irradiated with 365 nm UV light at $15\,\text{mW}\,\text{cm}^{-2}$).

Two important silole-based building blocks are BS [77, 78] and TS [79, 80]. The first is also known as silafluorene (SiF), in which a silicon atom replaces the carbon atom in the 9-position of fluorene; the second is the silicone-bridged bithiophene (Fig. 17). In both structures the LUMO energy level is lowered by a σ^*–π^* conjugation within the silole ring.

A series of diphenyl-, dithienyl-, and dipyridylsubstituted TSs and their trimethylsilyl derivatives have been reported by the Ohshita and Kunai group [81] (Fig. 18). TS with trimethylsilylpyridyl substituents TSPy_2 have shown good ET

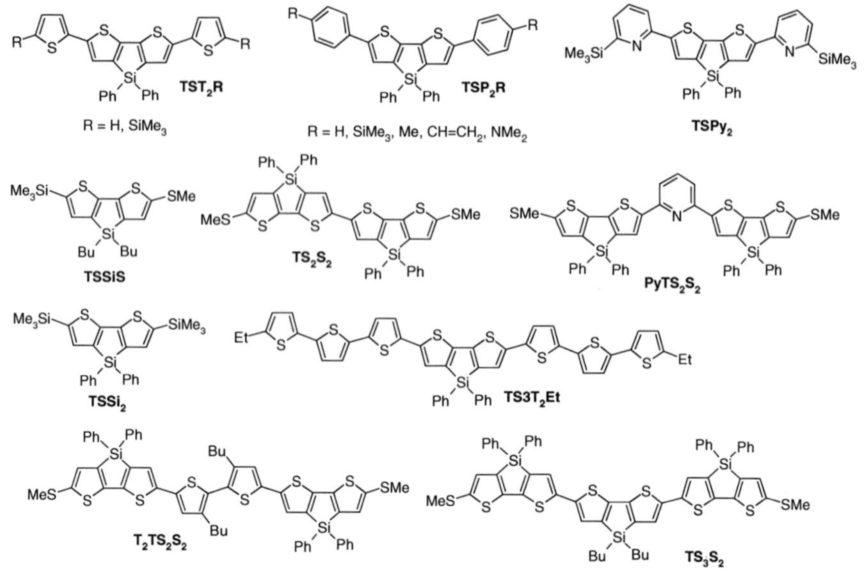

Fig. 18 Dithienosilole-based oligomers

properties in OLEDs: a device with the structure of ITO/TPD/Alq$_3$/DTSPy$_2$/Mg-Ag, where Alq$_3$ acted as the emitter and TPD (N,N'-diphenyl-N,N'-di(m-tolyl)biphenyl-4,4'-diamine) as hole-transporting material, emitted a strong green light with the maximum luminance of 16,000 cd m^{-2}. Lee et al. reported on the synthesis of TS derivatives containing *para*-substituted phenyl with methyl, vinyl, or dimethylamino groups [82]. TSP$_2$Me, TSP$_2$Vin, and TSP$_2$NMe$_2$ emit light with a maximum emission at 512, 517, and 560 nm, which correspond to green to yellowish-green light. At a voltage of 10 V the luminance of advanced multilayered OLED devices using these TS derivatives as emitters were found to be 680, 515, and 250 cd m^{-2}, respectively. The group of Ohshita and Kunai reported on a series of TS-containing oligomers with different conjugation lengths (2–8 conjugated aromatic rings, see Fig. 18). Attached electron-donating methylthio groups raise the HOMO energy, thus leading to oligomers with smaller HOMO–LUMO energy gaps [83]. Hence, compared to oligomers with the same π-conjugated chain length but without silole subunits, their optical absorption (395–498 nm) and the emission (498–576 nm) spectra maxima are shifted to longer wavelengths. Emission colors of these oligomers cover the whole visual spectral range from violet to red (Fig. 19) despite their moderate efficiency (Φ_F ranges from 0.02 for T$_2$TS$_2$S$_2$ to 0.22 for PyTS$_2$S$_2$). Solely TS3T$_2$Et, the oligomer containing eight conjugated aromatic rings, was found to be active as semiconductor in OTFT, vapor-deposited films as well as spin-coated films of which exhibiting a hole mobility of $\mu = 2.6 \times 10^{-5}$ cm^2 V^{-1} s^{-1} and $\mu = 1.2 \times 10^{-7}$ cm^2 V^{-1} s^{-1}, respectively.

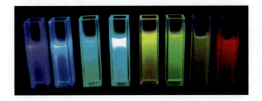

Fig. 19 Luminescence of the THF solutions of TS derivatives (from *left* to *right*): TSSi$_2$, TSSiS, PyTS$_2$S$_2$, TSPy$_2$, T$_2$TS$_2$S$_2$, 2TS$_2$S$_2$, TS3T$_2$Et, TS$_3$S$_2$ (for chemical formulas – see Fig. 18). Reproduced with permission of the American Chemical Society from [83]

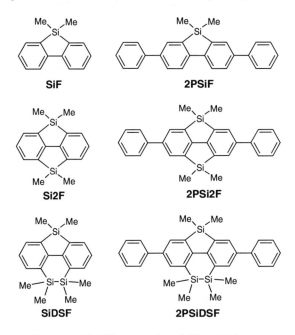

Fig. 20 Silafluorene oligomers with different number of silicon-bridge atoms

Shimizu et al. investigated silicon-bridge effects on photophysical properties of silafluorenes with different bridge structures (Fig. 20) [84]. 4,5-Dimethylsilylene- or 4,5-tetramethyldisilylene-bridged 9-silafluorenes Si2F and SiDSF were prepared by lithiation of 2,2′,6,6′-tetrabromobiphenyls followed by silylation with dichlorodimethylsilane or 1,2-dichloro-1,1,2,2-tetramethyldisilane, respectively. X-ray analysis of the silylene-bridged silafluorene revealed that the molecular framework was perfectly planar and four Si-C(methyl) σ-bonds were completely orthogonal to the plane. Additional silicon bridges connecting the 4,5-positions were found to induce a red shift in the absorption and fluorescence spectra compared to 9-silafluorenes (SiF). Density functional theory (DFT) calculations suggested that an introduction of the silicon bridges to 9-silafluorene increases the energy of the HOMO and LUMO levels simultaneously.

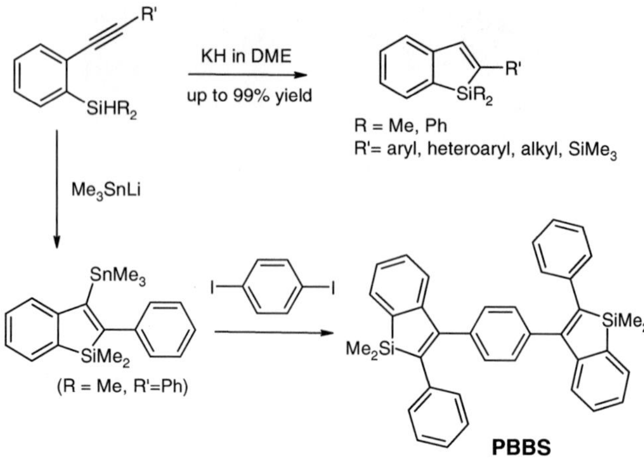

Fig. 21 Functionalized benzosiloles and their derivatives

Benzosilole derivatives, in which phenylene ring replaces a double bond in the silole structure, have also been recognized as prospective building blocks for the synthesis of oligomers with interesting electronic and photonic properties. The group of Nakamura and Tsuji reported a trimethylstannyllithium promoted cyclization of (o-alkynylphenyl)silane to a 3-stannylbenzosilole that proceeds via an addition to the triple bond followed by intramolecular cyclization in a cascade fashion. This intermediate can be functionalized with either electrophiles or nucleophiles to 2,3-disubstituted benzosiloles (Fig. 21). Phenylene-bis(benzosilole) PBBS shows an electron mobility up to $\mu = 6 \times 10^{-4}\,cm^2\,V^{-1}\,s^{-1}$ in an amorphous film, which makes this class of compounds promising for the use in organic light-emitting devices and OPVs [85]. Recent results from the same group showed an intramolecular cyclization of (2-alkynylphenyl)silanes in the presence of potassium hydride to obtain a variety of new 2-substituted benzosiloles in good up to excellent yields (Fig. 21). Some of these compounds showed a high fluorescence quantum yield both in solution and in the solid state [86].

Silicon-bridged biaryls (SBArs) can be synthesized conventionally by dilithiating the corresponding 2,2'-dihalobiaryls followed by reaction with dichlorosilanes. Recently Shimizu et al. reported on a novel and versatile approach to SBArs which involves Pd-catalyzed intramolecular direct arylation of readily available 2-(arylsilyl)aryl triflates (Fig. 22) [87]. This approach is applicable to the facile synthesis of not only symmetrical and asymmetrical functionalized 9-silafluorenes, but also SBArs containing heteroaromatic rings, such as furan, thiophene, and pyrrole. This synthetic approach was used to synthesize silicon-bridged 2-phenylindole (SBPI), which exhibits highly efficient blue fluorescence in the solution ($\Phi_F = 0.70$) and in the solid state ($\Phi_F = 0.90$–1.00) (Fig. 23).

In summary, silole-based oligomers present an interesting class of organosilicon σ–π conjugated compounds with unique electronic and optical properties. A low lying LUMO level can be further adjusted by appropriate combination with other

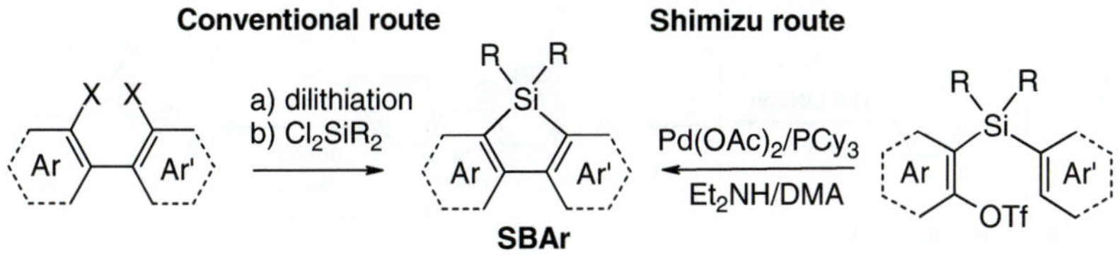

Conjugated Organosilicon Materials for Organic Electronics and Photonics   57

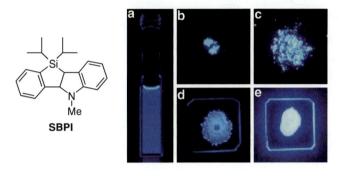

Fig. 22 Two approaches to silicon-bridged biaryls

Fig. 23 Silicon-bridged 2-phenylindole (SBPI) and its fluorescence images ($\lambda_{ex} = 365$ nm): (**a**) 1.9×10^{-5} M solution in cyclohexane; (**b**) microcrystal; (**c**) powder; (**d**) thin-film prepared by spin-coating from a toluene solution; (**e**) dispersed in PMMA film (reproduced with permission of John Wiley & Sons, Inc from [87])

π-conjugated (hetero)cycles. In combination with the unique AIE phenomena, this class of compounds yields an extremely high solid state PL efficiency. The highest potential for applications of silole-based oligomers seems to be in OLED structures, especially as blue light emitters.

2.4 Silicon Analogs of Oligo(p-Phenylenevinylene)s

Recently a new class of conjugated organosilicon oligomers has been reported: silicon analogs of oligo(p-phenylenevinylenes). Yamaguchi, Xu, and Tamao synthesized a homologous series of bis-silicon-bridged stilbenes [88] via an intramolecular reductive cyclization of bis(o-silyl)-diphenylacetylene (Scheme 4). Bis(o-silyl)-diphenylacetylenes are reacted with excess lithium naphthalenide to undergo a two-electron reduction at the acetylene moiety to produce a dianion intermediate. This dianion further undergoes a double cyclization in a 5-exo mode to yield bis-silicon-bridged stilbenes. This method can be successfully applied to the synthesis of tetrakis-silicon-bridged bis(styryl)benzenes. The obtained silicon-bridged π-conjugated systems show an intense fluorescence in the visible region which differs significantly from its carbon analog.

This synthesis methodology was expanded to the synthesis of ladder oligo(p-phenylenevinylene)s (LOPVs) and related π-electron systems, having annelated

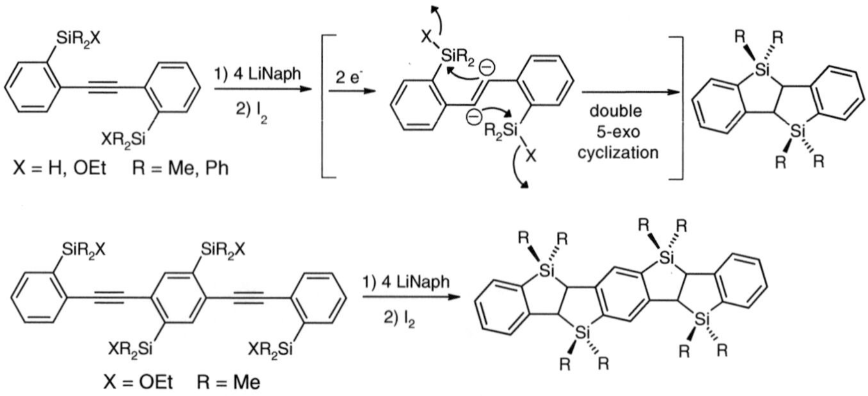

Scheme 4 Synthesis of bis-silicon-bridged stilbenes

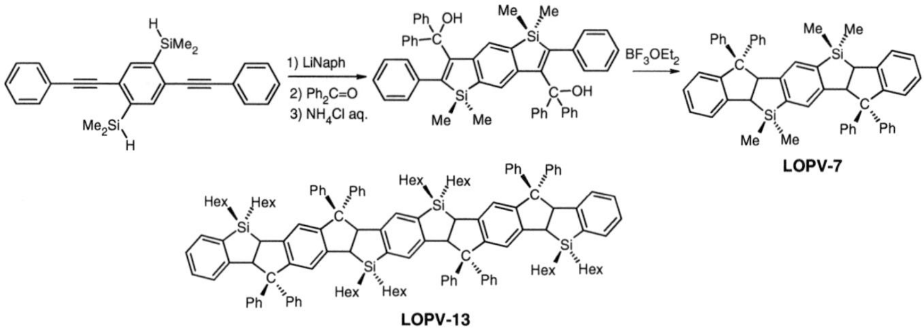

Scheme 5 Synthesis of ladder oligo(*p*-phenylenevinylene)s and the longest molecule obtained by this technique

π-conjugated structures with silicon and carbon bridges [89, 90]. In this case a combination of two cyclization reactions was applied, i.e., an intramolecular reductive cyclization of (*o*-silylphenyl)acetylene derivatives and an Friedel–Crafts-type cyclization (Scheme 5). This allowed the synthesis of a homologous series of ladder molecules up to LOPV-13, a system with 13 fused rings. The crystal structure of LOPV-13 proves a nearly flat π-conjugated framework with a length of ca. 2.9 nm. All obtained ladder π-electron systems show intense fluorescence in the visible region (λ_{max} = 443–523 nm) with high quantum yields up to Φ_F = 0.84, but relatively small Stokes shifts (18–23 nm).

Disilene (Si = Si) analogs of oligo(*p*-phenylene-vinylene)s were successfully synthesized by use of a 1,1,3,3,5,5,7,7-octaethyl-*s*-hydrindacen-4-yl (Eind) ligand (Fig. 24) [91]. Their X-ray crystal structures and spectroscopic data demonstrate that the π-conjugation effectively extends over the Si-OPV framework. Tetrasiladistyrylbenzene TSDSB exhibits an orange fluorescence even at room temperature both in solution and in the solid state, which is attributable to the effective extension of conjugation and a low tendency to aggregate. TSDSB is the first example of a disilene derivative with fluorescence even at room temperature.

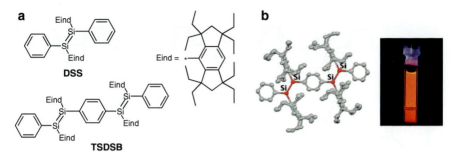

Fig. 24 (**a**) Disilene analogs of the oligo(*p*-phenylenevinylene)s (Si-OPVs): disilastilbene DSS and tetrasiladistyrylbenzene TSDSB. (**b**) Highly planar crystal structure of tetrasiladistyrylbenzene and its fluorescence – reproduced with permission of the American Chemical Society from [91]

Despite the few reports on silicon analogs of oligo(*p*-phenylenevinylenes) it appears that this emerging class of organosilicon materials exhibits promising optical properties for future photonic applications.

3 Branched Conjugated Organosilicon Oligomers

Albeit some of the linear oligomers described in Sect. 2 also have branched molecular shape, all of them have a 1D conjugation backbone that allows them to be considered as linear conjugated oligomers. This section will concentrate on oligomers having several conjugated units connected to three- or four-functional central silicon atom.

One of the first conjugated organosilicon oligomers was spiro-bis-septithiophene Spiro-Si-7T (Fig. 25) [92] This molecule is a branched analog of septithiophene Me_3Si–T7–Me_4SiMe_3 (see Sect. 2.1), in which two oligothiophenes connect via a C-Si isolated bridge and build a 90° torsion angle. This unique structure separates the two perpendicular chromophores and allows oxidation of each of the 7T units sequentially and independently of each other that proceeds via four distinct species: a mono radical cation, a bis-(radical cation), a radical cation/dication, and a bis-(dication). It has been proposed to use these structures in future to build molecular electronic devices for memory, logic, and amplification functions [93, 94], in which the oxidized (doped) and the neutral (nondoped) states of the oligothiophenes may serve as bit states "one" and "zero."

A series of four asymmetrically aryl-substituted 9,9′-spiro-9-silabifluorene (SSF) derivatives, 2,2′-di-*tert*-butyl-7,7′-diphenyl-9,9′-spiro-9-silabifluorene (PSSF), 2,2′-di-*tert*-butyl-7,7′-dipyridin-2-yl-9,9′-spiro-9-silabifluorene (PySSF), 2,2′-di-*tert*-butyl-7,7′-dibiphenyl-4-yl-9,9′-spiro-9-silabifluorene (BPSSF), and 2,2′-di-*tert*-butyl-7,7c-bis(2′,2″-bipyridin-6-yl)-9,9′-spiro-9-silabifluorene (BPySSF) have been reported by Lee et al. [95] (Fig. 26). These molecules were synthesized by the cyclization of the corresponding 2,2′-dilithiobiphenyls with silicon tetrachloride.

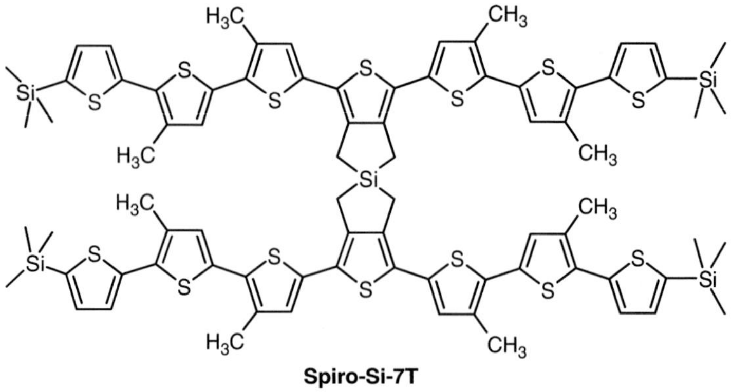

Fig. 25 Spiro-bis-septithiophene Spiro-Si-7T

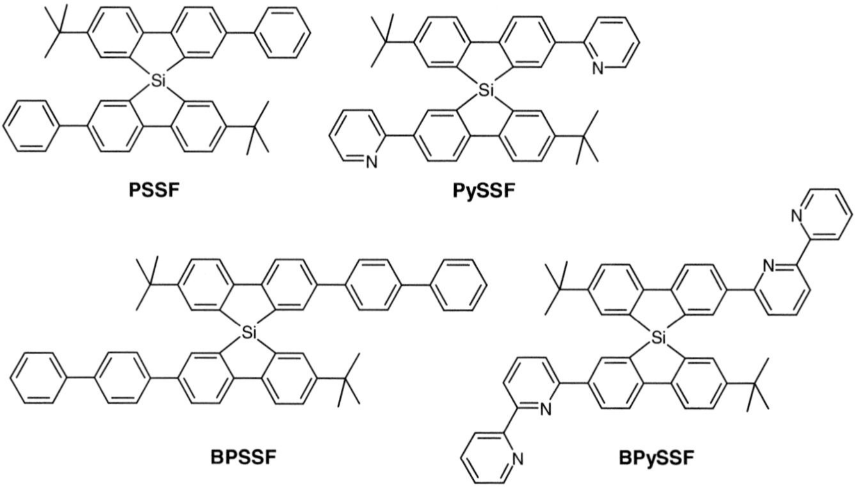

Fig. 26 Spiro-9-silabifluorene derivatives

These spiro-linked siloles form transparent and stable amorphous films with glass transition temperatures above 200°C. The absorption spectrum of each compound shows a significant bathochromic shift relative to the corresponding carbon analog as a result of the effective σ^*–π^* conjugation between the σ^* orbital of the exocyclic Si-C bond and the π^* orbital of the oligoarylene fragment. Solid state films exhibit an intense violet–blue emission with maxima at 398–415 nm and high absolute photoluminescence quantum yields ($\Phi_F = 0.30$–0.55).

As discussed in Sect. 2.1, luminescence efficiency of bi- and terthiophenes can be increased by silylene- and disilylelene substituents. This effect might be enhanced by a further branching of the chromophore to avoid aggregation. Ponomarenko and his group reported a significant increase of the luminescence quantum yield Φ_F in dilute solutions of a series of mono-, di-, tri-, and tetra(5′-hexyl-2,2′-bithiophene)silanes from 0.06 to 0.20 while the absorption and luminescence spectra only slightly change [96]. Interestingly, tri- and tetra-substituted bithiophenesilanes have almost the same Φ_F (0.19 and 0.20). These oligomers were synthesized in good to excellent yields from the lithium derivative of 5-hexyl-2,2′-bithiophene and the corresponding chlorosilanes similar to the synthesis reported before for a series of bithienylhydridesilanes by Lukevics et al. [97]. Recently a similar route was successfully applied for the synthesis of a series of tetrahedral arylethynyl substituted silanes [98].

An impressive luminescence efficiency and luminescent lifetime was reported by the Ishikawa group for a series of tri-arm star-like bithiophene–disilylene molecules 2T-DSi-stars, in which the synergy of both the branched structure and the disilylene substituents raised the efficiency up to $\Phi_F = 0.49$–0.75 (Fig. 27) [99, 100]. The highest efficiency of $\Phi_F = 0.78$ was reported for bithiophene–silylene star 2T-Si-star. These effects originate from the star-like structure as well as σ–π conjugation

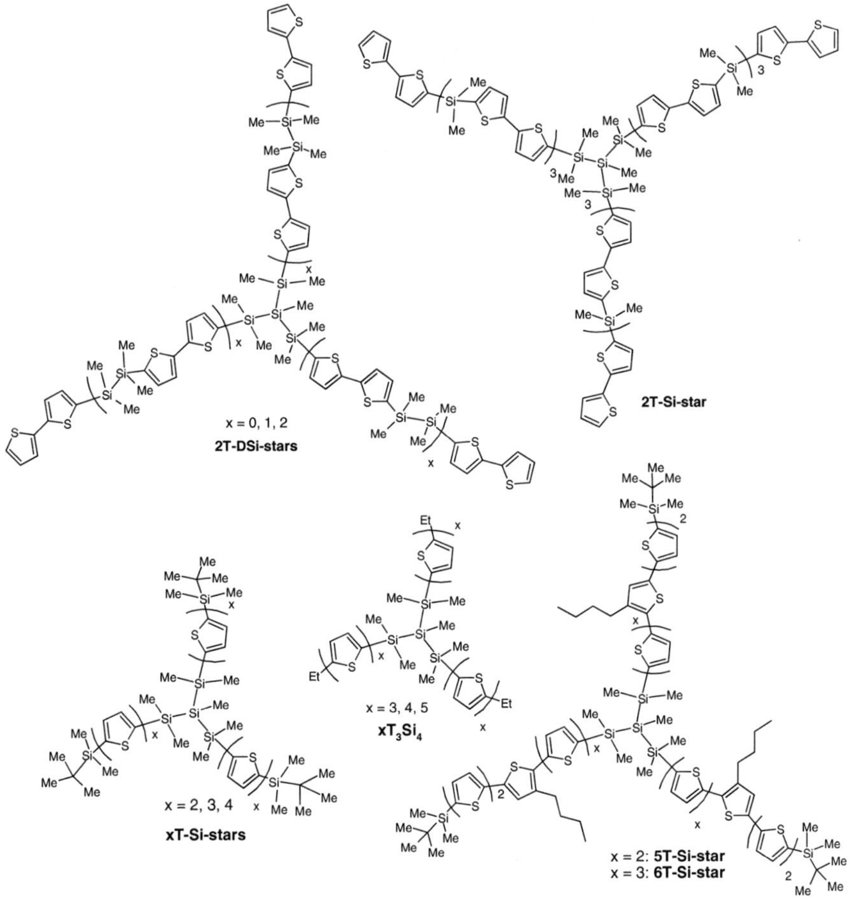

Fig. 27 Tri-substituted oligothiophenesilane nanosized star molecules

between the silicon atoms and bithiophene units by a significant decrease of the nonradiative deactivation rate. Ter-, quater-, quinque-, and sexithiophene-silanylene star molecules also showed increased Φ_F in comparison to their linear analogs, albeit not as dramatic as in the case of the 2T chromophore: (compare $\Phi_F(6T) = 0.36$, Φ_F (Me$_3$Si – T6 – Oct$_2$SiMe$_3$) = 0.25, and Φ_F (6T-Si-star)=0.61) [101].

Ohshita et al. has reported on ter-, quater- and quinquethiophenesilane trisubstituted stars xT$_3$Si$_4$ (x = 3–5, see Fig. 27) and compared them with linear dimers (Fig. 3) [21]. The quinquethiophenesilane star 5T$_3$Si$_4$ possesses the best semiconducting properties among the oligothiophenesilanes investigated: the OFETs prepared by vacuum sublimation shows a mobility of 6.4×10^{-2} cm^{2}V^{-1}s^{-1} and on/off ratio of 10^4. It was, however, hardly soluble in organic solvents, and therefore could not be processed to films by spin-coating.

The solubility issue could be solved for tetrasubstituted silanes, possessing tetrahedral symmetry. Lukevicz et al. has reported on the synthesis of a series of radial oligothienylsilanes having four bi-, ter-, or quaterthiophenesilane units attached to a silicon center via a short –SiMe$_2$C$_2$H$_4$– spacer [102]. Unfortunately, no optical or semiconducting properties for these materials have been reported. The group of Roncali has described several tetra(terthienyl)silanes (Fig. 28) among which Si(3T)$_4$ was investigated as 3D electro-active π-conjugated material [103] with the expectation that the rigid 3D structure might enhance the light absorption independent

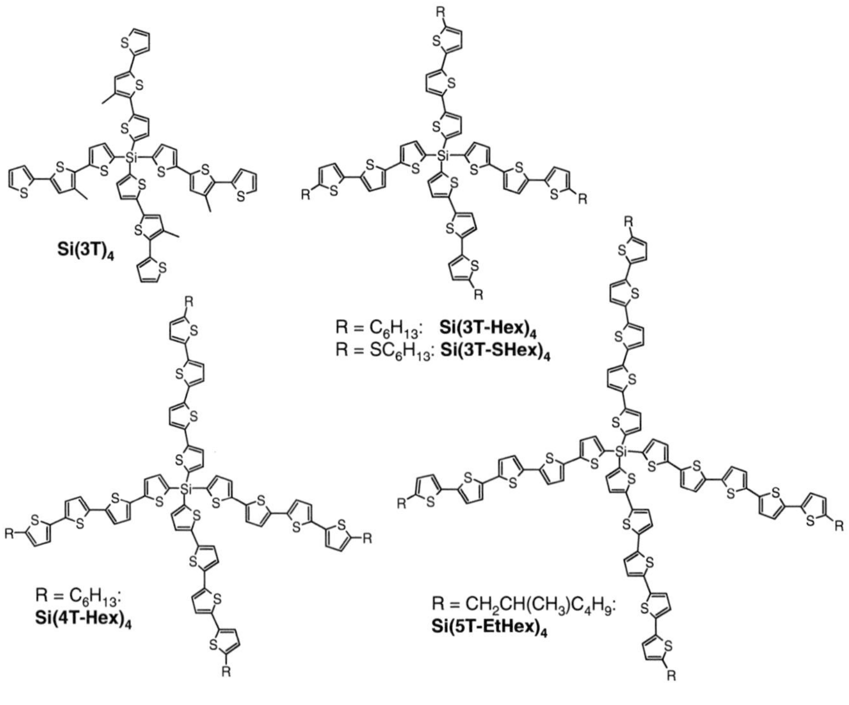

Fig. 28 Tetra(oligothienyl)silanes

of the layer packaging. It is known that linear oligothiophenes have almost perpendicular orientation to the substrate [104] that is useful for OFETs, but prevents their usage in OVPs. Consequently, Si(3T-Hex)$_4$ and Si(3T-SHex)$_4$ were used as donor materials for organic solar cells [105]. Optimal devices using Si(3T-Hex)$_4$ as donor material and PCBM as acceptor material showed an unsatisfactory low PCE of 0.30%; however, this value is five times higher than for a device with the linear oligomer 5-hexyl-2,2′ : 5′,2″-terthiophene Hex-3T. The solar cells efficiency could be improved by increasing the length of the oligothienyl units in corresponding silanes. The group of Ponomarenko recently prepared tetra(quaterthienyl)silane Si(4T-Hex)$_4$ and tetra(quinquethienyl)silane Si(5T-EtHex)$_4$ (Fig. 28) [106]. Heterojunction photovoltaic cells employing composites with C[70]PCBM ([6, 6]-phenyl C71-butyric acid methyl ester) yielded a PCE of 1.0% and 1.35%, respectively. The increasing efficiency in the series of 3D ter-, quater- and quinque thiophenesilanes was attributed to a more effective absorption of the solar light caused by shifting the absorption maxima in the series of oligomers from 390 nm to 414 nm and 437 nm, respectively.

If linear oligothiophenes attach to a silicon branching core via flexible aliphatic spacers, branched structures can be obtained, the ordering of which in layer is determined by the crystallization of the oligothiophene units rather than by the 3D structure of the molecules. Several examples of such materials have been prepared in our group (Fig. 29) [107, 108]. They combine the high crystallinity and semiconducting properties of oligothiophenes with good solubility and solution processability of branched structures – both are useful properties for organic electronics. The terminal groups have a significant influence on the orientation of these molecules along the surface depending on whether they are linear or branched. Molecules with linear end groups G0(Und-4T-Hex)$_4$ orient perpendicular to the surface (Fig. 30), which can be utilized in solution-processed OFETs with the mobility up to 2×10^{-2} cm^2 V^{-1} s^{-1} and on/off ratio of 10^5. It is noteworthy that devices made from such materials can be prepared under ambient conditions and their characteristics are stable after storage without any packaging layer for years. Molecules having branched 2-ethylhexyl

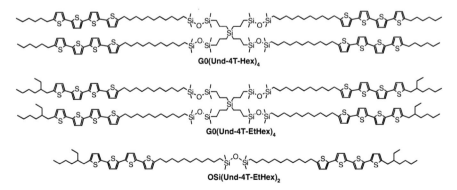

Fig. 29 Quaterthiophene-based flexible multipods bearing two (dimer D1) or four (tetrapods D2-D3) quaterthiophene arms with flexible carbosilane-siloxane cores

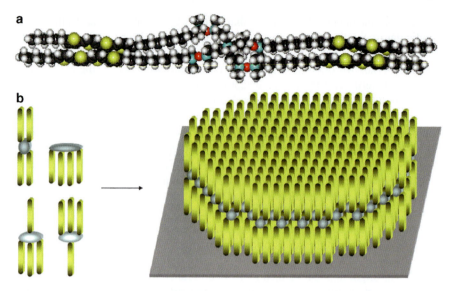

Fig. 30 (**a**) Molecular model of G0(Und-4T-Hex)$_4$ in its extended conformation. (**b**) Schematic representation of different conformations of this molecule as a core with four arms and its self-organization into a monolayer. Reproduced with permission of the American Chemical Society from [107]

end groups G0(Und-4T-EtHex)$_4$ or OSi(Und-4T-EtHex)$_2$ tend to orient parallel to the surface [109], which was used for the creation of photovoltaic cells. The most efficient OPV devices were made from the blend of the dimer OSi(Und-4T-EtHex)$_2$ with fullerene derivative C[70]PCBM, which showed a PCE of 0.9% [110].

Thus, connecting of π-conjugated units to branching silicon center could lead to the following consequences resulting from a decreased tendency to agglomerate: (1) increased luminescence efficiency; (2) improved solubility; (3) changed film morphology. Branching therefore is a versatile concept to modify the properties of conjugated oligomers and provides promising candidates for functional (semiconducting, photo-, or electroluminescent) layers in different organic electronic and photonic devices.

4 Conjugated Organosilicon Dendrimers

As discussed in Sect. 3, the σ-orbitals of silicon overlap with the π-system of the chromophore if the silicon is directly attached by a single bonds to the π system. Despite this σ–π-interaction, silicon does not facilitate conjugation between two π-systems connected by a silicon bridge. Therefore silicon does not extend the π-system which will have consequences for organosilicon dendrimers, in which π-conjugated units are connected either by silicon atoms which build branches in

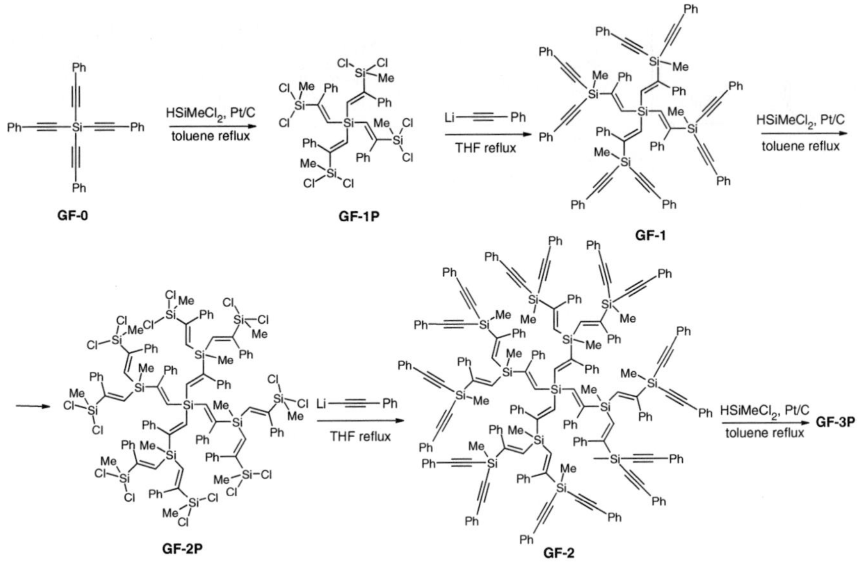

Scheme 6 Synthesis of phenylenevinylene–carbosilane dendrimers

the dendritic structure, or by nonconjugated spacer units. Thiophenes, phenylenes, phenylenevinylenes, and other subunits have been proven to be suitable structures for the construction of chromophores.

The first organosilicon dendrimer containing conjugated units was published by Kim [111]. In an elegant synthetic scheme utilizing a hydrosilylation of the phenylacetylene units of GF-0 with dimethylchlorosilane followed by reaction of terminal chlorosilane groups with lithium phenylacetylide, a series of dendrimers was constructed, which was followed up to the third generation of phenylenevinylene–carbosilane dendrimers with terminal phenyleneacetylene groups (Scheme 6). Unfortunately, few optical properties were reported, among them the absorption maximum and molar extinction coefficient. Independent of the generation the absorption maximum was observed at 287–289 nm. The molar extinction coefficient increased proportionally to the number of alkynyl groups incorporated in the dendrimer.

During recent years, tremendous scientific attention has been devoted towards thiophene-containing dendrimers: (oligo)thiophene groups have been incorporated as core units of the aromatic dendrimers [112, 113] or located at the periphery of organic [114, 115] or organophosphorus [116] dendrimers. Dendritic branches were made from quaterthienyl [117] or 2,3,5-thiophene [118] units. Polythiophene dendrimers have been reported as well [119, 120]. The latter polymers were found to have promising properties in OPV devices [121, 122]. Less attention has been devoted towards oligothiophene-containing organosilicon dendrimers.

Nakayama and Lin have reported on the first organosilicon dendrimer Si_5T_{16}, containing 16 2,5-thienylene rings, connected via tetrasubstituted silicon atoms [123]. Si_5T_{16} was synthesized by the reaction between a tetralithium

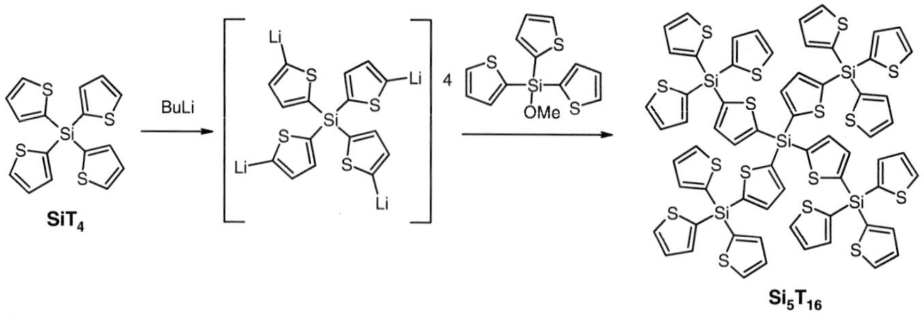

Scheme 7 Synthesis of the first thiophenesilane dendrimer

derivative of tetrakis(2-thienyl)silane with a fourfold molar amount of tris(2-thienyl)methoxysilane with the yield of 19% (Scheme 7). An attempt to access higher generation dendrimers by this synthetic scheme was unsuccessful which may be explained by a low solubility of the polylithium derivatives as well as lithium exchange reactions between products and reagents, both containing active protons in the 5 position of 2-thienyl substituents. Unfortunately neither optical nor electrical properties of this dendrimer have been reported, albeit they would potentially be interesting for organic electronic and photonic applications.

In 2005 Ponomarenko et al. started their work on a series of oligothiophene-silane monodendrons and dendrimers to exploit their optical properties (Fig. 31) [124–128, 130]. The first dendrimer generation G1(2-x) ($x = 1$–4) was synthesized by Kumada and Suzuki coupling reactions of corresponding precursors with moderate to good yields [124, 125]. Bithiophenesilane mono-dendrons of different generations Mn(2–2) ($n = 1$–3) were prepared by reaction of the corresponding chlorosilane with bithienyllithium reagents. This type of reaction was used for a convergent dendrimer synthesis [126] and for further modifications, i.e., to yield carboxyl-containing monodendrons Mn–COOH ($n = 1$–3) [127]. Mn–COOH forms stable and uniform Langmuir monolayers at the air–water interface at a modest surface pressure ($<10\,\text{mN m}^{-1}$), which can easily be transferred to a solid substrate. All dendrimers with bithienylsilane core and monothienylsilyl end groups G1(2–1) show luminescent spectrum with a λ_{max} at 385 nm and a quantum yield ($\Phi_F = 0.20$) similar to those obtained for branched tri- and tetrasubstituted bithiophenesilanes [96]. Independent of the generation, all bithiophenesilane dendrimers Gn(2–2) ($n = 1$–3) exhibit violet–blue light emissions with two almost identical emission maxima at 373 and 390 nm and a photoluminescence efficiency of (Φ_F) of 0.30. Bithiophenesilane dendrimers exhibit a significantly (five times) increased luminescence efficiency compared to its constituent luminophores (bithiophenesilanes with $\Phi_F = 0.06$). An explanation may be more favorable energy levels of the 2T chromophores caused by both the inductive effect of the silicon atom and the influence of a local field from the adjacent 2T units, arranged into star-shaped structures within the dendritic molecule [128]. Dendrimers with ter- and quaterthiophenesilane cores G1(3–2) and G1(4–2) as well as "butterfly-like" molecules B1(3–2) and B1(4–2) have shown a so-called "molecular antennae" effect [129, 130]. Energy

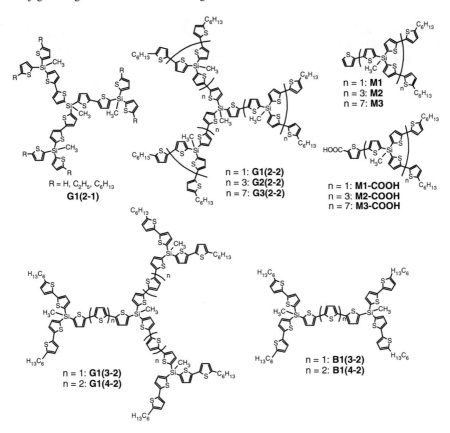

Fig. 31 Oligothiophenesilane monodendrons and dendrimers

captured in the outer segments is efficiently transferred to the inner fragments within the dendritic molecules and have been proposed as promising structures for photo- and electrooptical devices. While this effect has been known since 1994 [131], it took until 2009 to demonstrate G1(3-2) and B1(3-2) as the first organosilicon molecular antennas [129]. Dendrimers G1(3-2) and G1(4-2) emit green light in simple one layer OLED devices [132]. Although their efficiency was rather low, these results indicate that oligothiophenesilane dendrimers possess not only hole but also ET properties. This untypical behavior for oligothiophenes can be explained by a σ–π conjugation between the silicon atoms and the oligothiophene units lowering the dendrimer HOMO level.

More efficient OLEDs have been reported for phosphorescent tris-cyclometalated homoleptic Ir(III) complex Ir(TPSppy)$_3$ (TPSppy = 2 − (4′-(triphenylsilyl)biphenyl-3-yl)pyridine) with a silane-based dendritic substituent (Fig. 32) [133]. This dendrimer has shown high luminescence efficiency both in solution ($\Phi_F = 0.63$) and in film ($\Phi_F = 0.74$). A polymer-based triplet OLED with this Ir(III) doped emitter exhibits a remarkable efficiency of 32.8 cd A^{-1}, maximal brightness of

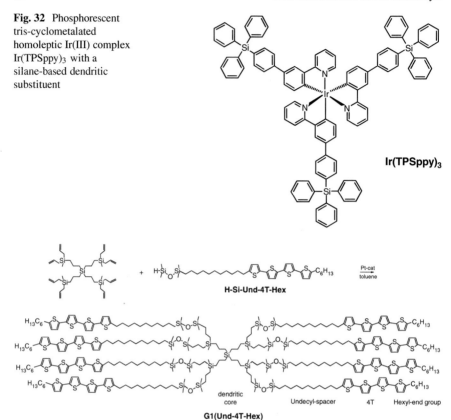

Fig. 32 Phosphorescent tris-cyclometalated homoleptic Ir(III) complex Ir(TPSppy)$_3$ with a silane-based dendritic substituent

Fig. 33 Synthesis and structural formula of the quaterthiophene-containing carbosilane dendrimer G1(Und-4T-Hex)

21,250 cd m^{-2} and maximal power efficiency of 18.7 lm W^{-1} at driving voltage of 5–10 V and doped ratio of 5–30%. These values exceed the performance of optimized devices with Ir(ppy)$_3$ (ppy = 2-phenylpyridine) – a similar complex without dendritic phenylenesilane ligands, measured in the same work.

Another type of conjugated organosilicon dendrimers, quaterthiophene-containing carbosilane dendrimers Gn(Und-4T-Hex), have been designed similar to star-shaped oligothiophenes using silicon atoms as branches and additionally flexible aliphatic spacer groups to link the oligothiophene chromophores [107]. They were prepared by a hydrosilylation reaction of polyallylcarbosilane dendrimers of generations 0 (tetraallylsilane), 1, 3, and 5 with a monofunctional quaterthiophene precursor H-Si-Und-4T-Hex [134]. An example showing the synthesis of the first generation dendrimer is shown in Fig. 33. All dendrimers containing the quaterthiophene chromophores and hexyl end groups were soluble in common organic solvents at an elevated temperature of 40–60°C, while at room temperature they formed gels. AFM measurements revealed three typical morphologies:

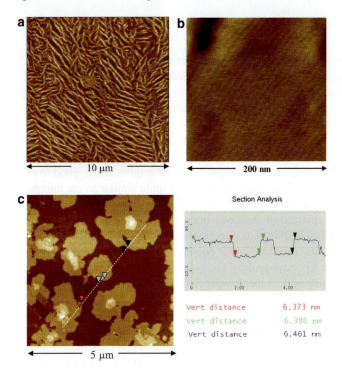

Fig. 34 AFM height images of the dendrimer thick films on graphite substrate: (**a**) G0(Und-4T-Hex) spin-coated from 1 mg ml^{-1} THF solution; (**b**) G1(Und-4T-Hex) from melt; (**c**) G3(Und-4T-Hex) from melt and its cross-section analysis along the line crossing the image [134]

mesowire-like structure, ordered lamellar domains, and smectic-like layered structures, which were common for all these dendrimers (Fig. 34). DSC measurements revealed a high degree of crystallinity, which significantly decreases for dendrimers of higher generations (G3 and G5). A unique feature of these materials is the formation of highly ordered thin films with smectic-like ordering due to π–π-stacking of quaterthiophene groups which might be favorable for OFETs. Indeed, 4T-containing dendrimers up to third generation exhibited field-effect characteristics with a maximum charge carrier mobility of 0.02 cm^2 V^{-1} s^{-1} (for the G0 dendrimer, decreasing with increasing generation), on/off ratios in the order of 10^5–10^6, and threshold voltages close to 0 V.

The few works published on conjugated organosilicon dendrimers indicate that the dendritic structure generally seems to help to yield increased luminescence efficiency, important for OLEDs. Flexibility in the molecular design and good processability from solution are prerequisites for cost efficient OFETs and OLEDs. The current results suggest that the first generation dendrimers already show the optimal effect, and thus the costly synthesis of higher generations will not be necessary.

5 Hyperbranched Conjugated Organosilicon Polymers

Hyperbranched polymers are highly branched macromolecules with 3D dendritic architecture [135]. They may be considered as simplified analogs of dendrimers, which often possess similar properties. In contrast to dendrimers, HB polymers can be produced in a one-pot reaction, leading to highly branched, albeit nonregular molecular structures.

In 1999 Yao and Son reported one of the first examples, an HB poly(2,5-silylthiophene)s HB-SiT-R, prepared by one-pot reaction between 2-bromo-5-(trimethoxysilyl)thiophene TMOS-T-Br and magnesium in good to excellent yields (Scheme 8) [136]. This structure is indeed quite similar to the thiophenesilane dendrimer of Nakayama [123], the main difference being the highly regular and almost defect-free structure of the dendrimer. The intermediate methoxy-substituted HB polymer HB-SiT-OMe was functionalized with Grignard reagents, leading to a series of air- and moisture-stable polymers HB-SiT-R with R being methyl-, vinyl-, 2-thienyl-, and phenyl-groups. The extent of σ–π conjugation in these polymers was examined using optical spectroscopy, which showed (compared to a non-substituted thiophene) a significantly red-shifted π–π* transition (at ~250 vs 230 nm) independent of the group R. This absorption originating from the HB-SiT-R polymer subunit indicates a significant degree of conjugation between the thiophene ring and the silicon atom, but not between the thiophene ring and the attached group R. A weak and broad absorption in the region of 300–320 nm was attributed to a charge transfer from thiophene to silicon, albeit it may also stem from a small fraction of bithiophene–silylene fragments – potential side products from the Grignard reaction of monothienyl reagents [137]. Therefore, the electronic effects of this 3D polymer structure may require more investigations, but at least the effects contributing to the π–π* absorptions appear to be well established. The λ_{max} and extinction coefficients for HB-SiT-R are similar to those obtained for linear polymers.

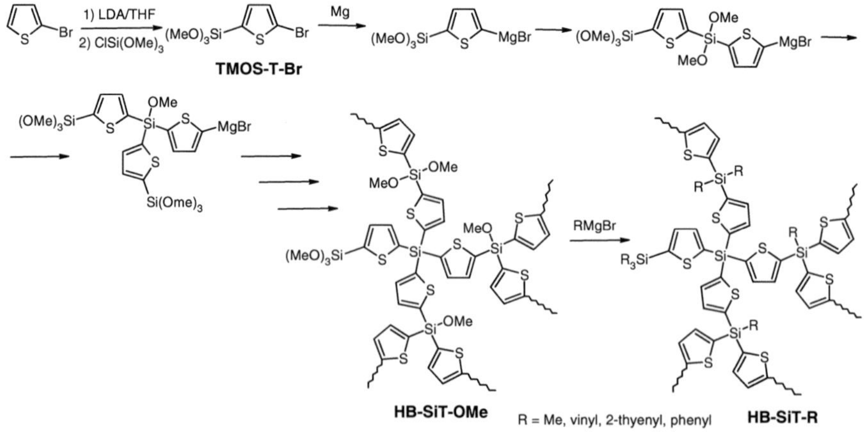

Scheme 8 Reaction sequence, leading to hyperbranched poly(2,5-silylthiophene)s

Two other types of HB organosilicon polymers containing conjugated units were synthesized by hydrosilylation reaction of AB$_2$- or AB$_3$-functional monomers [138, 139]. Poly(silylenevinylene) with ethynyl functional groups HB-SVE were prepared by hydrosilylation of AB$_2$-functional methyldiethynylsilane (MDES) (Scheme 9) [138]. HB poly(carbosilarylene)s HB-SPVs were prepared by hydrosilylation of AB$_3$-functional arylsilane monomers 1-(dimethylsilyl)-4-(trivinylsilyl)-benzene (DMS-1,4-Ph-TVS), 1-(dimethylsilyl)-3-(trivinylsilyl)-benzene (DMS-1,3-Ph-TVS), and 1,1,3,3-tetramethyl-1-(4′-(trivinylsilyl)-phenylene-1)-disiloxane (TMDS-1,4-Ph-TVS) (Scheme 10) [139]. The latter type of polymers contains σ–π conjugated disilylene-phenylene units separated by ethylene or disiloxane nonconjugated spacers. Unfortunately, in both cases no optical or electrical characterization of the HB polymers were reported, allowing no insight into their potential in the field of organic photonics and electronics.

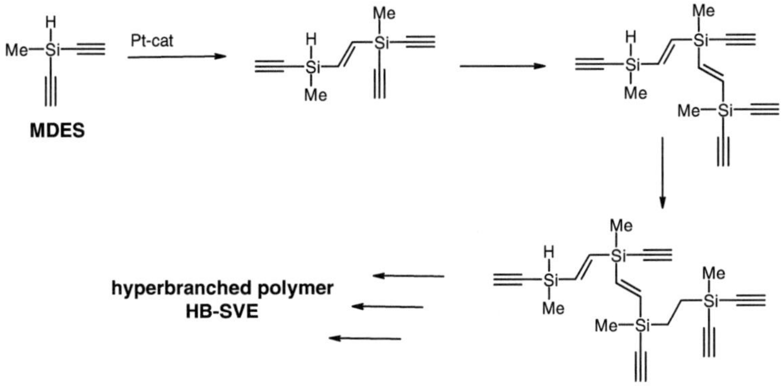

Scheme 9 Synthesis of hyperbranched poly(silylenevinylene) with ethynyl functionalization

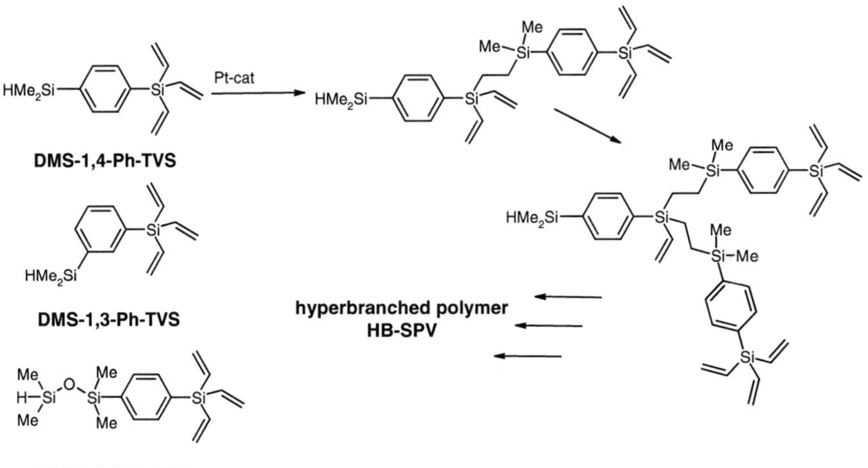

Scheme 10 Arylsilane monomers and an example of hyperbranched poly(carbosilarylene) formation from the first of them

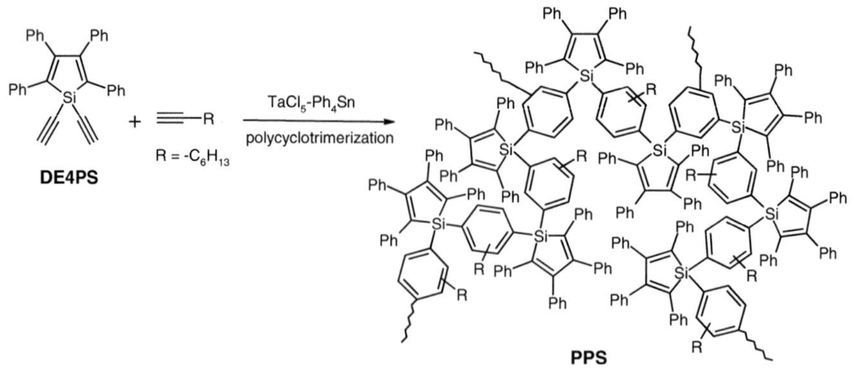

Scheme 11 Synthesis of hyperbranched poly(penylenesilolene)s

Very interesting features were reported by Chen et al. for HB poly(phenylenesilolene)s (PPS) [140]. These HB polymers were prepared in a one-pot synthesis by homo-polycyclotrimerization of 1,1-diethynyl-2,3,4,5-tetraphenylsilole (DE4PS) and its co-polycyclotrimerizations with 1-octyne catalyzed by $TaCl_5 - Ph_4Sn$ in high yields (Scheme 11). The silole units in PPS are interconnected by trisubstituted benzene units, which link the phenyl rings to form 3D structures. This unique molecular structure leads to a number of interesting properties: these polymers are readily processible (completely soluble), thermally stable (T_d up to ~400°C), contain extended conjugated chromophores ($\lambda_{max} \sim 520$ nm), exhibit nonlinear optical activity ($F_L \sim 180$ mJ cm^{-2}), and readily emit light upon excitation at low temperatures (cooling-enhanced emission, CEE). Their molecular weights, measured by GPC relative to a linear polystyrene standard, were measured to be in the range 3,530–5,820 Da (M_w), but this value might significantly underestimate the real molecular weight (the authors suggested the "real" molecular weight may be up to seven times higher) due to the HB and rigid structure of these polymers. All PPS polymers absorb in the visible region, while the monomeric siloles absorb in the UV spectral region ($\lambda_{max} = 378$ nm). The extended conjugation of these HB polymers originates possibly from the synergistic interplay of a σ–π conjugation of the silole rings with the π-orbitals of the chromophore and an electronic interaction of the aromatic rings via the silicon bridges. This extended electronic conjugation results in a nonlinear optical effect, an optical limitation.[2] In this regard the effect of PPSs has been found superior to C_{60}, the best-known optical limiter so far. In contrast to siloles or linear poly(silolylacetylene)s [141], solutions of PPSs are somewhat luminescent ($\Phi_F = 0.01$, this low value is however 10–100 higher than that for low molar weight siloles), but are completely inactive towards AIE. The luminescence of PPS polymers is, however, enhanced at low temperatures. This unique phenomenon of cooling-enhanced emission (CEE) was explained by the restricted intramolecular rotations of the phenyl rings around the axes of the

---

[2] An optical limitation is an optical transmittance that sharply decreases at high light levels and strongly attenuates the optical power of intense laser pulses.

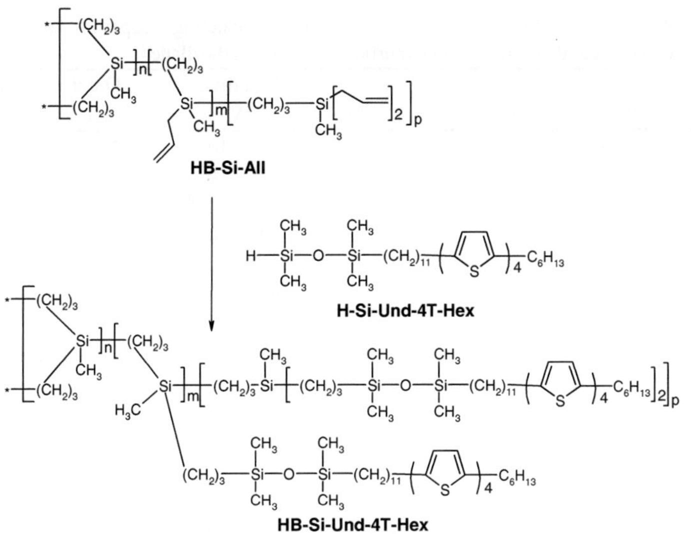

Scheme 12 Synthesis of hyperbranched polycarbosilanes with quaterthiophene end groups

single bonds linked to the silole cores at lowest temperatures. This effect may be regarded as a special type of thermochromism, offering a versatile mean for tuning the photoluminescence of the HB polymers.

A recent example of HB conjugated organosilicon polymers are HB polycarbosilanes with quaterthiophene end substituents prepared by the authors group [142]. They were received from nonconjugated HB polyalyllsilanes HB-Si-All modified with a silane-containing quaterthiophene precursor H-Si-Und-4T-Hex by hydrosilylation similar to the synthesis of carbosilane dendrimers Gn(Und-4T-Hex) [134] (Scheme 12). While dendrimers resemble a single species of defined molecules, HB polymers represent a distribution of molecules with different molecular weights and a large polydispersity. Samples of HB polymers with three different molecular weight distributions were compared to corresponding dendrimers: the initially obtained HB polycarbosilane was recrystallized to remove low molecular weight impurities (HB-1) and two samples with narrow molecular weight distribution were prepared by fractionation of the initial HB-Si-All (HB-2, HB-3). In OFETs all samples showed sufficient hole mobility, good on/off ratio, and good threshold voltage comparable to their dendritic analogs (Table 1). Sample HB-2 with the lowest molar weight and narrowest distribution showed the highest field-effect mobility ($\mu = 10^{-3} cm^2 V^{-1} s^{-1}$) comparable to G1(Und-4T-Hex).

To date, the information on HB conjugated organosilicon polymers is rather limited. While dendrimers and especially linear polymers have been extensively studied, many of their HB analogs are waiting for their synthesis. Some HB polymers have been made without performing a detailed analysis of their optical and electrical characteristics. Nevertheless, available data indicate that HB polymers indeed might serve as a more cost efficient alternative to dendrimers in electronic applications. In addition, some HB structures may also induce unique optical

Table 1 Molecular weights characteristics and semiconducting properties of Und-4T-Hex-containing carbosilane dendrimers and hyperbranched polycarbosilanes

Sample	M_n^a	M_w^a	M_w/M_n	Mobility, $cm^2V^{-1}s^{-1}$	On/off ratio	Threshold voltage, V
HB-1	13,600	35,600	2.62	1×10^{-4}	10^4	−2
HB-2	10,100	11,500	1.14	1×10^{-3}	10^5	−1
HB-3	21,700	24,800	1.14	2×10^{-4}	10^5	−3
G0	$2,997^b$	–	–	2.0×10^{-2}	10^6	0
G1	$6,301^b$	–	–	1.3×10^{-3}	10^6	1
G3	$16,279^b$	–	–	4.0×10^{-4}	10^6	−2

[a]Determined by GPC using polystyrene standards
[b]Molar weight of the dendrimers measured by MALDI–TOF (matrix assisted laser desorption ionization–time-of-flight mass spectrometry)

properties (i.e., CEE) different to those of their monomeric, linear, or dendritic analogs. We expect further progress in this field, striving for a comprehensive understanding of conjugated HB molecular structures and their contribution to electronics and photonics. This will result in a manifold of new structures and a detailed insight into electrical and optical properties of these compounds.

6 Linear Conjugated Organosilicon Polymers

Presently described linear conjugated organosilicon polymers can be divided into three large subgroups: (1) conjugated polymers with Si atoms in the side chains, which are directly linked to the conjugated backbone, (2) silanylene-containing polymers having Si atoms, which link the conjugated fragments to form a main chain, and (3) silol-containing polymers, possessing the special "silol"-units being a kind of silicon-containing chromophore.

6.1 Polymers with Silicon Atoms in the Side Chains

The predominant π-conjugated polymers with organosilicon substituents in the side chains are polyacetylenes [143–148], polythiophenevinylenes (PTV) [149], and poly(1,4-phenylene vinylene)s (PPV) [150–175].

Unlike unsubstituted polyacetylene, its silicon derivatives are air stable and highly soluble [143]. Since the discovery of poly[1-(trimethylsilyl)-1-propene] [144], silicon-substituted polyacetylenes are well-known for their extremely high gas permeation properties caused by a large free volume in the polymer films [145, 146]. However, organosilicon substituents, directly attached to a polyacetylene backbone, weaken the interchain interactions, which might cause inferior electrical properties. This can be a reason for the little work reported on the application of this type of polymers in organic electronics and photonics.

Soluble, air stable, and conductive copolymers were prepared by polymerization of (HC≡CSiMe$_2$)X (where X = single bond, –CH$_2$–, –CH$_2$CH$_2$–, –SiMe$_2$–, O) with WCl$_6$ or MoCl$_5$ catalysts [147]. Although the pristine polymers were not conductive, a conductivity of 10^{-2} S cm^{-1} was obtained by doping with (NEt$_2$)$_3$S$^+$SiF$_2$Me$_3^-$ (or NBu$_4^+$F$^-$) and I$_2$ (or SO$_3$). A series of 1,2,3,4,5-pentaphenylsilolyl-containing polyacetylenes 5PS-Ac, 5PS-9C-Ac, and 5PS-9C-PAc (Fig. 35) have been prepared by polymerization of the corresponding substituted acetylenes using NbCl$_5$- and WCl$_6$-Ph$_4$Sn catalysts [148]. The polymers were thermally stable up to 350°C, but practically non-luminescent in good solvents. Nevertheless, some polymers with nonanyloxy spacer between the pentaphenylsilol pendant groups and polyacetylene backbone (5PS-9C-Ac and 5PS-9C-Pac) showed aggregation- or cooling-induced emission. These effects were explained by the restricted intramolecular rotation or twisting of the silole chromophores introduced by aggregation due to incomplete solution (by using poor solvents or at low temperature). A multilayer electroluminescent device using 5PS-9C-PAc as an active layer emitted blue light of 496 nm with a maximum brightness of 1,118 cd m^{-2}, and showed a current efficiency of 1.45 cd A^{-1} with an external quantum yield of 0.55%.

The synthesis of long alkylsilyl-substituted poly(thienylenevinylene) PSiTV (Fig. 36) via heteroaromatic dehydrohalogenation polymerization was reported by Shim et al. [149]. The polymer was determined to have a molecular weight of

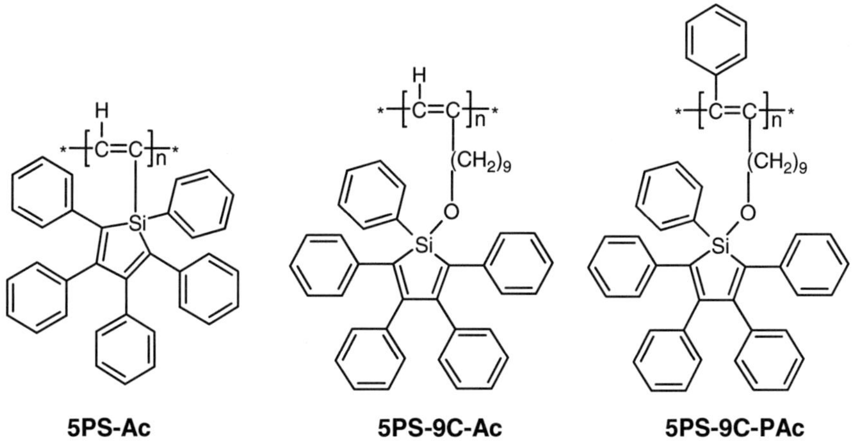

Fig. 35 1,2,3,4,5-Pentaphenylsilolyl-containing polyacetylenes

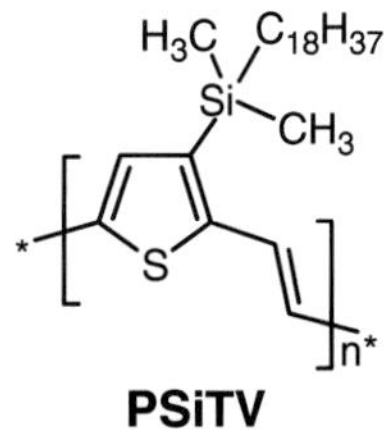

Fig. 36 Silyl-substituted polythiophenevinylene PSiTV

26.4 kDa (M_n, polydispersity of 5.5) and was completely soluble in common organic solvents (e.g., chloroform, THF, xylene), due to the attached dimethyloctadecylsilyl side chains. PSiTV was thermally stable up to 340°C (5% weight loss) and had a T_g of approximately 148°C. The maximum absorption occurs at 563 nm with an onset at 788 nm that corresponds to a promising rather low bandgap of 1.57 eV.

Work in the area of π-conjugated organosilicon substituted polymers has mainly been devoted to the synthesis and investigation of silicon-substituted poly(1,4-phenylenevinylene)s (Si-PPV). The first Si-PPV derivative, poly[2-(3-*epi*-cholestanol)-5-(dimethylthexylsilyl)-1,4-phenylenevinylene] (CS-PPV), was reported by Wudl et al. (Fig. 37) [150]. The silicon group in this polymer enlarges the band gap and changes the polymer's absorption peak from 510 nm (orange, reported for the carbon analog bis(3-*epi*-cholestanoxy)PPV) to 450 nm (yellow). In a non-optimized single layer OLED device with Al electrodes CS-PPV doped with an electron transporting molecular dopant, 2-(4-biphenylyl)-5-(4-*tert*-butylphenyl)-1,3,4-oxadiazole (PBD), emitted green light with a power efficiency of 0.3% [151].

The simplest representative of this class of polymers is poly(2-trimethylsilyl-1,4-phenylenevinylene) (TMS-PPV), first reported by Shim et al., which also shows green electroluminescent emission [152] The electrical conductivity of TMS-PPV doped with I_2 was measured to be 2×10^{-2} cm^{-1}. This is quite different from unsubstituted PPV, which cannot be doped with iodine. Substitution of one

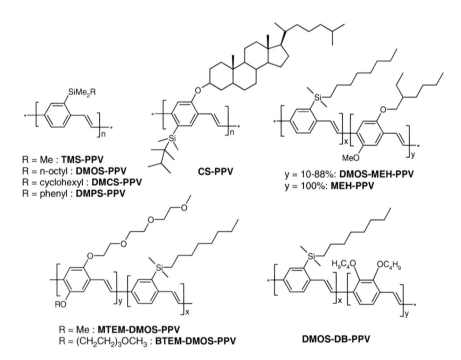

Fig. 37 Silyl-substituted poly(1,4-phenylene vinylene)s (PPV)

methyl group in TMS-PPV by an n-octyl group led to poly(2-dimethyloctylsilyl-1,4-phenylenevinylene) DMOS-PPV – another green polymeric emitter [153]. Films of DMOS-PPV were found to be highly fluorescent, with a quantum efficiency exceeding 0.60, an efficiency significantly higher than the value for PPV and MEH-PPV (0.27 and 0.15, respectively) [154]. Single layer electroluminescent devices (layer structure: ITO/DMOS-PPV/Ca or Al) exhibit an emission maximum at 520 nm with an internal quantum efficiency in the range of 0.2–0.3% [155] In double layer devices with an additional PBD electron conducting and hole blocking layer the internal quantum efficiency increased up to 2% [156]. However, as emitter in OLEDs the homopolymer DMOS-PPV has a disadvantage of a high turn-on voltage (15 V). Statistical copolymers of DMOS-MEH-PPV (Fig. 37) have been prepared to overcome this problem, which indeed showed a reduced turn-on voltage in OLED devices (6–7 V), however, accompanied by a decreased PL and EL efficiency with increasing ratio of DMOS to MEH [157]. Further improvements were achieved in DMOS-MTEM-PPV and DMOS-BTEM-PPV copolymers, containing silyl-substituted DMOS-PPV units as well as ion-transporting 2-methoxy-5-(trimethoxyethoxy)-1,4-phenylenevinylidene (MTEM-) or 2,5-bis(trimethoxyethoxy)-1,4-phenylenevinylidene (BTEM-)units. In so-called light-emitting electrochemical cells (LECs) mobile ions move during charge transfer from and into the polymer layer [158]. Both copolymers yielded LECs with low turn-on voltages (down to 2.5 V) and an increased efficiency (up to 0.5lm W^{-1}).

Cyclohexylsilyl- and phenylsilyl-substituted PPV derivatives, poly[2-dimethylcyclohexylsilyl-1,4-phenylene vinylene] (DMCS-PPV) and poly[2-dimethylphenyl-silyl-1,4-phenylene vinylene] (DMPS-PPV) (Fig. 37), as well as disubstituted poly[2,5-bis(dimethylcyclohexylsilyl)-1,4-phenylene vinylene] (BDMCS-PPV) and poly[2,5-bis(dimethylphenylsilyl)-1,4-phenylene vinylene] (BDMPS-PPV) (Fig. 38), were synthesized from bromine functionalized precursors and Gilch dehydrohalogenation polyaddition [159]. The disubstituted polymers BDMCS-PPV and BDMPS-PPV yielded insoluble thin films from soluble polymer precursor materials by a thermal conversion mechanism. Monosilylsubstituted DMCS-PPV and DMPS-PPV exhibited good solubility, good film-forming properties, and high molecular weights. These polymers feature a thermal stability which outperforms most other PPV derivatives including alkylsilyl-substituted PPVs, with higher glass transition temperature ($Tg = 125$–$127°C$), and remarkably high PL efficiencies both in solution ($\Phi_F = 0.86$–0.88) and as film ($\Phi_F = 0.82$–0.83). Simple single layer OLEDs fabricated from DMCS-PPV and DMPS-PPV (layer structure: ITO/polymer/Al) showed EL maxima at 510 and 515 nm, respectively, with an external EL quantum efficiency up to 0.03%. An additional poly(vinylcarbazole) (PVK) HTL increased the EL quantum efficiencies up to a maximum of 0.08%.

Silyldisubstituted PPV [poly(2,5-bis(trimethylsilyl)-1,4-phenylenevinylene)] (BTMS-PPV, Fig. 38) was prepared via two different precursor polymers, a water-soluble sulfonium and an organic soluble thiophenoxy precursor polymer [160]. In a single layer OLED device fabricated with BTMS-PPV (from the thiophenoxy precursor) as emitter showed a emission peak at about 545 nm, although with a turn-on voltage of 20 V and an external quantum efficiency of

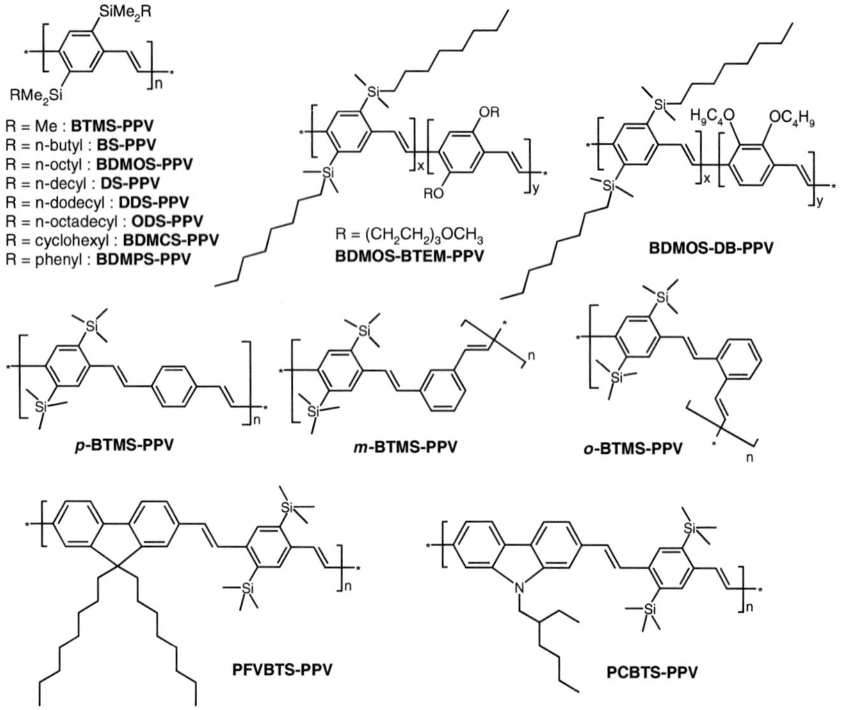

Fig. 38 Silyl-disubstituted poly(1,4-phenylene vinylene)s (PPV)

6.0×10^{-4}% in air. Silyl disubstituted PPV with longer alkylsilyl groups – poly (2,5-bis (dimethylbutylsilyl)-1,4-phenylenevinylene) BS-PPV [161, 162], poly(2,5-bis(dimethyloctyl-silyl)-1,4-phenylenevinylene) BDMOS-PPV [163], poly(2,5-bis(dimethyldecylsilyl)-1,4-phenylenevinylene DS-PPV [162], poly(2,5-bis (dimethyldodecylsilyl)-1,4-phenylene-vinylene DDS-PPV [162], and poly(2,5-bis(dimethyloctadecylsilyl)-1,4-phenylenevinylene ODS-PPV [162, 164], as well as unsymmetrically disilylsubstituted PPVs - were also reported [165]. Disubstituted BDMOS-PPV were found to have similar properties as their monosubstituted analog DMOS-PPV, both emitting light in the green region. However, BDMOS-PPV is thermotropic liquid crystalline between 160 and 180°C and emits polarized light from the film. Apparently the polymer is both hole and electron conducting, since a single layer device emits light at positive as well as negative bias. The statistical copolymer poly[2,5-bis(dimethyloctylsilyl)-1,4-phenylenevinylene)]-co-[2,5-bis(trimethoxyethoxy)-1,4-phenylenevilylene] BDMOS-BTEM-PPV exhibits an optical absorption and luminescence, in between the two homopolymers [166]. Both polymers were tested in OLED and LEC devices. Statistical copolymers of BDMOS units with 2,3-bis(dibutoxy)-1,4-phenylenevilylene units (BDMOS-DB-PPV) showed an EL efficiency up to $0.72 \, \text{cd} \, \text{A}^{-1}$ with a maximum luminance of $1,384 \, \text{cd} \, \text{m}^{-2}$ at 12 V and a turn-on voltage of 4.0 V, which was superior to its monosilylsubstituted analog, the DMOS-DB-PPV copolymer [167, 168].

Tuning the color of silyl-disubstituted PPVs is possible by introduction of *o*-, *m*-, or *p*-phenylenevinylene units into the copolymers having different effective conjugation lengths. Shim et al. reported on synthesis of poly[*o*(*m*,*p*)-phenylene-vinylene-*alt*-2,5-bis(trimethylsilyl)-*p*-phenylenevinylene], *o*(*m*,*p*)-PBTMS-PPV (Fig. 38) by the Wittig condensation polymerization of diphosphonium salts with dialdehyde monomers such as terephthaldicarboxaldehyde, isophthalaldehyde, and phthalicdicarboxaldehyde [169, 170]. Their photoluminescence spectra have peaks at 485, 470, and 440 nm for *p*-PBTMS-PPV, *o*-PBTMS-PPV, and *m*-PBTMS-PPV, respectively. The electroluminescence spectra of *o*-PBTMS-PPV and *p*-PBTMS-PPV exhibited EL emission at 470 and 490 nm, respectively, with a threshold voltage of 8–9 V. The emission wavelength of *o*-PBTMSPPV corresponds to a pure blue light which obviously is a consequence of the reduced π-conjugation length effected by the ortho-linkage and trimethylsilyl substituent. It is noteworthy that blue EL emission from PPV derivatives is quite exceptional.

Another approach to tune the emission color of PPV derivatives was attempted by introduction of fluorene or carbazole units into the PPV backbone. Thus, poly[*N*-ethylhexyl-3,6-carbazolevinylene-*alt*-2,5-bis-(trimethylsilyl)-*p*-phenylene-vinylene] (PCBTS-PPV) and poly[9,9-*n*-dihexyl-2,7-fluorenediylvinylene-*alt*-2,5-bis(trimethylsilyl)-*p*-phenylenevinylene] (PFBTS-PPV) have been synthesized by a Wittig polycondensation reaction [171]. These resulting polymers have shown a photoluminescence with maximum peaks at 480 and 495 nm corresponding to blue and greenish-blue emissions, respectively. The shift to shorter wavelengths compared to other PPV-based copolymers seems to be caused by the electron-donating effect of the silyl group. In films, PFBTS-PPV and PCBTS-PPV showed remarkably high PL efficiencies ($\Phi_F = 0.64$ and 0.81, respectively). As single layer OLEDs (layer structure: ITO/polymer/Al) the polymers emit blue and greenish-blue light with emission maxima of 480 nm (PCBTS-PPV) and 500 nm (PFBTS-PPV). Compared to MEH-PPV the relative EL quantum efficiencies of PCBTS-PPV and PFBTS-PPV were found to be 13 and 32 times higher.

A series of PPV derivatives containing a dimethyldodecylsilylphenyl unit as a pendant group was reported by the Jin group (Fig. 39) [172–174]. All polymers were completely soluble in common organic solvents, had high thermal stability (up to 400°C), and were used as emitting layer in OLEDs. Poly[2-(4-dimethyldodecylsilylphenyl)-1,4-phenylenevinylene] (*p*-SiPhPPV), synthesized via

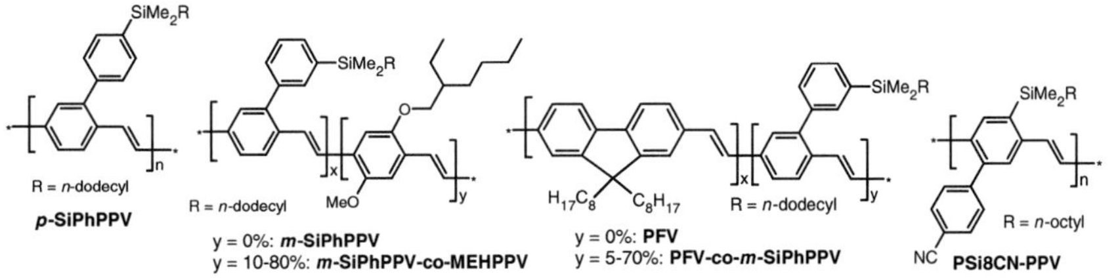

Fig. 39 Alkylsilylphenyl-containing PPV homo- and co-polymers used in OLEDs

Gilch polymerization, had a very high molecular weight of 300 kDa (M_n, polydispersity of about 3). In a single layer OLED device it showed a strong green emission at 524 nm with a maximum brightness of 5,900 cd m^{-2} (at 17 V) [172]. Asymmetric homo- and co-polymers: poly[2-(3′-dimethyldodecylsilylphenyl)-1,4-phenylene vinylene] (m-SiPhPPV) and poly[2-(3′-dimethyldodecylsilylphenyl)-1,4-phenylene vinylene-co-2-methoxy-5-(2′-ethylhexyloxy)-1,4-phenylene vinylene] (m-SiPhPPV-co-MEHPPV) were prepared similarly [173]. These polymers also had high molecular weights with narrow weight distributions. The copolymers showed better optical and EL properties than those of the homopolymers m-SiPhPPV and MEH-PPV. While devices made from the homopolymer m-SiPhPPV need a rather high turn-on voltage (14 V), the replacement of just 10% of the monomeric units by MEHPPV units in the copolymer reduced the turn-on voltage to 5.5 V, while 25% reduced to 2.3 V. Single layer light-emitting devices fabricated from m-SiPhPPV-co-MEHPPV emitted orange–red light ($\lambda_{max} = 588 - 595$ nm) with maximum brightness and an external luminance efficiency up to 19,180 cd m^{-2} or 2.9 lm W^{-1}. A series of similar copolymers, poly[9,9-di-n-octylfluorenyl-2,7-vinylene]-co-(2-(3-dimethyldodecylsilyl-phenyl)-1,4-phenylene vinylene)] (PFV-co-m-SiPhPV), emitted green light with turn-on voltages in the range of 4.5–6.0 V, maximum brightness up to 9,691 cd m^{-2} at 16 V, and a luminance efficiency up to 3.27 cd A^{-1} [174]. Thin films of PFV, m-SiPhPV, and PFV-co-m-SiPhPV were found to exhibit photoluminescence quantum yields between 21 and 42%, exceeding those of MEH-PPV.

Poly(p-phenylenevinylene) derivatives with an electron-withdrawing cyanophenyl group on the polymer backbone, poly[2-dimethyloctylsilyl-5-(4′-cyanophenyl)-1,4-phenylenevinylene] (PSi8CN-PPV, Fig. 39), were synthesized via the Gilch polymerization [175]. They showed very high glass transition temperatures (above 180°C). The presence of the electron-withdrawing cyanophenyl group lowered the HOMO and LUMO energy levels of PSi8CN-PPV relative to common PPV derivatives. OLEDs (layer structure: ITO/PEDOT/PSi8CN-PPV/LiF/Al) emitted light with a maximum at 513 nm, corresponding to green light with a CIE coordinate of (0.330, 0.599) close to standard green (0.30, 0.60), a maximum external quantum efficiency of 0.67%, and a maximum brightness of 2,900 cd m^{-2}.

The optical properties of silicon-substituted PPV have been summarized in Table 2. As can be seen from these data, these materials often outperform their carbon analogs (PPV, MEH-PPV, PFV), especially with photoluminescence quantum efficiency which reaches 83% in films. Their absorption and emission spectra as well as HOMO–LUMO levels can be tuned by appropriate substitution or copolymerization using a range of comonomers, leading to highly efficient green, blue, or orange–red emitters. The high solubility of such polymers is apparently caused by the twisting of the silicon-containing phenyl ring away from the plane of the conjugated PPV backbone and proves silicon side groups as a valid concept to introduce wet processability into classes of polymers with promising optical and electrical properties.

Table 2 Optical properties of PPV and its silyl-substituted derivatives

Polymer	λ_{max} (UV, nm) Solution[a]	λ_{max} (UV, nm) Film	λ_{max} (PL, nm)[b] Solution[a]	λ_{max} (PL, nm)[b] Film	PL efficiency (Φ) Solution[a]	PL efficiency (Φ) Film	λ_{max} (EL, nm)	E_g (eV, (UV/nm))[c]	HOMO (eV)	LUMO (eV)	Reference
PPV	–	426	–	–	–	0.27	–	2.5	−5.0/−5.1	−2.5	[152, 154, 161]
MEH-PPV	–	510	–	592	0.14	0.15	590	2.1	−4.94	−2.82	[154, 159, 161]
						0.18				−3.00	[173]
										−2.73	[175]
TMS-PPV		420	–	510 (550, 590)	–	–	–	–	−5.1	–	[152]
DMOS-PPV	420	414/420	–	520 (563)	–	0.60	520	2.43 (510)	−5.0	–	[153, 155]
CS-PPV	–	450	–	515 (550)	–	–	530	2.48 (500)	−5.7	–	[150, 151]
BDMOS-PPV	–	436	–	513	–	0.60	545	–	–	–	[160]
ODS-PPV	436	447	494 (527)	510 (526)	–	–	–	2.38 (520)	−5.66	−2.54	[164]
	435	445	–	510 (536)	0.89	–	–	2.42 (513)	−6.52	−2.62	[162]
BS-PPV	434	434	494 (526)	512 (543)	0.72	–	–	2.43 (510)	−5.59	−2.56	[161]
					0.87			2.45 (506)	−5.54	−2.60	[162]
DS-PPV	435	439	501 (535)	520 (548)	0.86	0.53	–	2.42 (512)	−5.52	−2.62	[162]
DDS-PPV	436	448	494 (523)	511 (540)	0.87	0.56	–	2.43 (510)	−5.56	−2.58	[162]
BDMCS-PPV	–	413	–	519 (555)	–	0.53	518 (542)	–	–	–	[159]
BDMPS-PPV	–	370	–	525 (560)	–	0.56	525 (556)	–	–	–	[159]
DMCS-PPV	414	420	483 (515)	510 (545)	0.88	0.82	510	–	−5.64	−2.58	[159]
DMPS-PPV	418	422	485 (517)	513 (547)	0.86	0.83	515	–	−5.55	−2.64	[159]
o-PBTMS-PPV	–	340	–	470	–	–	470	–	–	–	[169, 170]
m-PBTMS-PPV	–	330	–	440	–	–	–	–	–	–	[169, 170]
p-PBTMS-PPV	–	380	–	485	–	–	490	–	–	–	[169, 170]
PCBTS-PPV	–	355	–	480	–	0.64	480	–	–	–	[171]
PFBTS-PPV	–	385	–	495	–	0.81	500	–	–	–	[171]

(continued)

Table 2 (continued)

Polymer	λ_{max} (UV, nm)[a] Solution	λ_{max} (UV, nm)[a] Film	λ_{max} (PL, nm)[b] Solution[a]	λ_{max} (PL, nm)[b] Film	PL efficiency (Φ) Solution[a]	PL efficiency (Φ) Film	λ_{max} (EL, nm)	E_g (eV, (UV/nm))[c]	HOMO (eV)	LUMO (eV)	Reference
p-SiPhPPV	427	427	486	524	–	–	524	2.44	−5.43	−2.99	[172]
m-SiPhPPV	–	435	–	525	–	0.35	528	2.38	−5.30	−2.99	[173]
m-SiPhPPV-co-MEH-PPV (3:1)	–	433	–	518	–	–	525 (562)		−5.18	−2.80	[174]
	–	464	–	594	–	0.21	595		−5.21	−3.15	[173]
PFV-co-m-SiPhPV (1:1)	–	419	–	518	–	0.42	525 (562)	2.39	−5.15	−2.76	[174]
PFV	–	409	–	468, 500	–	0.29	471 (503, 534)	2.63	−5.26	−2.63	[174]
PSi8CNPV	427	434	496 (527)	513 (547)	0.62	–	513	2.47 (501)	−5.72	−2.75	[175]

Notes: UV – absorption, PL – photoluminescence, E_g – optical band gap
[a]Measured in chloroform solutions
[b]Values in parentheses represent shoulder peaks
[c]Determined from the absorption edge (value in parentheses) of the UV-vis spectrum

6.2 Silanylene-Containing Polymers

There is increasing interest in the polymers composed from alternating organosilicon and π-electron units in the backbone due to their unique electronic structure [176, 177]. Typical examples of silanylene-containing polymers are presented in Fig. 40. The main synthetic pathways to such polymers include: (1) Wurtz-type coupling of bis(chlorosilyl) compounds [178, 179], (2) polycondensation of dichlorosilanes with organodilithium derivatives [180, 181], and (3) transition metal-catalyzed polycondensation of magnesium- [182, 183], zinc- [184], or tin-derivatives [185] of diarylsilanes. The first two routes lead to polymers with one to three linked conjugated arylene units within the polymeric chain. The third route allows preparation of polymers with longer chromophores, which show promising properties in organic electronic devices.

A series of mono- and disilylene copolymers with di-, ter-, or quaterphenylene was reported by Ishikawa and his group (MS2P, DSmP, and MS2P3 in Fig. 40) [182] from which the disilanylene-substituted polymers DSmP were found to be photoactive. Thin polymer films degraded during irradiation in air by scission of the Si–Si bonds and degradation products including silanol and siloxy units were formed. The photoactivities of the disilanylene-substituted polymers in solution decrease with increasing extension of the π-electron system. Cyclic voltammogram (CV) measurements proved that the Si–Si orbital plays an important role in the first oxidation step of disilanylene–oligophenylene polymer films [186]. Monosilylene-copolymers MS2P and MS2P3 were found to be inert toward UV irradiation.

Poly[(1,2-tetraethyldisilanylene)-9,10-diethynylanthracene] (DSDEA) was evaluated as hole-transport material in OLEDs with Alq$_3$ as an electron transporting emitter layer [187]. These devices exhibited a maximum $\eta_{EL} = 0.2\%$. Although this is approximately a magnitude higher than a single layer of Alq$_3$ in absence of hole-transport material, it is very much below the state of the art, triplet emitter

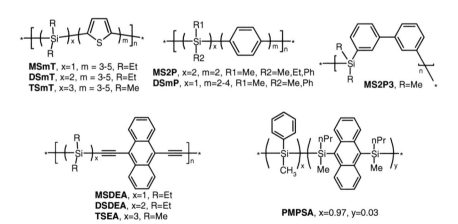

Fig. 40 Typical examples of silanylene-containing polymers

OLEDs reaching an external quantum yield up to 20% and more [188]. The low η_{EL} values of the organosilicon-based devices have been attributed to the poor electron-blocking properties of the organosilicon polymers which allows electrons to migrate from the emitter into the hole-transport layer and combine without light emission. Subsequently a series of similar mono-, di-, and trisilanylene copolymers composed of alternating 9,10-diethynylanthracene and organosilicon units were prepared by coupling reactions of 9,10-di(lithioethynyl)anthracene with dichloromono-, di-, and trisilanes, $Cl(SiR_2)_mCl$ ($m = 1$–3) by the same group [189] and evaluated in double-layer-type OLEDs employing these polymers as hole-transport materials in combination with Alq_3 as electron transporting-emitter. Increasing the number of silicon atoms in the polymer unit from $m = 1$ to 3 leads to an increased turn-on voltage and decreases the maximum current density of the devices. The highest luminance of $1,300 \, cd \, m^{-2}$ was measured for the silanylene-diethynylanthracene polymer MSDEA (Fig. 40).

A poly(methylphenylsilane) containing 3% anthracene units in the polymer backbone (PMPSA) was synthesized by a Wurtz-type coupling reaction of 9,10-bis(methylpropylchlorosilyl)anthracene and methylphenyldichlorosilane [190]. This polymer exhibited light blue photoluminescence with 87% quantum efficiency with a strong photoluminescence emission at 500 nm originating from the anthracene units, and a weak one at 350 nm from the polysilane chain. Intramolecular energy transfer from the polysilane σ^* orbital to the anthracene π^* orbital enables the emission at 500 nm not only by π–π^* excitation of the anthracene unit at 420 nm, but also by σ–σ^* excitation of the polysilane fragments at 250–350 nm.

Poly(silylene)oligothiophenes MSmT were first reported by Corriou and coworkers [180]. Copolymers with smaller oligothienyl blocks ($m = 1 - 3$) were prepared by a coupling reaction of corresponding dilithioderivatives of oligothiophenes and dichlorosilanes in 51–80% yield. Copolymers with longer oligothienyl blocks ($m = 3 - 5$) were synthesized by Pd-catalyzed coupling of zinc derivatives of dithienylsilanes with corresponding aryldibromides in 66–95% yield. Two series of copolymers were obtained with quite low molecular weights (M_w between 1,930 and 5,940). Upon doping with $NOBF_4$ polymer films showed a maximum conductivity of 10^{-2} to $10^{-1} \, S \, cm^{-1}$.

Ishikawa and his group prepared a series of copolymers with alternating mono-, di-, or trisilanylene units and 2,5-oligothienylene groups $[(SiR_2)_xT_m]_n$ with R = Me, Et, $x = 1 - 3$, $m = 2 - 5$ by a $NiCl_2(dppe)$-catalyzed Grignard coupling ($NiCl_2(dppe)$ – 1,2-bis(diphenylphosphino)ethane nickel(II) chloride) [183, 191]. DSmT copolymers were readily obtained in 75–97% yield and a moderate molecular weight of 17–53 kDa (M_w, polydispersity of 2.1–4.8) [183]. Other copolymers were obtained with similar properties. Irradiation of the DSmT copolymers with UV light resulted in cleavage of the silicon–silicon bonds, but with an increasing number of thienyl units the sensibility towards UV irradiation decreased. With an increasing number of thienyl units the UV absorption and emission maxima shifts to lower energies, and is little affected by the length of the silicon chain. Doping the polymers by exposition of films to $FeCl_3$ vapor, moderate conductivities of $1.3 \times 10^{-4} - 2.3 \times 10^{-1} \, S \, cm^{-1}$ were observed. In general the conductivities tended

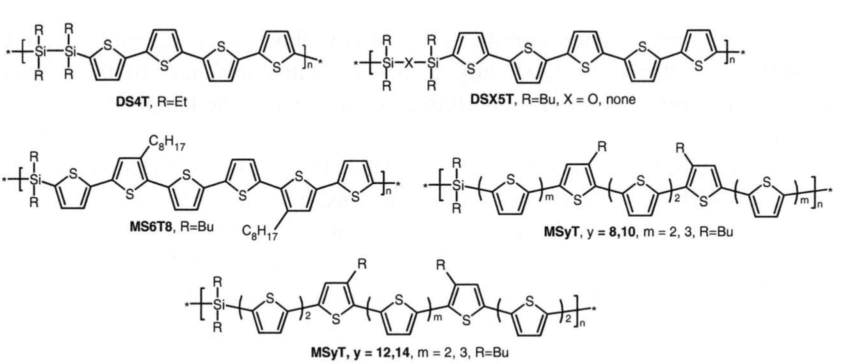

Fig. 41 Chemical structures of copolymers of silylene and 2,5-oligothienylenes used in EL and OFET devices

to increase with the number of thienyl units, and decreased with the length of the silicon chain. Conductive films made by electrochemical doping showed similar results and tendencies [192].

In 1993 the evaluation of poly[(silylene)- and (disilanylene)oligothienylenes] in an OLED was started by Hadziioannou and coworkers [193]. MS6T8 used as emitter (Fig. 41) in a simple device with ITO as the hole- and electron-injecting electrodes emits orange–red light ($\lambda_{max} = 612$ nm). More sophisticated double-layer-type OLEDs were tested using MSmT ($m = 3$–5) and DS4T copolymers as the hole-transporting materials in combination with the emitter Alq$_3$ [191]. Both, a decreasing number of thienyl units and a decreasing length of the silicon chain increased the turn-on voltage and reduced the maximum current density of the device. The highest luminance of 2,000 cd m^{-2} at 12 V and the lowest threshold voltage at 5 V was obtained from a device based on the DS4T copolymer.

DS5T and its disiloxyl analog DSO5T (Fig. 41) with reasonable molecular weights of 76–94 kDa (M_w with a polydispersity of 2.5–3.8) were used as hole-transport materials in combination with Alq$_3$ as emitter in a double-layer OLED setup [194]. The OLEDs showed the expected green light from Alq$_3$ emission after reaching the turn-on voltage. DS5T exhibited a slightly lower turn-on voltage compared to DSO5T, while a similar maximum luminance of 4,000 cd m^{-2} was obtained for both polymers.

Ohshita, Kunai, and coworkers synthesized monosilanylene-oligothienylene alternating polymers **MSyT** with increasing conjugation length of the oligothiophene chromophore (T = 2,5-thienylene or 3-butyl-2,5-thienylene, $y = 8, 10, 12,$ and 14, R = Bu, see Fig. 41) and tested these polymer in OLEDs and OFETs [185]. The polymers were obtained by Stille coupling reactions with a molecular weight of 7.3–14.2 kDa (M_w, polydispersity of 1.3–1.7). The onset voltages for the OLED (layer structure: ITO/**MSyT**/Alq$_3$/Mg:Ag) decreased with increasing conjugation in the chromophore with a maximum luminance of 900 cd m^{-2} for $y = 12$ before higher voltage introduce irreversible damages. The external quantum efficiencies of the devices with MSyT were about 0.1–0.2% at the maximum luminance.

Field-effect charge carrier mobilities in **MSyT** films were determined to be $(3.4$–$6.9) \times 10^{-5}$ cm^2 V^{-1} s^{-1}, slightly increasing with the length of the oligothiophene unit from $y = 8$ to 12, and almost not affected by the length of the silicon atom chain.

Extensive electrochemical studies of mono- [195] and disilanylene [196–198] copolymers with oligothienylenes proved the sensitivity of the Si–Si bond towards electrochemical oxidation. In the CV of films made from MSmT ($m = 4, 5$) two couples of redox peaks were found at a potential in the range of 0–1.2 V. Combined with doping level and in situ spectroelectrochemical measurements the results suggest a mechanism in which in a first step the polymer repeating unit is oxidized to the radical cation followed by the oxidation to the dication. From in situ UV-vis-NIR measurement it was concluded that the radical cation forms π-dimers via a coupling reaction [195]. Poly[(tetraethyldisilanylene)oligo(2,5-thienylene)] derivatives (DSmT; $m = 3$–5) have been successfully doped by electrochemical oxidation. Band-gap energies of 2.52, 2.65, 2.82, and 3.27 eV were measured for DS5T, DS4T, DS3T, and DS2T respectively. Doped films of DS5T, DS4T, and DS3T exhibited electrical conductivities in the order of 10^{-3} to 10^{-4} S cm^{-1} with BF$_4$$^-$ as counter ion while the work functions increased from 5.1 to ca. 5.5 eV during doping [196]. The instability of Si–Si bonds towards electrochemical oxidation in DSmT ($m = 3$–5) was demonstrated by its cleavage at an electrochemical potential as low as 0.5 V (vs Ag/Ag+), resulting in the dissolution of oligothiophene-like species. The decomposition products were oxidized to form a new polymer film on the surface of an original DSmT film and the new composite polymer film is subsequently doped [197].

Sotzing et al. exploited the electrochemical instability of the copolymers containing silylene and thiophene in an ingenious way [199, 200]. They prepared a series of copolymers of 3,4-ethylenedioxythiophenes and their derivatives with dialkylsilylene units, processed films from solution, and converted the films electrochemically to yield an insoluble highly conductive film of poly(3,4-ethylenedioxythiophene) PEDOT (Scheme 13). The precursor polyarylsilanes, PSEDT, were prepared via step-growth polymerization of dialkyldichlorosilane and dilithiated EDOT derivatives with a molecular weight of 51 and 85 kDa (M_w and polydispersities of 2.1 and 1.7) for PSEDT(a) and PSEDT(c), respectively. Thin films of precursor polymers PSEDT were converted into the conjugated polymer PSEDTx via solid state electrodesilylation under conditions in which polymers PSEDT did not dissolve. Albeit the conversion is not quantitative and residual silylene groups remain in the polymer PSEDTx, the conductivity of PSEDTx(a) was determined to be 20 S cm^{-1} which is in accordance with the conductivity of conventionally electrodeposited PEDOT (16 S cm^{-1}).

Ohshita et al. synthesized a series of alternating copolymers of 2,6-diaryldithienosilole and organosilicon units [201, 202]. These polymers were obtained in 33–71% yield by nickel-catalyzed coupling of 2,6-dibromodithienosilole with di-Grignard reagents prepared from the corresponding bis(bromoaryl)disilanes or monosilane (Scheme 14) resulting polymer with limited molecular weights of 9.4–15.9 kDa (M_w, polydispersity 1.4–2.6). With units of four conjugated aryl rings

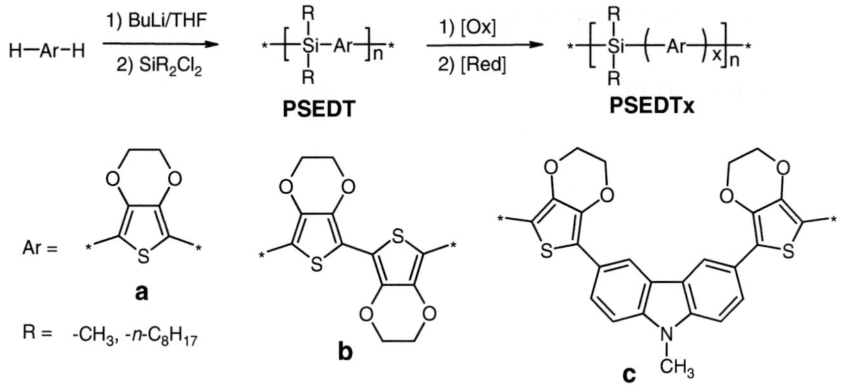

Scheme 13 Schematic representation of synthesis of soluble 3,4-ethylenedioxythiophene containing poly(arylsilane) PSEDT and its electrochemical conversion into highly conducting conjugated polymer PSEDTx

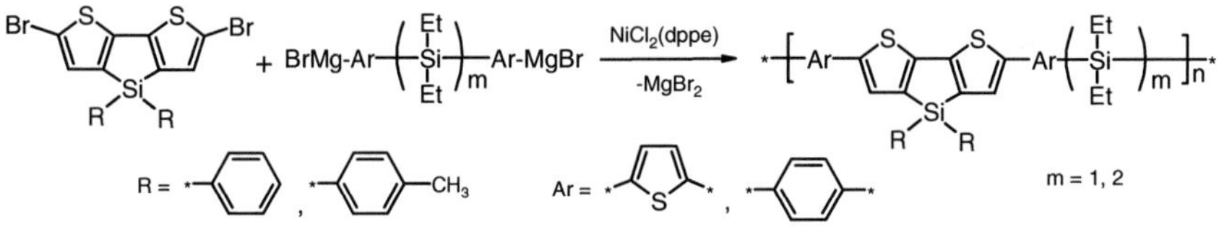

Scheme 14 Synthesis of alternating copolymers of 2,6-diaryldithienosilole and organosilicon units

separated by mono- or disilyl bridges, these polymers strongly absorb light in the UV-vis region and exhibit emission bands at 422–444 and 482–541 nm, respectively. The emission of the Si-bridged polymers is remarkably red-shifted in comparison to the corresponding di- and monosilanylene-quaterthienylene polymers DS4T and MS4T (Fig. 40), indicating the existence of expanded π-conjugation due to the silole ring. Their CV display first oxidation peaks at 0.95–1.00 V vs. SCE (saturated calomel electrode). Upon doping films with FeCl$_3$ vapor, they became conducting with a low conductivity of 3.3×10^{-5} to 8.7×10^{-3} S cm^{-1}. Although copolymers containing disilylene units showed an order of magnitude higher conductivity, their hole-transporting properties appeared to be lower than polymers containing monosilylene units which was evidenced by a study of hole-transport material in OLEDs. Introduction of the silole ring in the copolymer structure introduced some electron transporting properties to the polymers, which were higher compared to MS4T or PVK, but significantly lower compared to typical electron transporting material, such as Alq$_3$.

Silanylene-containing polymers have shown some (semi)conductive and luminescence properties, which, however, were often well below the values reported for other classes of conjugated polymers. In our view, the most interesting feature of such polymers is redox- and UV-light instability of silanylene and especially disilanylene units, which might be utilized in patterning of the functional electronic layers, necessary to produce complex organic electronics circuits, or for preparation of (insoluble) conductive polymer films from soluble silanylene precursors.

6.3 Silol-Containing Polymers

The incorporation of silole unit opens a wide field of research to modify the electronic structure of known carbon-based polymeric chromophores. Chromophores like the carbon analog of TS and 2,7-fluorene have been successfully used as emitters in OLEDs and semiconductors in OTFT and have been combined with additional structural units to adjust electronic properties. Replacement of the bridging carbon atom in such structural units by silicon has a strong impact on the electronic properties: interaction between the silicon σ-orbitals and the π-system of the carbon chromophore will lift the HOMO and lower the LUMO energy and therefore will decrease the band gap. Silicon modified chromophores may be combined with their carbon analogs and other successfully used structural units like thiophene, bithiophene and phenylene to adjust further the energy levels of HOMO and LUMO. Incorporation of benzothiadiazole (BT) and other structural units with nonbonding electrons will introduce new energy levels to form donor–acceptor type polymers with even smaller band gaps (Fig. 42).

The TS-based homopolymer, poly(4,4-di-*n*-hexyldithienosilole) (TS6) as well as its copolymers with mono- and bithiophene subunits, poly(4,4-di-*n*-hexyldithienosilole-*alt*-(bi)thiophene) (TS6T1, TS6T2), were almost simultaneously reported by Facchetti and Marks [203, 204] and Ohshita [205] (Fig. 43). Facchetti and Marks examined in detail the consequences of introducing TS and BS structures into a thiophene polymer backbone by comparison of the new BS-based homopolymer, poly(9,9-di-*n*-octyldibenzosilole) (BS8) and copolymers with mono- and bithiophene subunits, poly(9,9-di-*n*-octyldibenzosilole-*alt*-(bi)thiophene) (BS8T1, BS8T2), with their carbon analogs F8T1 and F8T2. Thiophene-based copolymers were prepared by Stille coupling, while for phenylene-based copolymers Suzuki coupling reactions proved to be suitable. The molecular weight of thiophene-based copolymers was around 35 kDa (M_w, polydispersity 2.9–3.7) and around 120 kDa

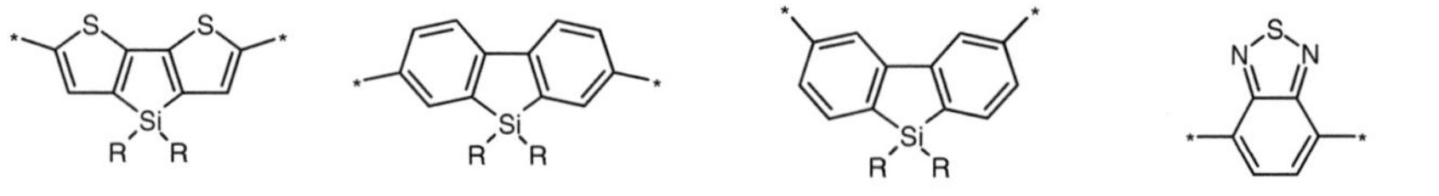

4,4-dialkyldithienosilole (**TS**)   9,9-dialkyl-2,7-dibenzosilole (**BS**)   9,9-dialkyl-3,6-dibenzosilole (**PS**)   2,1,3-benzothiadiazole (**BT**)

Fig. 42 General structural units often used in silole-containing copolymers design

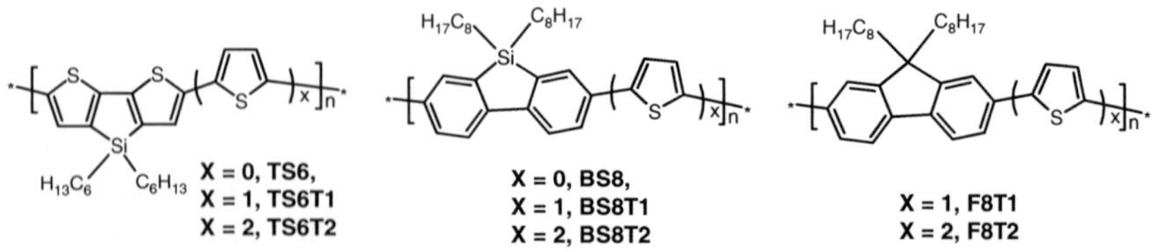

X = 0, TS6,
X = 1, TS6T1
X = 2, TS6T2

X = 0, BS8,
X = 1, BS8T1
X = 2, BS8T2

X = 1, F8T1
X = 2, F8T2

Fig. 43 Dithienosilole- and dibenzosilole-based homopolymers, their mono- and bithiophene copolymers and carbon analogs

for phenylene-based copolymers. Charge carrier mobilities turned out to be higher for thiophene-based copolymers with a maximum of $0.08\,\text{cm}^2\,\text{V}^{-1}\,\text{s}^{-1}$ for copolymer TS6T2 and a quite acceptable on/off ratio of 10^4–10^6 (unaligned films under ambient conditions). Non-encapsulated OTFTs were proven to be highly stable in air, obviously due the absence of alkyl side groups on the thiophene ring in contrast to regioregular poly(3-hexylthiophene) (P3HT). All silole-containing copolymers studied are apparently of excellent thermal stability, no significant weight loss being measured below 400°C by TGA. The absorption maxima values of BS-based copolymers were red-shifted by 40–50 nm vs fluorene-based analogs F8T1 and F8T2, demonstrating Si σ^*-orbital overlap with the π-conjugated chromophore.

Ohshita and his coworkers synthesized poly(dithienosilole-2,6-diyl)s with a molecular weight of 2.7–13.2 kDa (M_w, polydispersity 1.1–1.8 after precipitation) using palladium-catalyzed oxidative homocoupling of 2,6-bis(tributylstannyl) dithienosiloles with $CuCl_2$ (Scheme 15a) [205]. Compared to the corresponding silole-free polythiophenes, the absorption peak of these polymers is red-shifted by about 100 nm. Alternating copolymers with molecular weights of 5.3–12.8 kDa (M_w, polydispersity 1.3–1.7 after precipitation) were prepared by palladium-catalyzed cross-coupling reactions using 2,6-dibromodithienosiloles and distannylthiophene or bithiophene as starting materials (Scheme 15b, c).

The polymers were studied in OLEDs as emitters and as hole-transport materials in combination with the electron conducting emitter Alq_3. Used as emitter in a single layer device they emit red light while in combination with Alq_3 they act as

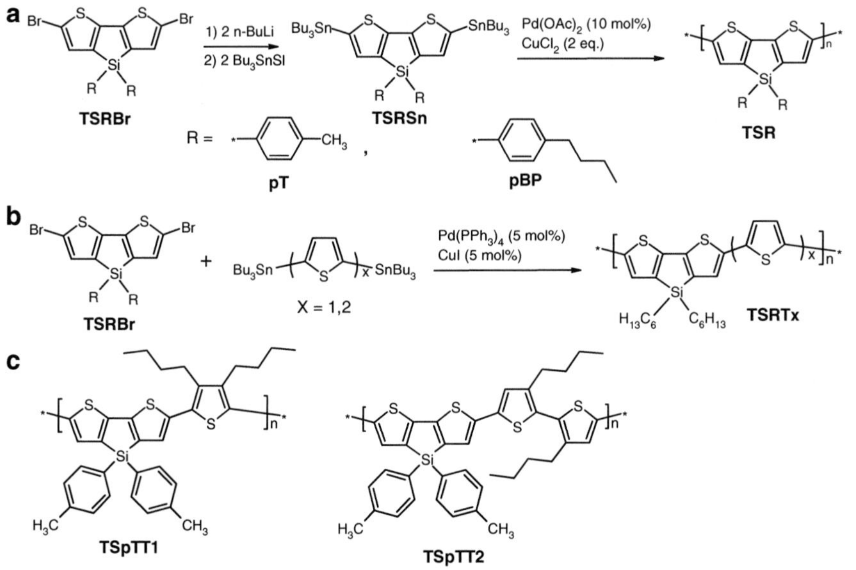

Scheme 15 (a) Synthesis of poly(dithienosilole-2,6-diyl)s by oxidative homocoupling, (b) synthesis of alternate dithienosilol copolymers by Stille cross-coupling reactions, and (c) structures of the most soluble copolymers

hole-transport material with a maximum luminance of 500 cd m^{-2} at 13 V (obtained for copolymer TSpTT2). In this class of polymers, solely TSpBP was found to work as semiconductor in OTFT with a low mobility of 10^{-7} cm^2 s^{-1}.

In 2005 Holmes et al. prepared poly(9,9-dihexyl-2,7-dibenzosilole) BS6 by Suzuki copolymerization of dibromo and bis(boronate) 2,7-disubstituted BS monomers. The monomers were synthesized by the selective *trans*-lithiation of 4,4'-dibromo-2,2'-diiodobiphenyl followed by silylation with dichlorodihexylsilane [206]. BS6 was obtained in 93% yield with molecular weight of 220 kDa (M_w) and broad polydispersity (approximately 7). The polymer revealed similar optical properties in thin films as the corresponding polyfluorene (PF) with an absorption maximum at 390 nm and a luminescence maximum at 425 nm and an optical band gap of 2.93 eV (PL efficiency 0.62). Preliminary studies with single layer light-emitting OLEDs confirmed the emission maxima at 431 and 451 nm and a similar efficiency as PF8. Films of polymer BS6 exhibit a glass temperature of 149°C and are apparently resistant to thermal stress: annealing at 250°C for 16 h in air with ambient light did not change its photoluminescence spectrum significantly (Fig. 44). This is in clear contrast to its carbon analog PF8, the photoluminescence spectrum of which degraded completely after 4 h at a temperature of 200°C.

Two random copolymers based on poly(di-*n*-hexylfluorene-*co*-4,4-diphenyldithienosilole) (PF-TS) were synthesized by Jen and coauthors by Suzuki coupling reaction (Scheme 16) [207]. The molecular weight of the polymers obtained reached 146 kDa (M_w, polydispersity 2.8) for PF9-TS, and 76 kDa (M_w, polydispersity 2.2) for PF19-TS. Both copolymers were thermally stable to thermal decomposition temperature up to 400°C. PF9-TS exhibit a slightly higher T_g of 100°C than PF19-TS (95°C), likely due to the higher content of rigid diphenyldithienosilole units in PF9-TS.

Apparently even the presence of a small fraction of electron-deficient TS moiety in the copolymers increases their electron affinity significantly. Subsequently they facilitate charge recombination by acting as charge-trapping sites for both holes and electrons. By exploiting a Förster energy transfer from the higher energy fluorene segments to the lower-energy TS-containing segments it is possible to tune the color of emitted light to longer wavelengths. In an OLED emitting

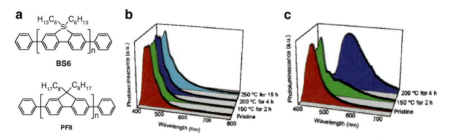

Fig. 44 (a) Structures of poly(9,9-dihexyl-2,7-dibenzosilole) BS6 and poly(9,9-dihexyl-2,7-fluorene). PL spectra of BS6 (b) and PF8 (c) films after annealing at different temperatures. Reproduced with permission of the American Chemical Society from [206]

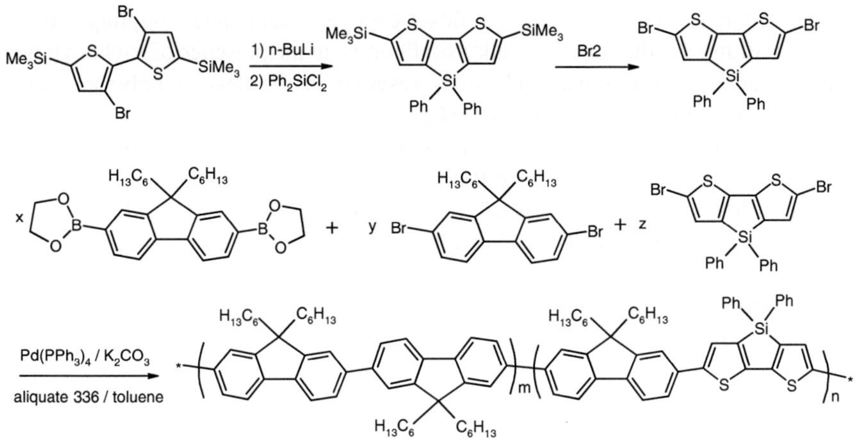

PF9-TS: x = n, y = 0.9n, z = 0.1n
PF19-TS: x = n, y = 0.95n, z = 0.05n

Scheme 16 Synthetic scheme for 4,4-diphenyldithienosilole monomer and silole-containing copolymers by Jen and coauthors [207]

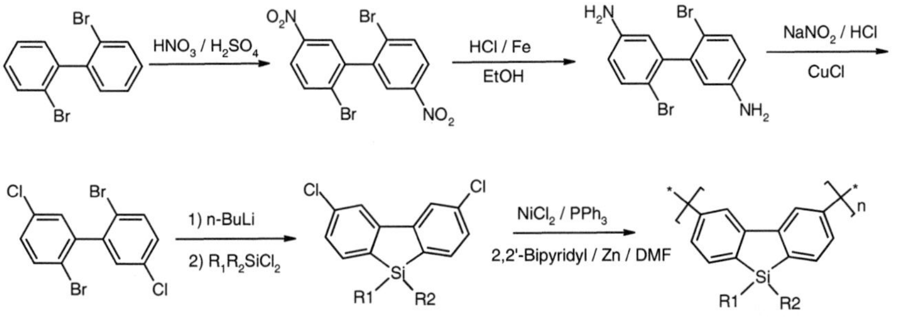

BrCl2P

R1 = R2 = C_6H_{13}: **SFC6**
R1 = CH_3, R2 = $C_{12}H_{25}$: **SFC1C12**

R1 = R2 = C_6H_{13}: **PSFC6**
R1 = CH_3, R2 = $C_{12}H_{25}$: **PSFC1C12**
R1 = R2 = C_8H_{17}: **PSFC8**

Scheme 17 Synthesis route of poly(3,6-silafluorene)s (PSF)

green light the low turn-on voltage of 4.6 V together with a maximum brightness of more than 25,900 cd m^{-2}, and a maximum external quantum efficiency of 1.64% was attributed to improved charge injection and recombination when PF9-TS was used as emitter in combination with in situ polymerized hole-transport material bis-tetraphenylenebiphenyldiamine-perfluorocyclobutane (BTPD-PFCB).

Cao et al. synthesized poly(9,9′-alkyl-3,6-silafluorenes as an ultraviolet-emitting polymers with a wide band gap of 4.0 eV [208]. Since the silole ring in the silafluorene undergoes a ring-opening reaction during bromination with standard bromination reagents and during oxidative polymerization, e.g., with $FeCl_3$, the authors had to design a new synthetic route (Scheme 17). The starting monomers, 3,6-dichloro-9,9′-alkylsilafluorenes (SFC6 and SFC1C12), were synthesized from 2,2′-dibromobiphenyl via 2,2′-dibromo-5,5′-dichlorobiphenyl (BrCl2P) following the procedure shown in Scheme 17. Nickel-catalyzed Yamamoto coupling reaction of the resulting 3,6-dichloro-9,9'-alkylsilafluorene in the presence of triphenylphosphine, zinc, 2,2'-bipyridine, and NiCl₂ resulted in the desired polymer with a molecular weight of around 10 kDa (M_n, polydispersity of around 1.6). Poly(3,6-silafluorene) reveals an excellent thermal stability with a decomposition temperature of 442°C for PSFC6 and 425°C for PSFC1C12, even higher than poly(2,7-fluorene) (385°C) [209] and a stability towards oxygen and light similar to PFs. Replacement of carbon in position 9 of PF by silicon obviously has little impact on rigidity of the polymer chain.

The optical properties of polymers PSFC6 and PSFC1C12 were found to be similar, both exhibiting an absorption peak at 250–300 nm and an absorption edge of around 310 nm in solid films and solutions. All absorptions are significantly blue-shifted compared to poly(9,9-dialkyl-2,7-fluorene)s [210] and poly(9,9-dialkyl-2,7-silafluorene with an absorption maximum of 380 nm and absorption onset at 420 nm [206]. A reduced conjugation of the 3,6-linkage compared to 2,7-linkage in silafluorenes may account for the change in optical properties. Poly(3,6-silafluorenes) PSFC6 and PSFC1C12 show a maximal photoemission at 355 and 360 nm, in solution and as solids, with a luminescence quantum efficiency of 0.25 and 0.30. Unlike most other conjugated polymers, the PL emission of polysilafluorene in the solid state shows almost no red shift compared to spectra in solution.

Almost simultaneously to Cao et al., Holmes et al. described the preparation of the 3,6-disubstituted dibenzosilole monomers 4, 5, and 10 by two alternative routes (Scheme 18) [211]. The first method employs selective transmetalation of 2,2'-diiodo-5,5'-dibromobiphenyl (3) followed by silacyclization (as used for the corresponding 2,7-dibenzosilole) [206], while the second exploits the displacement of methyl groups from a dibenzosilole 8. The latter route demonstrates the versatility of the alkyl-displacement on silicon reaction to introduce solubilising substituents to the bridging silicon, keeping in mind the limited commercial availability of disubstituted dichlorosilanes.

Suzuki copolymerization of monomers 4 and 5a with phenyl group end-capping yielded poly(9,9-dioctyl-3,6-dibenzosilole) (PSFC8) in 93% yield with similar molecular weight as obtained by Cao et al. (M_n 11 kDa, polydispersity 2.1). The polymer has a sufficiently high triplet energy (2.55 eV), higher than that of

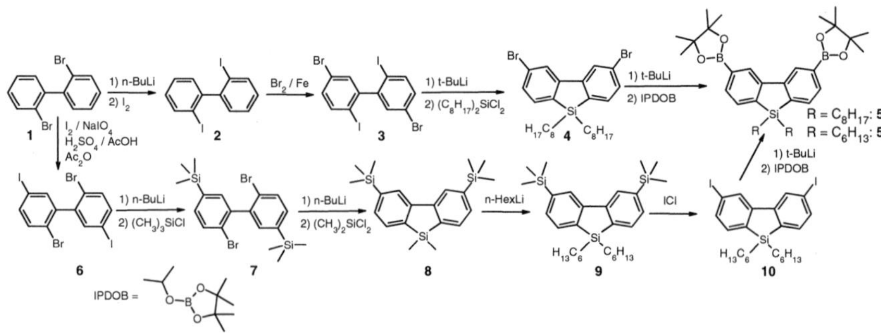

Scheme 18 Synthesis of 3,6-disubstituted dibenzosilole monomers

commonly used PFs (2.1 eV) [212] and comparable with that of polycarbazoles (2.6 eV) [213], to act as a host for green electrophosphorescent emitters. From CV measurements the LUMO energy of PSFC8 was determined as −2.15 eV (cf. PVK at −2.0 eV), indicating a withdrawal of electrons from the phenyl rings by σ^*–π^* conjugation, which is often observed in molecular and polymeric siloles [58, 214]. The HOMO level calculated from the optical band gap (3.5 eV) was estimated to be −5.65 eV (cf. PVK at −5.8 eV) [215]. A lower HOMO energy of polymer PSFC8 compared to PVK is expected to result in a lower hole-injection barrier at the PEDOT:PSS interface. The HOMO and LUMO energies of PSFC8 facilitate both electron and hole injection in OLEDs. In a triplet layer structure a complete energy transfer occurred from the polymer to the triplet dopant fac-tris[2-(2-pyridyl-κN)-5-methylphenyl]iridium(III) (Ir(mppy)$_3$) (12) with a triplet energy of 2.4 eV and resulted in a at a low device turn-on voltage.

Another series of copolymers (PSiFF) based on 3,6-silafluorene and 2,7-fluorene were synthesized by Cao and coauthors via the Suzuki reaction (Fig. 45a) [216]. The content of the 3,6-silafluorene subunits was between 5 and 50 mol%. The molecular weights ranged from 19 to 39 kDa (M_n, polydispersity 1.4–2.5). All copolymers exhibited good thermal stability with degradation temperatures of 395–418°C.

Both in solution and as film, poly(3,6-silafluorene-co-2,7-fluorene) showed only a single absorption peak which shifts to shorter wavelengths with increasing 3,6-silafluorene content in the copolymers (381 nm for the PFO homopolymer, 283 nm for the PSiF homopolymer). Parallel to the absorption spectra, the PL spectra of the PSiFF copolymers in solution and as films significantly shift to the blue with increasing 3,6-silafluorene content in the copolymers (431 nm for the PFO homopolymer to 420, 416, and 400 nm for copolymers). Absolute PL efficiencies of copolymer films start with 0.67 for the poly(9,9-dihexyl)fluorine homopolymer,

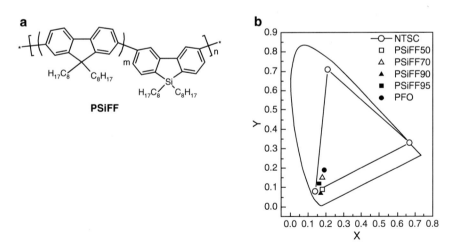

Fig. 45 (a) Chemical structure of 3,6-silafluorene and 2,7-fluorene copolymers PSiFF. (b) CIE coordinates of the devices fabricated from the PSiFF copolymers. Reproduced by permission of The Royal Society of Chemistry [216]

reach a maximum at the copolymer containing 10% of 3,6-silafluorene units (PSiFF90) (0.84), and then drop quickly with increasing 3,6-silafluorene content to 0.38 (alternating copolymer with 50% of 3,6-silafluorene units) and 0.14 for the poly(3,6-(9,9-dihexyl)silafluorene) homopolymer.

The incorporation of 3,6-silafluorene units into PF suppresses long-wavelength emission, and significantly improves the color purity in OLEDs. Fluorene and 3,6-silafluorene contribute to the conjugation of the chromophore and the emission is remarkably blue-shifted with increasing 3,6-silafluorene content. In OLEDs (layer structure: ITO/PEDOT:PSS/PVK/polymer/Ba/Al) PSiFF90 reached η_{EL} of 3.34% and a luminous efficiency of $2.02\,cd\,A^{-1}$ at a brightness of $326\,cd\,m^{-2}$. The CIE coordinates of (0.16, 0.07) almost match the NTSC specification for blue color (0.14, 0.08) (Fig. 45b). Moreover, the incorporation of 3,6-silafluorene into the PF main chain significantly improved the spectral stability during annealing. Poly(3,6-silafluorene-co-2,7-fluorene) therefore might be a promising blue emitter with good color purity.

A copolymer derivative P36–27SiF90, containing 10% of 3,6-silafluorene and 90% of 2,7-silafluorene units, was accessed via Suzuki polycondensation (Fig. 46a) [217] and obtained with a molecular weight of 47 kDa (M_n, polydispersity 2.5). OLEDs with P36–27SiF90 as emitter (layer structure ITO/PEDOT:PSS/PVK/polymer/Ba/Al) reached $\eta_{EL} = 1.95\%$, a luminous efficiency of $1.69\,cd\,A^{-1}$, and a maximal brightness of $6{,}000\,cd\,m^{-2}$. As in poly(3,6-silafluorene-co-2,7-fluorene) polymers, the silafluorene subunit successfully suppresses undesired long-wavelength green emission and produces a dominating emission at 500–600 nm (Fig. 46b, c). The total absence of vulnerable C-9 carbon in the main chain, which is easily oxidized by photo- and/or electro-oxidation, enhanced the oxidative stability ever further.

A series of soluble conjugated random and alternating copolymers (PFO-TST) derived from 9,9-dioctylfluorene (FO) and 1,1-dimethyl-3,4-diphenyl-2,5-bis(2'-thienyl)silole (TST) (Fig. 47a) were synthesized by Cao and his group [218] using palladium(0)-catalyzed Suzuki coupling. Random copolymers exhibited a PFO-segment-dominated UV absorption peak at 385 nm and absorption at 490 nm originating from the narrow band gap TST-segment. In contrast to the random copolymer, the alternating copolymer exhibited a broad absorption band which obviously results from the mixed configuration dominated by the TST segment. In films with a low

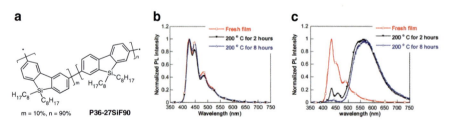

Fig. 46 Copolymer of 3,6-silafluorene and 2,7-silafluorene P36–27SiF90 (**a**). PL spectra of (**b**) P36–27SiF90 and (**c**) PFO in films with thermal annealing in air after different time at 200°C. Reproduced with permission of John Wiley and Sons Inc. from [217]

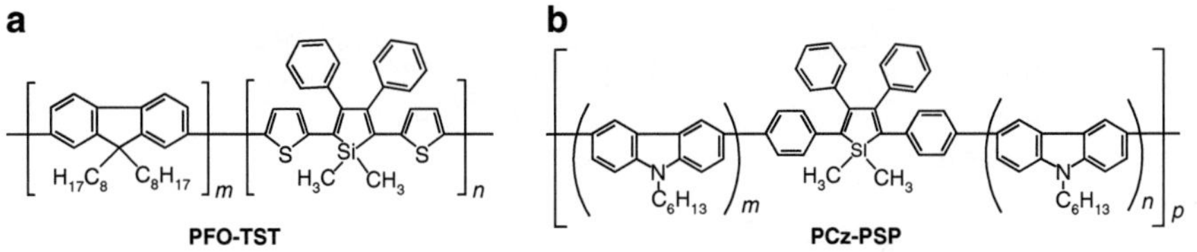

Fig. 47 Structural formulas of (**a**) 9,9-dioctylfluorene (FO) and 1,1-dimethyl-3,4-diphenyl-2,5-bis(2′-thienyl)silole (TST) copolymers PFO-TST [218] and (**b**) N-hexyl-3,6-carbazole (Cz) and 1,1-dimethyl-2,3,4,5-tetraphenylsilole (PSP) copolymers PCz-PSP [219]

TST content the excitation energy completely transferred from the PFO to the TST segment. In OLEDs (layer structure ITO/PEDOT/PVK/copolymer/Ba/Al) the emitted light from the copolymers was found to be red-shifted. In bulk-heterojunction photovoltaic cells the PFO-TST alternating copolymer proved their suitability as electron donor when combined with PCBM as electron acceptor to yield an energy conversion efficiency of 2.01%. In OTFT the field-effect hole mobility of the same copolymer was moderate (4.5×10^{-5} cm^2 V^{-1} s^{-1}).

Chen and coauthors incorporated N-hexyl-3,6-carbazole (Cz) and 1,1-dimethyl-2,3,4,5-tetraphenylsilole (PSP) to yield random and alternating copolymers (PCz-PSP) (Fig. 47b) by Suzuki coupling reactions [219]. The molecular weights were around 11–17 kDa (M_w, polydispersity between 1.3 and 1.7). The HOMO levels of the copolymers were found to be between −5.15 and −5.34 eV. Single layer OLEDs (layer structure ITO/copolymer/Ba/Al) using a copolymer with a content of 20% PSP segments exhibited a maximum η_{EL} of 0.77%. In OTFT, hole mobilities of the copolymers used as semiconductor decreased with the content of PSP segments with a maximum mobility of 9.3×10^{-6} cm^2 V^{-1} s^{-1}.

In a search for highly efficient red and green emitters, the Cao group synthesized via Suzuki coupling a series of 2,7-silafluorene copolymers with electron rich comonomers like 4,7-di(4-hexyl-2-thienyl)-2,1,3-benzothiadiazole (DHTBT) and 2,1,3-benzothiadiazole (BT) which narrow the bandgap of PSiF-DHTBT10 and PSiF-BT10 copolymers with a 10% molar ratio (Scheme 19) [220]. The molecular weights were determined to be 40 and 17 kDa (M_n, polydispersity 2.8 and 2.5) Both copolymers exhibited good thermal stability with degradation temperatures above 400°C and glass transition temperatures of 72–93°C. Compared to their PF analogs, PSiF-DHTBT10 and PSiF-BT10 show a higher PL emission. OLEDs (layer structure ITO/PEDOT:PSS/PVK/polymer/Ba/Al) with these polymers as emitters showed a maximum η_{EL} of 2.89% (PSiF-DHTBT10) and 3.81% (PSiF-BT10) as well as current efficiency of 2.0 cd A^{-1} (PSiF-DHTBT10) and 10.6 cd A^{-1} (PSiF-BT10). The CIE coordinates of both polymers ((0.67, 0.33) and (0.38, 0.57), respectively) are quite promising for future practical use, since they are close to pure red and green colors.

Fluoro-substituted silole-containing polymers were recently prepared by Suzuki polycondensation reaction using 2,5-dihydroxyboryl-1,1-dimethyl-3,4-bis(3-fluorophenyl)-silole or 1,3-dibromo-5-fluoro-benzene as fluoro-containing monomers and 2,7-dibromo-9,9-dioctyl-fluorene or 2,5-dihydroxyboryl-1,1-dimethyl-3,4-bis(phenyl)-silole as co-monomers [221]. With a polymerization

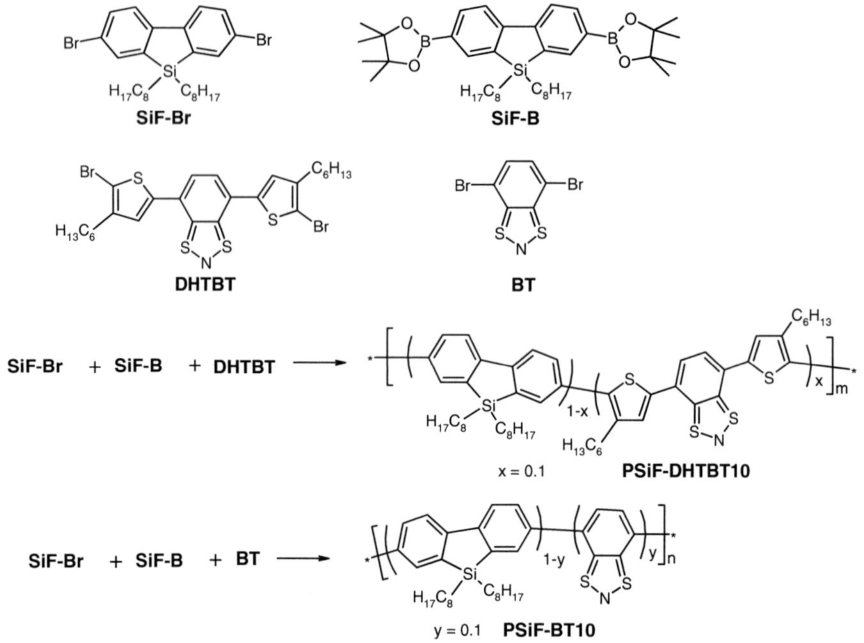

Scheme 19 Synthesis routes to PSiF-DHTBT10 and PSiF-BT10 copolymers

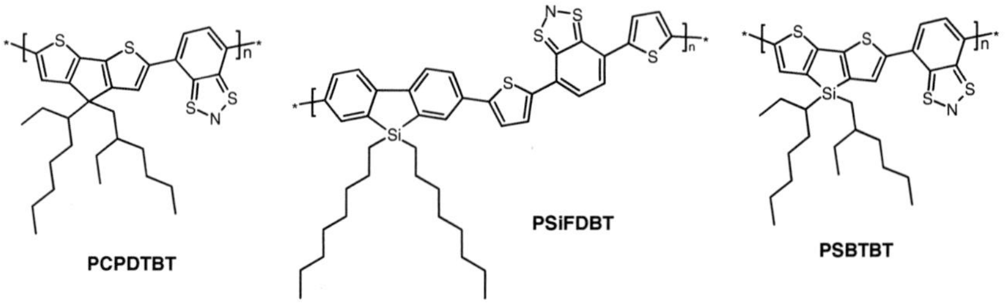

Fig. 48 Structures of PCPDTBT, PSiFDBT, and PSBTBT copolymers

degree ranging between 4 and 8, more or less oligomers were obtained with a weak green PL in the solid state. Cyclic voltammetry, visible absorption spectroscopy, and DFT calculations showed that the fluoro substituents withdraw electrons sufficiently to lower both the HOMO and the LUMO orbital in the oligomer.

The combination of 2,7-silafluorene (SiF) with the electron rich 4,7-di(2-thienyl)-2,1,3-benzothiadiazole (DBT) allowed preparation of an alternating copolymer PSiFDBT with a molecular weight of 79 kDa (M_n, polydispersity 4.2) (Fig. 48) [222]. Besides an excellent thermal stability (up to 430°C) PSiFDBT compared to its carbon analog PFDTBT exhibited a slightly lower optical band gap (1.82 vs 1.92 eV), and an approximately 20 nm red-shifted absorption peak. The shift to longer wavelengths allows absorption of longer wave lengths from solar radiation, and indeed, bulk junction solar sells with PCBM as the electron acceptor yielded a remarkable PCE up to 5.4% with an open-circuit voltage (V_{oc}) of 0.90 V, a short-circuit current (I_{sc}) of 9.5 mA cm^{-2}, and a fill factor (FF) of 50.7%. In contrast to poly-3-hexylthiophene (P3HT), the commonly used electron acceptor for OPV, the high efficiency was reached without post annealing or addition of additives to control the film morphology. A broad spectrum of absorption, a sufficient hole mobility, and low HOMO are clues to reach a high performance as OPV material. Field-effect transistors fabricated from PSiF-DBT showed a hole mobility of $\sim 1 \times 10^{-3}$ cm^2 V^{-1} s^{-1}, which is nearly ten times higher than that of PFO-DBT ($\sim 3 \times 10^{-4}$ cm^2 V^{-1} s^{-1}). The HOMO level of PSiF-DBT was estimated to be -5.39 eV.

The introduction of 4,7-bis(2-thienyl)-2,1,3-benzothiadiazole units into dithieno[3,2-b:2',3'-d]silole backbone as alternating copolymers will also reduce the band gap and increase the interchain interaction which in comparison to the nonmodified silole improve the energy conversion efficiency in the bulk-heterojunction organic solar cells by a factor of three [223]. Copolymers based on dithieno[3,2-b:2',3'-d]silole and 2,1,3-benzothiadiazole were also efficient as active layer of OPV [224, 225]. Poly[(4,4'-bis(2-ethylhexyl)dithieno[3,2-b:2',3'-d]silole)-2,6-diyl-*alt*-(2,1,3-benzothiadiazole)-4,7-diyl] (PSBTBT) with branched 2-ethylhexyl silicon substituents was of specific interest for wet processing due to its high solubility (Fig. 48). Scheme 20 depicts its synthesis from dichlorobis(2-ethylhexyl)silane **12**, which is synthesized by a two-step reaction route from tetrachlorosilane with a total yield of 40–45% (~ 65% for the first step and ~ 70% for the second step) and the substituted bithiophene **13** to yield 4,4'-dialkyldithieno[3,2-b:2',3'-d]silole **14** in 70% yield [226]. The polymer PSBTBT was prepared by a Stille coupling reaction from the monomer **16**, which was synthesized from **14** by commonly used methods. The molecular weight of the polymer was found to be 18 kDa (M_n, polydispersity 1.2). The polymer is stable up to 250°C in air. The HOMO and LUMO levels of PSBTBT were determined at -5.05 and -3.27 eV with an optical band gap of 1.45 eV, similar to its carbon analog PCPDTBT (Fig. 48). In OTFT the

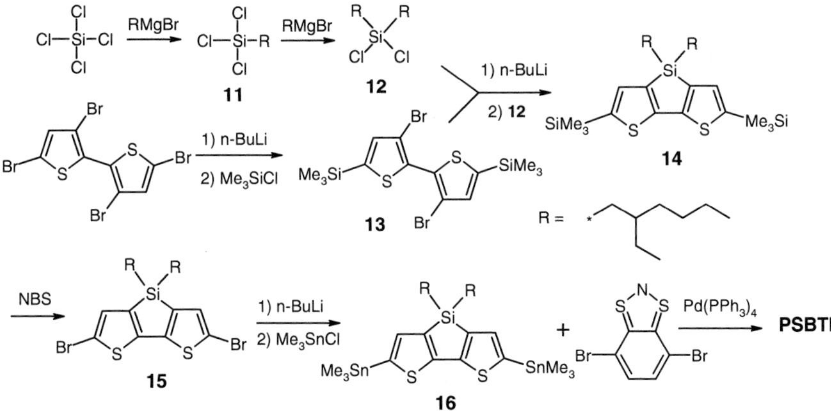

Scheme 20 Synthesis route of the polymer PSBTBT

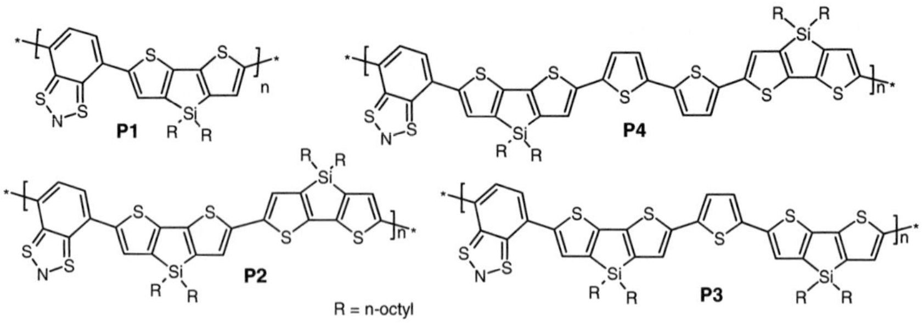

Fig. 49 Structures of DTS-BT copolymers P1, P2, P3, and P4 [228]

hole mobility value of 3×10^{-3} cm^2 V^{-1} s^{-1} was found to be three times higher than that for PCPDTBT [227]. These characteristics indicate its suitability as active material for OPV, and indeed bulk junction solar cells (layer structure: ITO/PEDOT-PSS/PSBTBT:PCBM/Ca/Al) exhibited a PCE of 4.7% (average of 100 devices, maximum: 5.1% with V_{oc} 0.68 V, J_{sc} 12.7 mA cm^{-2}, and FF 55%).

Recently Müllen and Reynolds described another successful combination of silole and benzothiazole donor–acceptor copolymers: the copolymer of dithieno[3,2-b:2′,3′-d]silole (DTS) and 2,1,3-benzothiadiazole (BT) (Fig. 49) [228]. In a series of DTS-BT copolymers (P1, P2, P3, and P4) two polymers (P2, P3) were found with unusually broad homogeneous spectral absorptions (in the range of 400–800 nm). Stille coupling reactions were used to obtain the unsymmetrical copolymers P1, P3, and P4, while the symmetrical P2 was polymerized by oxidative coupling with FeCl$_3$ followed by reduction with hydrazine. All polymers exhibited moderate molecular weights 10–20 kDa (M_n) and large polydispersities (3.2–3.8). Hole mobilities in bottom-contact OFETs increased from 10^{-6} cm^2 V^{-1} s^{-1} for polymer P1 to 10^{-2} cm^2 V^{-1} s^{-1} for polymer P4 which may be explained by an increasing trend to crystallize due to additional, unsubstituted thienyl monomer units. The unsubstituted bithiophene group in P4 creates effective hopping sites for the charge carriers, reduces the concentration of solubilizing groups, and enhances the backbone planarity. As required for solar cell applications, P4 absorbs light homogeneously across the entire visible spectrum.

In summary, a manifold of silole-containing polymers have been synthesized and studied. Especially polymers with TS and BS building blocks present an emerging class of functional materials, the electronic and optical properties of which open good perspectives for creation of highly thermal and environmentally stable organic electronics devices with outstanding performance.

7 Conclusions and Outlook

The analysis of different conjugated organosilicon structures revealed that silicon may be incorporated into the structures of oligomeric, dendritic, HB, and polymeric semiconducting molecules by many ways to achieve quite different effects.

A number of brilliant examples exploit the electronic interaction of silicon with the π-systems to adjust HOMO and LUMO energy levels and modify (usually lower) the band gap. The incorporation of silicon may be combined with the attachment of other groups with similar function to achieve synergistic effects. Attachment of silicon-containing bulky groups will hinder aggregation, therefore raise the solubility, and have an impact on electronic properties associated with the formation of aggregates.

Many silicon-containing organic semiconductors define the state of the art in their target application like organic electronic and photonic devices. Among them are silicon-modified oligoacenes (TES-ADT-F2) with a hole mobility greater than $1.0\,cm^2\,V^{-1}\,s^{-1}$ processed from solutions [53] and $6\,cm^2\,V^{-1}\,s^{-1}$ made from single crystals [54]. Oligothiophenes with silicon-containing anchor groups for the first time yielded SAMFETs with a mobility of $0.04\,cm^2\,V^{-1}\,s^{-1}$ and on/off ratio of 1×10^8 at $40\,\mu m$ channel length devices [25] as well as functional integrated circuits containing over 300 SAMFETS working simultaneously [28]. Other organosilicon semiconductors based on siloles-containing oligomers and polymers yielded highly efficient blue OLEDs with external quantum yield 8% and the power efficiency of $20\,lm\,W^{-1}$ [70] or superior thermal and devices stability [206, 217]. Improved solar cells were prepared from silafluorene copolymers and many other examples having a PCE up to 5.4% [222]. Most of the organosilicon materials considered above possess high thermal, electrical and environmental stability lacking today in conventional organic electronics. Bearing these achievements we foresee an increasing interest in this class of compounds which might result in numerous publications and successful innovation in the nearest future.

References

1. Forrest SR (2004) The path to ubiquitous and low-cost organic electronic appliances on plastic. Nature 428:911–918
2. Klauk H (ed) (2006) Organic electronics: materials, manufacturing and applications. Wiley-VCH, Weinheim
3. Facchetti A (2007) Semiconductors for organic transistors. Mater Today 10:28–38
4. Braun D (2002) Semiconducting polymer LEDs. Mater Today 5:32–39
5. Brabec C, Dyakonov C, Scherf U (eds) (2008) Organic photovoltaics. Wiley-VCH, Weinheim
6. Mayer AC, Scully SR, Hardin BE, Rowell MW, McGehee MD (2007) Polymer-based solar cells. Mater Today 10:28–33
7. Lloyd MT, Anthony JE, Malliaras GG (2007) Photovoltaics from soluble small molecules. Mater Today 10:34–41
8. Sokolov AN, Roberts ME, Bao Z (2009) Fabrication of low-cost electronic biosensors. Mater Today 12:12–20
9. Sun Y, Liu Y, Zhu D (2005) Advances in organic field-effect transistors. J Mater Chem 15: 53–65
10. Murphy AR, Fréchet JMJ (2007) Organic semiconducting oligomers for use in thin film transistors. Chem Rev 107:1066–1096
11. Sauvajol JL, Lère-Porte JP, Moreau JJE (1997) Silicon-containing thiophene oligomers and polymers: synthesis, characterization and properties. In: Nalwa NS (ed) Conductive polymers. Handbook of organic conductive molecules and polymers, vol 2. Wiley, New York

12. Herrema JK, Hutten PF, Gill RE, Wildeman J, Wieringa RH, Hadziioannou G (1995) Tuning of the luminescence in multiblock alternating copolymers. 1. synthesis and spectroscopy of poly [(silanylene)thiophenels. Macromolecules 28:8102–8116
13. Tour JM, Wu R (1992) Synthesis and UV-visible properties of soluble α-thiophene oligomers. Monomer to octamer. Macromolecules 25:1901–1907
14. Hapiot P, Gaillon L, Audebert P, Moreau JJE, Lère-Porte JP, Wong Chi Man M (1995) Solvent effects on the polymerization kinetics of some alfa-sylulated thiophene oligomers. Special influence of the α-silyl group. Synth Met 72:129–134
15. Lere-Porte JP, Moreau JJE, Torreilles C, Bouachrine M, Sauvajol JL, Serein-Spirau F (1999) Oxidative polymerisation of silyl monomers. Applications and limits. Synth Met 101:15–16
16. Barbarella G, Ostoja P, Maccagnani P, Pudova O, Antolini L, Casarini D, Bongini A (1998) Structural and electrical characterization of processable bis-silylated thiophene oligomers. Chem Mater 10:3683–3689
17. Halik M, Klauk H, Zschieschang U, Schmid G, Radlik W, Ponomarenko S, Kirchmeyer S, Weber W (2003) High-mobility organic thin-film transistors based on α,α'-didecyloligothiophenes. J Appl Phys 93:2977–2981
18. Halik M, Klauk H, Zschieschang U, Schmid G, Ponomarenko S, Kirchmeyer S, Weber W (2003) Relationship between molecular structure and electrical performance of oligothiophene organic thin film transistors. Adv Mater 15:917–922
19. Yassar A, Garnier F, Deloffre F, Horowitz G, Ricard L (1994) Crystal structure of α,ω-bis(triisopropylsylyl)-sexithiophene: unusual conjugated chain distortion induced by interchain steric effects. Adv Mater 6:660–663
20. Kim DH, Ohshita J, Kosuge T, Kunugi A, Kunai A (2006) Synthesis of silicon-bridged oligothiophenes and applications to thin film transistors. Chem Lett 35:266–267
21. Ohshita J, Izumi Y, Kim DH, Kunai A, Kosuge T, Kunugi Y, Naka A, Ishikawa M (2007) Applications of silicon-bridged oligothiophenes to organic FET materials. Organometallics 26:6150–6154
22. Facchetti A, Mushrush M, Yoon MH, Hutchison GR, Ratner MA, Marks TJ (2004) Building blocks for n-type molecular and polymeric electronics. perfluoroalkyl- versus alkyl-functionalized oligothiophenes (nT; n = 2–6). Systematics of thin film microstructure, semiconductor performance, and modeling of majority charge injection in field-effect transistors. J Am Chem Soc 126:13859–13874
23. Meyer-Friedrichsen T, Elschner A, Keohan F, Lövenich W, Ponomarenko SA (2009) Conductors and semiconductors for advanced organic electronics. Proc SPIE 7417:741704
24. Ponomarenko SA, Borshchev OV, Setayesh S, Smits ECP, Mathijssen SGJ, Pleshkova AP, Meyer-Friedrichsen T, Kirchmeyer S, Muzafarov AM, de Leeuw DM (2010) Synthesis of monochlorosilyl derivatives of dialkyloligothiophenes for self-assembling monolayer field-effect transistors. Organometallics (submitted)
25. Mathijssen SGJ, Smits ECP, van Hal PA, Wondergem HJ, Ponomarenko SA, Moser A, Resel R, Bobbert PA, Kemerink M, Janssen RAJ, de Leeuw DM (2009) Monolayer coverage and channel length set the mobility in self-assembled monolayer field-effect transistors. Nat Nanotechnol 4:674–680
26. Gholamrezaie F, Mathijssen SGJ, Smits ECP, Geuns TCT, van Hal PA, Ponomarenko SA, Cantatore E, Blom PWM, de Leeuw DM (2010) Ordered semiconducting self-assembled monolayers on polymeric surfaces applied in organic integrated circuits. Nano Lett (accepted)
27. Mottaghi M, Lang P, Rodriguez F, Rumyantseva A, Yassar A, Horowitz G, Lenfant S, Tondelier D, Vuillaume D (2007) Adv Funct Mater 17:597–604
28. Smits ECP, Mathijssen SGJ, van Hal PA, Setayesh S, Geuns TCT, Mutsaers KAHA, Cantatore E, Wondergem HJ, Werzer O, Resel R, Kemerink M, Kirchmeyer S, Muzafarov AM, Ponomarenko SA, de Boer B, Blom PWM, de Leeuw DM (2008) Bottom up organic integrated circuits. Nature 455:956–959
29. Anthony JE (2006) Functionalized acenes and heteroacenes for organic electronics. Chem Rev 106:5028–5048

30. Landis CA, Parkin SR, Anthony JE (2005) Silylethynylated anthracene derivatives for use in organic light-emitting diodes. Jpn J Appl Phys 44:3921–3922
31. Odom SA, Parkin SR, Anthony JE (2003) Tetracene derivatives as potential red emitters for organic LEDs. Org Lett 5:4245–4248
32. Karatsu T, Hazuku R, Asuke M, Nishigaki A, Yagai S, Suzuri Y, Kita H, Kitamura A (2007) Blue electroluminescence of silyl substituted anthracene derivatives. Org Electron 8:357–366
33. Kelley TW, Muyres DV, Baude PF, Smith TP, Jones TD (2003) High performance organic thin film transistors. Mater Res Soc Symp Proc 771:169-179
34. Allen CFH, Bell A (1942) Action of Grignard reagents on certain pentacenequinones, 6,13-diphenylpentacene. J Am Chem Soc 64:1253–1260
35. Anthony JE, Brooks JS, Eaton DL, Parkin SR (2001) Functionalized pentacene: improved electronic properties from control of solid-state order. J Am Chem Soc 123:9482–9483
36. Anthony JE, Eaton DL, Parkin S.R (2002) A road map to stable, soluble, easily crystallized pentacene derivatives. Org Lett 4:15–18
37. Sheraw CD, Jackson TN, Eaton DL, Anthony JE (2003) Functionalized pentacene active layer organic thin-film transistors. Adv Mater 15:2009–2011
38. Park SK, Jackson TN, Anthony JE, Mourey DA (2007) High mobility solution processed 6,13-bis(triisopropyl-silylethynyl) pentacene organic thin film transistors. Appl Phys Lett 91:063514
39. Troisi A, Orlandi G, Anthony JE (2005) Electronic interactions and thermal disorder in molecular crystals containing cofacial pentacene units. Chem Mater 17:5024–5031
40. Lobanova Griffith O, Gruhn NE, Anthony JE, Purushothaman B, Lichtenberger DL (2008) Electron transfer parameters of triisopropylsilylethynyl-substituted oligoacenes. J Phys Chem C 112:20518–20524
41. Lloyd MT, Mayer AC, Tayi AS, Bowen AM, Kasen TG, Herman DJ, Mourey DA, Anthony JE, Malliaras GG (2006) Photovoltaic cells from a soluble pentacene derivative. Org Electron 7:243–248
42. Miller GP, Briggs J, Mack J, Lord PA, Olmstead MM, Balch AL (2003) Fullerene–acene chemistry: single-crystal X-ray structures for a [60]fullerene–pentacene monoadduct and a cis-bis[60]fullerene adduct of 6,13-diphenylpentacene. Org Lett 5:4199–4202
43. Sakamoto Y, Suzuki T, Kobayashi M, Gao Y, Fukai Y, Inoue Y, Sato F, Tokito S (2004) Perfluoropentacene: high-performance p-n junctions and complementary circuits with pentacene. J Am Chem Soc 126:8138–8140
44. Inoue Y, Sakamoto Y, Suzuki T, Kobayashi M, Gao Y, Tokito S (2005) Organic thin-film transistors with high electron mobility based on perfluoropentacene. Jpn J Appl Phys 44:3663–3668
45. Swartz CR, Parkin SR, Bullock JE, Anthony JE, Mayer AC, Malliaras GG (2005) Synthesis and characterization of electron-deficient pentacenes. Org Lett 7:3163–3166
46. Wolak MA, Melinger JS, Lane PA, Palilis LC, Landis CA, Delcamp J, Anthony JE, Kafafi ZH (2006) Photophysical properties of dioxolane-substituted pentacene derivatives dispersed in tris(quinolin-8-olato)aluminum(III). J Phys Chem B 110:7928–7937
47. Wolak MA, Delcamp J, Landis CA, Lane PA, Anthony J, Kafafi Z (2006) High-performance organic light-emitting diodes based on dioxolane-substituted pentacene derivatives. Adv Funct Mater 16:1943–1949
48. Chiang CL, Wu MF, Dai CC, When YS, Wang JK, Chen CT (2005) Red-emitting fluorenes as efficient emitting hosts for non-doped, organic red-light-emitting diodes. Adv Funct Mater 15:231–238
49. Laquindanum JG, Katz HE, Lovinger AJ (1998) Synthesis, morphology, and field-effect mobility of anthradithiophenes. J Am Chem Soc 120:664–672
50. Payne MM, Parkin SR, Anthony JE, Kuo CC, Jackson TN (2005) Organic field-effect transistors from solution-deposited functionalized acenes with mobilities as high as 1 cm^2/vs. J Am Chem Soc 127:4986–4987
51. Dickey KC, Smith TJ, Stevenson KJ, Subramanian S, Anthony JE, Loo YL (2007) Establishing efficient electrical contact to the weak crystals of triethylsilylethynyl anthradithiophene. Chem Mater 19:5210–5215

52. Lloyd MT, Mayer AC, Subramanian S, Mourey DA, Herman DJ, Bapat AV, Anthony JE, Malliaras GG (2007) Efficient solution-processed photovoltaic cells based on an anthradithiophene/fullerene blend. J Am Chem Soc 129:9144–9149
53. Subramanian S, Park SK, Parkin SR, Podzorov V, Jackson TN, Anthony JE (2008) Chromophore fluorination enhances crystallization and stability of soluble anthradithiophene semiconductors. J Am Chem Soc 130:2706–2707
54. Jurchescu OD, Subramanian S, Kline RJ, Hudson SD, Anthony JE, Jackson TN, Gundlach DJ (2008) Organic single-crystal field-effect transistors of a soluble anthradithiophene. Chem Mater 20:6733–6737
55. Platt AD, Day J, Subramanian S, Anthony JE, Ostroverkhova O (2009) Optical, fluorescent, and (photo)conductive properties of high-performance functionalized pentacene and anthradithiophene derivatives. J Phys Chem C 113:14006–14014
56. Payne MM, Parkin SR, Anthony JE (2005) Functionalized higher acenes: hexacene and heptacene. J Am Chem Soc 127:8028–8029
57. Payne MM, Odom SA, Parkin SR, Anthony JE (2004) Stable, crystalline acenedithiophenes with up to seven linearly fused rings. Org Lett 6:3325–3328
58. Yamaguchi S, Tamao K (1998) Silole-containing σ- and π-conjugated compounds. J Chem Soc Dalton Trans:3693–3702, Doi: http://dx.doi.org/10.1039/a804491k
59. Tamao K, Yamaguchi S, Shiro M (1994) Oligosiloles: first synthesis based on a novel endo-endo mode intramolecular reductive cyclization of diethynylsilanes. J Am Chem Soc 116:11715–11722
60. Yamaguchi S, Jin RZ, Tamao K, Shiro M (1997) Silicon-catenated silole oligomers: oligo(1,1-silole)s. Organometallics 16:2486–2488
61. Kanno K, Ichinohe M, Kabuto C, Kira M (1998) Synthesis and structure of a series of oligo[1,1-(2,3,4,5-tetramethylsilole)]s. Chem Lett 27:99–100
62. Tamao K, Yamaguchi S, Ito Y, Matsuzaki Y, Yamabe T, Fukushima M, Mori S (1995) Silole-containing π-conjugated systems. 3. Series of silole-thiophene cooligomers and copolymers: synthesis, properties, and electronic structures. Macromolecules 28:8668–8675
63. Tamao K, Ohno S, Yamaguchi S (1996) Silole–pyrrole co-oligomers: their synthesis, structure and UV-VIS absorption spectra. Chem Commun:1873–1874
64. Tamao K, Uchida M, Izumizawa T, Furukawa K, Yamaguchi S (1996) Silole derivatives as efficient electron transporting materials. J Am Chem Soc 118:11974–11975
65. Murata H, Kafafi ZH, Uchida M (2002) Efficient organic light-emitting diodes with undoped active layers based on silole derivatives. Appl Phys Lett 80:189–191
66. Palilisa LC, Mäkinen AJ, Uchida M, Kafafi ZH (2003) Highly efficient molecular organic light-emitting diodes based on exciplex emission. Appl Phys Lett 82:2209–2214
67. Yamaguchi S, Endo T, Uchida M, Izumizawa T, Furukawa K, Tamao K (2000) Toward new materials or organic electroluminescent devices: synthesis, structures, and properties of a series of 2,5-diaryl-3,4-dippenylsiloles. Chem Eur J 6:1683–1692
68. Lee SH, Jang BB, Kafafi ZH (2005) Highly fluorescent solid-state asymmetric spirosilabifluorene derivatives. J Am Chem Soc 127:9071–9078
69. Braye EH, Hübel W, Caplier I (1961) New unsaturated heterocyclic systems I. J Am Chem Soc 83:4406–4413
70. Luo J, Xie Z, Lam JWY, Cheng L, Chen H, Qiu C, Kwok HS, Zhan X, Liu Y, Zhu D, Tang BZ (2001) Aggregation-induced emission of 1-methyl-1,2,3,4,5-pentaphenylsilole. Chem Commun 1740–1741
71. Yu G, Yin S, Liu Y, Chen J, Xu X, Sun X, Ma D, Zhan X, Peng Q, Shuai Z, Tang B, Zhu D, Fang W, Luo Y (2005) Structures, electronic states, photoluminescence, and carrier transport properties of 1,1-disubstituted 2,3,4,5-tetraphenylsiloles. J Am Chem Soc 127:6335–6346
72. Chen J, Law CCW, Lam JWY, Dong Y, Lo SMF, Williams ID, Zhu D, Tang BZ (2003) Synthesis, light emission, nanoaggregation, and restricted intramolecular rotation of 1,1-substituted 2,3,4,5-tetraphenylsiloles. Chem Mater 15:1535–1546
73. Son HJ, Han WS, Chun JY, Lee CJ, Han JI, Ko J, Kang SO (2007) Spiro-silacycloalkyl tetraphenylsiloles with a tunable exocyclic ring: preparation, characterization, and device application of 1,1'-silacycloalkyl-2,3,4,5-tetraphenylsiloles. Organometallics 26:519–526

74. Zeng Q, Li Z, Dong Y, Di C, Qin A, Hong Y, Ji L, Zhu Z, Jim CKW, Yu G, Li Q, Li Z, Liu Y, Qin J, Tang BZ (2007) Fluorescence enhancements of benzene-cored luminophors by restricted intramolecular rotations: AIE and AIEE effects. Chem Commun 70–72
75. Tang BZ, Zhan X, Yu G, Lee PPS, Liu Y, Zhu D (2001) Efficient blue emission from siloles. J Mater Chem 11:2974–2978
76. Mi B, Dong Y, Li Z, Lam JWY, Häußler M, Sung HHY, Kwok HS, Dong Y, Williams ID, Liu Y, Luo Y, Shuai Z, Zhu D, Tang BZ (2005) Making silole photovoltaically active by attaching carbazolyl donor groups to the silolyl acceptor core. Chem Commun 3583–3585
77. Gilman H, Gorsich RD (1955) A silicon analog of 9,9-diphenylfluorene. J Am Chem Soc 77:6380–6381
78. Gilman H, Gorsich RD (1958) Cyclic organosilicon compounds. I. Synthesis of compounds containing the dibenzosilole nucleus. J Am Chem Soc 80:1883–1886
79. Ohshita J, Nodono M, Watanabe T, Ueno Y, Kunai A, Harima Y, Yamashita K, Ishikawa M (1998) J Organomet Chem 553:487–491
80. Ohshita J, Nodono M, Kai H, Watanabe T, Kunai A, Komaguchi K, Shiotani M, Adachi A, Okita K, Harima Y, Yamashita K, Ishikawa M (1999) Synthesis and optical, electrochemical, and electron-transporting properties of silicon-bridged bithiophenes. Organometallics 18:1453–1459
81. Ohshita J, Kai H, Takata A, Iida T, Kunai K, Ohta N, Komaguchi K, Shiotani M, Adachi A, Sakamaki K, Okita K (2001) Effects of conjugated substituents on the optical, electrochemical, and electron-transporting properties of dithienosiloles. Organometallics 20:4800–4805
82. Lee IS, Kim SJ, Kwak YW, Choi MC, Park JW, Ha CS (2008) Synthesis of 2,6-diaryl-4,4-diphenyldithienosiloles and their luminescent properties. J Ind Eng Chem 14:344–349
83. Kim DH, Ohshita J, Lee KH, Kunugi Y, Kunai A (2006) Synthesis of π-conjugated oligomers containing dithienosilole units. Organometallics 25:1511–1516
84. Shimizu M, Tatsumi H, Mochida K, Oda K, Hiyama T (2008) Silicon-bridge effects on photophysical properties of silafluorenes. Chem Asian J 3:1238–1247
85. Ilies L, Tsuji H, Sato Y, Nakamura E (2008) Modular synthesis of functionalized benzosiloles by tin-mediated cyclization of (o-alkynylphenyl)silane. J Am Chem Soc 130:4240–4241
86. Ilies L, Tsuji H, Nakamura E (2009) Synthesis of benzo[b]siloles via KH-promoted cyclization of (2-alkynylphenyl)silanes. Org Lett 11:3966–3968
87. Shimizu M, Mochida K, Hiyama T (2008) Modular approach to silicon-bridged biaryls: palladium-catalyzed intramolecular coupling of 2-(arylsilyl)aryl triflates. Angew Chem Int Ed 47:9760–9764
88. Yamaguchi S, Xu C, Tamao K (2003) Bis-silicon-bridged stilbene homologues synthesized by new intramolecular reductive double cyclization. J Am Chem Soc 125:13662–13663
89. Xu C, Wakamiya A, Yamaguchi S (2005) Ladder oligo(p-phenylenevinylene)s with silicon and carbon bridges. J Am Chem Soc 127:1638–1639
90. Yamaguchi S, Xu C, Yamada H, Wakamiya A (2005) Synthesis, structures, and photophysical properties of silicon and carbon-bridged ladder oligo(p-phenylenevinylene)s and related π-electron systems. J Organomet Chem 690:5365–5377
91. Fukazawa A, Li Y, Yamaguchi S, Tsuji H, Tamao K (2007) Coplanar oligo(p-phenylenedisilenylene)s based on the octaethyl-substituted s-hydrindacenyl groups. J Am Chem Soc 129:14164–14165
92. Guay J, Diaz A, Wu R, Tour JM (1993) Electrochemical and electronic properties of neutral and oxidized soluble orthogonally fused thiophene oligomers. J Am Chem Soc 115:1869–1874
93. Aviram A (1988) Molecules for memory, logic, and amplification. J Am Chem Soc 110:5687–5692
94. Tour JM, Wu R, Schumm JS (1991) Extended orthogonally fused conducting oligomers for molecular electronic devices. J Am Chem Soc 113:7064–7066
95. Lee SH, Jang BB, Kafafi ZH (2005) Highly fluorescent solid-state asymmetric spirosilabifluorene derivatives. J Am Chem Soc 127:9071–9078

96. Shumilkina EA, Borschev OV, Ponomarenko SA, Surin NM, Pleshkova AP, Muzafarov AM (2007) Synthesis and optical properties of linear and branched bithienylsilanes. Mendeleev Commun 17:34–36
97. Lukevics E, Ryabova V, Arsenyan P, Belyakov S, Popelis J, Pudova O (2000) Bithienylsilanes: unexpected structure and reactivity. J Organomet Chem 610:8–15
98. Schwarzer A, Schilling IC, Seichter W, Weber E (2009) Synthesis and X-ray crystal structures of new tetrahedral arylethynyl substituted silanes. Silicon 1:3–12
99. Tang H, Zhu L, Harima Y, Yamashita K, Lee KK, Naka A, Ishikawa M (2000) Strong fluorescence of nano-size star-like molecules. J Chem Soc Perkin Trans 2:1976–1979
100. Ishikawa M, Lee KK, Schneider W, Naka A, Yamabe T, Harima Y, Takeuchi T (2000) Synthesis and properties of nanosize starlike silicon compounds. Organometallics 19:2406–2407
101. Ishikawa M, Teramura H, Lee KK, Schneider W, Naka A, Kobayashi H, Yamaguchi Y, Kikugawa M, Ohshita J, Kunai A, Tang H, Harima Y, Yamabe T, Takeuchi T (2001) Nanosized, starlike silicon compounds. synthesis and optical properties of tris[(tert-butyldimethylsilyl)oligothienylenedimethylsilyl] methylsilanes. Organometallics 20:5331–5341
102. Arsenyan P, Pudova O, Popelis J, Lukevics E (2004) Novel radial oligothienyl silanes. Tetrahedron Lett 45:3109–3111
103. Roncali J, Thobie-Gautier C, Brisset H, Favart JF, Guy A (1995) Electro-oxidation of tetra(terthienyl)silanes: towards 3D electroactive π-conjugated systems. J Electroanal Chem 381:257–260
104. Garnier F, Yassar A, Hajlaoui R, Horowitz G, Deloffre F, Servet B, Ries S, Alnot P (1993) Molecular engineering of organic semiconductors: design of self-assembly properties in conjugated thiophene oligomers. J Am Chem Soc 115:8716–8721
105. Roquet S, de Bettignies R, Leriche P, Cravino A, Roncali J (2006) Three-dimensional tetra(oligothienyl)silanes as donor material for organic solar cells. J Mater Chem 16:3040–3045
106. Kleimyuk EA, Luponosov YN, Troshin PA, Khakina EA, Moskvin YL, Egginger M, Peregudova SM, Babenko SD, Razumov VF, Sariciftci NS, Muzafarov AM, Ponomarenko SA (2010) Three dimensional quater- and quinquethiophensilanes as promising electron donor materials for bulk heterojunction photovoltaic cells. J Mater Chem (submitted)
107. Ponomarenko SA, Tatarinova EA, Muzafarov AM, Kirchmeyer S, Brassat L, Mourran A, Moeller M, Setayesh S, de Leeuw DM (2006) Star-shaped oligothiophenes for solution-processible organic electronics: flexible aliphatic spacers approach. Chem Mater 18:4101–4108
108. Kirchmeyer S, Meyer-Friedrichsen T, Elschner A, Gaiser D, Lövenich W, Jonas F, Ponomarenko SA, Jang J (2008) Materials for organic electronics: conductors and semiconductors designed for wet processing. Proc SPIE 7054:705402
109. Mourran A, Defaux M, Luponosov YN, Ponomarenko SA, Muzafarov AM, Moeller M (2010) Film-formation of quaterthiophene derivatives and its multipods having branched 2-ethylhexyl end-groups. Thin Solid Films (submitted)
110. Troshin PA, Ponomarenko SA, Luponosov YN, Khakina EA, Egginger M, Meyer-Friedrichsen T, Elschner A, Peregudova SM, Buzin MI, Razumov VF, Sariciftci NS, Muzafarov AM (2010) Efficient solution-processible organic solar cells utilizing quaterthiophene-based multipods as electron donor materials. Solar Energy Materials & Solar Cells (submitted)
111. Kim C, Kim M (1998) Synthesis of carbosilane dendrimers based on tetrakis (phenylethynyl)silane. J Organomet Chem 563:43–51
112. Apperloo JJ, Janssen RAJ, Malenfant PRL, Fréchet JMJ (2000) Concentration-dependent thermochromism and supramolecular aggregation in solution of triblock copolymers based on lengthy oligothiophene cores and poly(benzyl ether) dendrons. Macromolecules 33:7038–7043
113. Adronov A, Malenfant PRL, Fréchet JMJ (2000) Synthesis and steady-state photophysical properties of dye-labeled dendrimers having novel oligothiophene cores: a comparative study. Chem Mater 12:1463–1472

114. Wang F, Kon AB, Rauh RD (2000) Synthesis of a terminally functionalized bromothiophene polyphenylene dendrimer by a divergent method. Macromolecules 33:5300–5302
115. Deng S, Locklin J, Patton D, Baba A, Advincula RC (2005) Thiophene dendron jacketed poly(amidoamine) dendrimers: nanoparticle synthesis and adsorption on graphite. J Am Chem Soc 127:1744–1751
116. Sebastian RM, Caminade AM, Majoral JP, Levillain E, Huchet L, Roncali J (2000) Electrogenerated poly(dendrimers) containing conjugated poly(thiophene) chains. Chem Commun 507–508
117. Zhang Y, Zhao C, Yang J, Kapiamba M, Haze O, Rothberg LJ, Ng MK (2006) Synthesis, optical, and electrochemical properties of a new family of dendritic oligothiophenes. J Org Chem 71:9475–9483
118. Mitchell WJ, Kopidakis N, Rumbles G, Ginley DS, Shaheen SE (2005) The synthesis and properties of solution processable phenyl cored thiophene dendrimers. J Mater Chem 15:4518–4528
119. Xia C, Fan X, Locklin J, Advincula RC (2002) A first synthesis of thiophene dendrimers. Org Lett 4:2067–2070
120. Xia C, Fan X, Locklin J, Advincula RC, Gies A, Nonidez W (2004) Characterization, supramolecular assembly, and nanostructures of thiophene dendrimers. J Am Chem Soc 126:8735–8743
121. Ma CQ, Mena-Osteritz E, Debaerdemaeker T, Wienk MM, Janssen RAJ, Bäuerle P (2007) Functionalized 3D oligothiophene dendrons and dendrimers: novel macromolecules for organic electronics. Angew Chem Int Ed 46:1679–1683
122. Ma CQ, Fonrodona M, Schikora MC, Wienk MM, Janssen RAJ, Bäuerle P (2008) Solution-processed bulk-heterojunction solar cells based on monodisperse dendritic oligothiophenes. Adv Funct Mater 18:3323–3331
123. Nakayama J, Lin JS (1997) An organosilicon dendrimer composed of 16 thiophene rings. Tetrahedron Lett 38:6043–6046
124. Ponomarenko SA, Muzafarov AM, Borshchev OV, Vodopyanov EA, Demchenko NV, Myakushev VD (2005) Synthesis of bithiophenesilane dendrimer of the first generation. Russ Chem Bull 3:684–690
125. Borshchev OV, Ponomarenko SA, Surin NM, Kaptyug MM, Buzin MI, Pleshkova AP, Demchenko NV, Myakushev VD, Muzafarov AM (2007) Bithiophenesilane dendrimers: synthesis and thermal and optical properties. Organometallics 26:5165–5173
126. Luponosov YN, Ponomarenko SA, Surin NM, Muzafarov AM (2008) Facile synthesis and optical properties of bithiophenesilane monodendrons and dendrimers. Org Lett 10:2753–2756
127. Gunawidjaja R, Luponosov YN, Huang F, Ponomarenko SA, Muzafarov AM, Tsukruk VV (2009) Photoluminescence and molecular ordering of functionalized bithiophenesilane monodendrons. Langmuir 25:9270–9284
128. Surin NM, Borshchev OV, Luponosov YN, Ponomarenko SA, Muzafarov AM (2010) Spectral-luminescent properties of oligothiophenesilane dendritic macromolecules. Russ J Phys Chem A (accepted)
129. Luponosov YN, Ponomarenko SA, Surin NM, Borshchev OV, Shumilkina EA, Muzafarov AM (2009) The first organosilicon molecular antennas. Chem Mater 21:447–455
130. Borshchev OV, Ponomarenko SA, Shumilkina EA, Luponosov YN, Surin NM, Muzafarov AM (2010) Branched oligothiophenesilanes with effective non-radiative energy transfer between the fragments. Russ Chem Bull (4) (accepted)
131. Xu Z, Moore JS (1994) Design and synthesis of a convergent and directional molecular antenna. Acta Polymer 45:83–87
132. Borschev OV (2007) Oligothiophenesilane dendrimers of the first generation: synthesis, optical and thermal properties. PhD Thesis, Moscow
133. You Y, An C, Lee D, Kim J, Park SY (2006) Silicon-containing dendritic tris-cyclometalated Ir(III) complex and its electrophosphorescence in a polymer host. J Mater Chem 16:4706–4713

134. Ponomarenko SA, Tatarinova EA, Meyer-Friedrichsen T, Kirchmeyer S, Setayesh S, de Leeuw DM, Magonov SN, Muzafarov AM (2007) Solution processible quaterthiophene-containing carbosilane dendrimers. Polym Mater Sci Eng 96:298–299
135. Gao C, Yan D (2004) Hyperbranched polymers: from synthesis to applications. Prog Polym Sci 29:183–275
136. Yao J, Son DY (1999) Hyperbranched poly(2,5-silylthiophenes). The possibility of σ-π conjugation in three dimensions. Organometallics 18:1736–1740
137. Ponomarenko S, unpublished results
138. Xiao Y, Wong RA, Son DY (2000) Synthesis of a new hyperbranched poly(silylenevinylene) with ethynyl functionalization. Macromolecules 33:7232–7234
139. Yoon K, Son DY (1999) Syntheses of hyperbranched poly(carbosilarylenes). Macromolecules 32:5210–5216
140. Chen J, Peng H, Law CCW, Dong Y, Lam JWY, Williams ID, Tang BZ (2003) Hyperbranched poly(phenylenesilolene)s: synthesis, thermal stability, electronic conjugation, optical power limiting, and cooling-enhanced light emission. Macromolecules 36:4319–4327
141. Chen J, Xie Z, Lam JWY, Law CCW, Tang BZ (2003) Silole-containing polyacetylenes. synthesis, thermal stability, light emission, nanodimensional aggregation, and restricted intramolecular rotation. Macromolecules 36:1108–1117
142. Kirchmeyer S, Ponomarenko S, Muzafarov A (2008) Macromolecular compounds with a core-shell structure. US patent 7,420,645
143. Masuda T, Higashimura T (1989) Synthesis and properties of silicon-containing polyacetylenes. In: Zeigler JM, Fearon FWG (eds) Silicon-based polymer science. Advances in chemistry, vol 224, pp 641–661, chapter doi: 10.1021/ba-1990-0224.ch035
144. Masuda T, Isobe E, Higashimura T, Takada K (1983) Poly[1-(trimethylsilyl)-1-propyne]: a new high polymer synthesized with transition-metal catalysts and characterized by extremely high gas permeability. J Am Chem Soc 105:7473–7474
145. Savoca AC, Surnamer AD, Tien CF (1993) Gas transport in poly(silylpropynes): the chemical structure point of view. Macromolecules 26:6211–6216
146. Yampolskii YP, Korikov AP, Shantarovich VP, Nagai K, Freeman BD, Masuda T, Teraguchi M, Kwak G (2001) Gas permeability and free volume of highly branched substituted acetylene polymers. Macromolecules 34:1788–1796
147. Kusumota T, Hiyama T. (1988) Polymerization of monomers containing two ethynyldimethylsilyl groups. Chem Lett 1149–1152
148. Chen J, Xie Z, Lam JWY, Law CCW, Zhong B (2003) Tang silole-containing polyacetylenes. synthesis, thermal stability, light emission, nanodimensional aggregation, and restricted intramolecular rotation. Macromolecules 36:1108–1117
149. Lee YB, Shim HK, Ko SW (2003) Silyl-substituted poly(thienylenevinylene) via heteroaromatic dehydrohalogenation polymerization. Macromol Rapid Commun 24:522–526
150. Höger S, McNamara JJ, Schricker S, Wudl F (1994) Novel silicon-substituted, soluble poly(phenylenevinylene)s: enlargement of the semiconductor bandgap. Chem Mater 6:171–173
151. Zhang C, Höger S, Pakbaz K, Wudl F, Heeger AJ (1994) Improved efficiency in green polymer light-emitting diodes with air-stable electrodes. J Electron Mater 23:453–458
152. Hwang DH, Shim HK, Lee JI, Lee KS (1994) Synthesis and properties of multifunctional poly(2-trimethylsilyl-1,4-phenylenevinylene): a novel, silicon-substituted, soluble PPV derivative. J Chem Soc Chem Commun 2461–2462
153. Kim ST, Hwang DH, Li XC, Grüner J, Friend RH, Holmes AB, Shim HK (1996) Efficient green electroluminescent diodes based on poly(2-dimethyloctylsilyl-1,4-phenylenevinylene). Adv Mater 8:979–982
154. Greenham NC, Samuel IDW, Hayes GR, Philips RT, Kessener YARR, Moratti SC, Holmes AB, Friend RH (1995) Measurement of absolute photoluminescence quantum efficiencies in conjugated polymers. Chem Phys Lett 241:89–96
155. Hwang DH, Kim ST, Shim HK, Holmes AB, Moratti SC, Friend RH (1996) Green light-emitting diodes from poly(2-dimethyloctylsilyl-1,4-phenylenevinylene). Chem Commun 2241–2242

156. Kim ST, Hwang DH, Holmes AB, Friend RH, Shim HK (1997) Green electroluminescent characteristics of poly(2-dimethyloctylsilyl-1,4-phenylenevinylene). Synth Met 84:655–656
157. Hwang DH, Kim ST, Shim HK, Holmes AB, Moratti SC, Friend RH (1997) Highly efficient green light-emitting diodes with aluminium cathode. Synth Met 84:615–618
158. Pei Q, Yu G, Zhang C, Yang Y, Heeger AJ (1995) Polymer light-emitting electrochemical cells. Science 269:1086–1088
159. Ahn T, Ko SW, Lee J, Shim HK (2002) Novel cyclohexylsilyl- or phenylsilyl-substituted poly(p-phenylene vinylene)s via the halogen precursor route and gilch polymerization. Macromolecules 35:3495–3505
160. Hwang DH, Kang IN, Lee JI, Do LM, Chu HY, Zyung T, Shim HK (1998) Synthesis and properties of silyl-substituted PPV derivative through two different precursor polymers. Polymer Bull 41:275–283
161. Chen ZK, Wang LH, Kang ET, Huang W (1999) Intense green light from a silyl-substituted poly(p-phenylenevinylene)-based light-emitting diode with air-stable cathode. Phys Chem Chem Phys 1:3789–3792
162. Chen ZK, Huang W, Wang LH, Kang ET, Chen BJ, Lee CS, Lee ST (2000) Family of electroluminescent silyl-substituted poly(p-phenylenevinylene)s: synthesis, characterization, and structure-property relationships. Macromolecules 33:9015–9025
163. Chu HY, Hwang DH, Do LM, Chang JH, Shim HK, Holmes AB, Zyung T (1999) Electroluminescence from silyl-disubstitutd PPV derivative. Synth Met 101:216–217
164. Wang LH, Chen ZK, Kang ET, Meng H, Huang W (1999) Synthesis, spectroscopy and electrochemistry study on a novel di-silyl substituted poly(p-phenylenevinylene). Synth Met 105:85–89
165. Geneste F, Fischmeister C, Martin RE, Holmes AB (2001) Ortho-methallation as a key step to the synthesis of silyl-substituted poly(p-phenylenevinylene). Synth Met 121:1709–1710
166. Rost H, Chuah BS, Hwang DH, Moratti SC, Holmes AB, Wilson J, Morgado J, Halls JJM, de Mello JC, Friend RH (1999) Novel luminescent polymers. Synth Met 102:937–938
167. Martin RE, Geneste F, Riehn R, Chuah BS, Cacialli F, Holmes AB, Friend RH (2001) Efficient electroluminescent poly(p-phenylenevinylene) copolymers for application in LEDs. Synth Met 119:43–44
168. Martin RE, Geneste F, Chuah BS, Fischmeister C, Ma Y, Holmes AB, Riehn R, Cacialli F, Friend RH (2001) Versatile synthesis of various conjugated aromatic homo- and copolymers. Synth Met 122:1–5
169. Ahn T, Jang MS, Shim HK, Hwang DH, Zyung T (1999) Blue electroluminescent polymers: control of conjugation length by kink linkages and substituents in the poly(p-phenylenevinylene)-related copolymers. Macromolecules 32:3279–3285
170. Shim HK, Song SY, Ahn T (2000) Efficient and blue light-emitting polymers composed of conjugated main chain. Synth Met 111/112:409–412
171. Ahn T, Song SY, Shim HK (2000) Highly photoluminescent and blue-green electroluminescent polymers: new silyl- and alkoxy-substituted poly(p-phenylenevinylene) related copolymers containing carbazole or fluorene groups. Macromolecules 33:6764–6771
172. Lee JH, Yu HS, Kim W, Gal YS, Park JH, Jin SH (2000) Synthesis and characterization of a new green-emitting poly(phenylenevinylene) derivative containing alkylsilylphenyl pendant. J Polym Sci A Polym Chem 38:4185–4193
173. Jin SH, Jang MS, Suh HS, Cho HN, Lee JH, Gal YS (2002) Synthesis and characterization of highly luminescent asymmetric poly(p-phenylene vinylene) derivatives for light-emitting diodes. Chem Mater 14:643–665
174. Jin SH, Jung HH, Hwang CK, Koo DS, Shin WS, Kim YI, Lee JW, Gal YS (2005) High electroluminescent properties of conjugated copolymers from poly[9,9-dioctylfluorenyl-2,7-vinylene]-co-(2-(3-dimethyldodecylsilylphenyl)-1,4-phenylene vinylene)] for light-emitting diode applications. J Polym Sci A Polym Chem 43:5062–5071
175. Ko SW, Jung BJ, Ahn T, Shim HK (2002) Novel poly(p-phenylenevinylene)s with an electron-withdrawing cyanophenyl group. Macromolecules 35:6217–6223

176. Ishikawa M, Ohshita J (1997) Silicon and germanium containing conductive polymers. In: Nalwa NS (ed) Conductive polymers. Handbook of organic conductive molecules and polymers vol 2. Wiley, New York
177. Ohshita J, Kunai A (1998) Polymers with alternating organosilicon and π-conjugated units. Acta Polym 49:379–403
178. Nate K, Ishikawa M, Ni H, Watanabe H, Saheki Y (1987) Photolysis of polymeric organosilicon systems. 4. Photochemical behavior of poly[p-(disilanylene)phenylene]. Organometallics 6:1673–1679
179. Ohshita J, Kanaya D, Ishikawa M, Koike T, Yamanaka T (1991) Polymeric organosilicon systems. 10. Synthesis and conducting properties of poly[2,5-(disilanylene)thienylenes]. Macromolecules 24:2106–2107
180. Chichart P, Corriu RJP, Moreau JJE, Garnier F, Yassar A (1991) Selective synthetic routes to electroconductive organosilicon polymers containing thiophene units. Chem Mater 3:8–10
181. Yi SH, Nagase J, Sato H (1993) Synthesis and characterization of soluble organosilicon polymers containing regularly repeated thiophene or terthiophene units. Synth Met 58:353–365
182. Ohshita J, Watanabe T, Kanaya D, Ohsaki H, Ishikawa M, Ago H, Tanaka K, Yamabe T (1994) Polymeric organosilicon systems. 22. Synthesis and photochemical properties of poly[(disilanylene)oligophenylylenes] and poly[(silylene)biphenylylenes]. Organometallics 13:5002–5012
183. Kunai A, Ueda T, Horata K, Toyoda E, Nagamoto I, Ohshita J, Ishikawa M, Tanaka K (1996) Polymeric organosilicon systems. 26. Synthesis and photochemical and conducting properties of poly[(tetraethyldisilanylene)oligo(2,5-thienylenes)]. Organometallics 15:2000–2008
184. Yi SH, Ohashi S, Sato H, Nomori H (1993) Syntheses and electrical properties of organosilicon polymers containing thiophene and anthraquinone units bull. Chem Soc Jpn 66:1244–1247
185. Ohshita J, Kim DH, Kunugi Y, Kunai A (2005) Synthesis of organosilanylene-oligothienylene alternate polymers and their applications to EL and FET materials. Organometallics 24:4494–4496
186. Ohshita J, Sugimoto K, Kunai A, Harima Y, Yamashita K (1999) Electrochemical and optical properties of poly[(disilanylene)oligophenylenes], peculiar behavior in the solid state. J Organomet Chem 580:77–81
187. Adachi A, Manhart SA, Okita K, Kido J, Ohshita J, Kunai A (1997) Multilayer electroluminescent device using organosilicon polymer as hole transport layer. Synth Met 91:333–334
188. He G, Pfeiffer M, Leo K, Hofmann M, Birnstock J, Pudzich R, Salbeck J (2004) High-efficiency and low-voltage p-i-n electrophosphorescent organic light-emitting diodes with double-emission layers. Appl Phys Lett 85:3911–3913
189. Manhart SA, Adachi A, Sakamaki K, Okita K, Ohshita J, Ohno T, Hamaguchi T, Kunai A, Kido J (1999) Synthesis and properties of organosilicon polymers containing 9,10-diethynylanthracene units with highly hole-transporting properties. J Organomet Chem 592:52–60
190. Suzuki H, Satoh S, Kimata Y, Kuriyama A (1995) Synthesis and properties of poly(methylphenylsilane) containing anthracene units. Chem Lett 451–452
191. Ohshita J, Takata A, Kai H, Kunai A, Komaguchi K, Shiotani M, Adachi A, Sakamaki K, Okita K, Harima Y, Kunugi Y, Yamashita K, Ishikawa M (2000) Synthesis of polymers with alternating organosilanylene and oligothienylene units and their optical, conducting, and hole-transporting properties. Organometallics 19:4492–4498
192. Kunugi Y, Harima Y, Yamashita K, Ohshita J, Kunai A, Ishikawa M (1996) Electrochemical anion doping of poly[(tetraethyldisilanylene) oligo(2,5-thienylene)] derivatives and their p-type semiconducting properties. J Electroanal Chem 414:135–139
193. Malliaras GG, Hadziioannou G, Herrema JK, Wildeman J, Wieringa RH, Gill RE, Lampoura SS (1993) Tuning of the photo- and electroluminescence in multi-block copolymers of poly[(silanylene)thiophene]s via exciton confinement. Adv Mater 5:721–723
194. Ohshita J, Yoshimoto K, Hashimoto M, Hamamoto D, Kunai A, Harima Y, Kunugi Y, Yamashita K, Kakimoto M, Ishikawa M (2003) Synthesis of organosilanylene–pentathienylene alternating polymers and their application to the hole-transporting materials in double-layer electroluminescent devices. J Organomet Chem 665:29–32

195. Tang H, Zhu L, Harima Y, Kunugi Y, Yamashita K, Ohshita J, Kunai A (2000) Optical study on electrochemical and chemical doping of polymers of oligothienyls bridged by monosilyl. Electrochim Acta 45:2771–2780
196. Kunugi Y, Harima Y, Yamashita K, Ohshita J, Kunai A, Ishikawa M (1996) Electrochemical anion doping of poly[(tetraethyldisilanylene) oligo(2,5-thienylene)] derivatives and their p-type semiconducting properties. J Electroanal Chem 414:135–139
197. Harima Y, Zhu L, Tang H, Yamashita K, Takata A, Ohshita J, Kunai A, Ishikawa M (1998) Electrochemical cleavage of a Si–Si bond in poly[(tetraethyldisilanylene) oligo(2,5-thienylene)] films. Synth Met 98:79–81
198. Tang H, Zhu L, Harima Y, Yamashita K, Ohshita J, Kunai A, Ishikawa M (1999) Electrochemistry and spectroelectrochemistry of poly[(tetraethyldisilanylene)quinque (2,5-thienylene)]. Electrochim Acta 44:2579–2587
199. Bokria JG, Kumar A, Seshadri V, Tran A, Sotzing GA (2008) Solid-state conversion of processable 3,4-ethylenedioxythiophene (EDOT) containing poly(arylsilane) precursors to π-conjugated conducting polymers. Adv Mater 20:1175–1178
200. Sotzing GA (2007) Conductive polymers from precursor polymers, method of making, and use thereof. US Patent Application US20070191576
201. Ohshita J, Nodono M, Watanabe T, Ueno Y, Kunai A, Harima Y, Yamashita K, Ishikawa M (1998) Synthesis and properties of dithienosiloles. J Organomet Chem 55:487–491
202. Ohshita J, Nodono M, Takata A, Kai H, Adachi A, Sakamaki K, Okita K, Kunai A (2000) Synthesis and properties of alternating polymers containing 2,6-diaryldithienosilole and organosilicon units. Macromol Chem Phys 201:851–857
203. Usta H, Lu G, Facchetti A, Marks TJ (2006) Dithienosilole- and dibenzosilole-thiophene copolymers as semiconductors for organic thin-film transistors. J Am Chem Soc 128:9034–9035
204. Lu G, Usta H, Risko C, Wang L, Facchetti A, Ratner MA, Marks TJ (2008) Synthesis, characterization, and transistor response of semiconducting silole polymers with substantial hole mobility and air stability. experiment and theory. J Am Chem Soc 130:7670–7685
205. Ohshita J, Kimura K, Lee KH, Kunai A, Kwak YW, Son EC, Kunugi Y (2007) Synthesis of silicon-bridged polythiophene derivatives and their applications to EL device materials. J Polym Sci A Polym Chem 45:4588–4596
206. Chan KL, McKiernan MJ, Towns CR, Holmes AB (2005) Poly(2,7-dibenzosilole): a blue light emitting polymer. J Am Chem Soc 127:7662–7663
207. Liu MS, Luo J, Jen AKY (2003) Efficient green-light-emitting diodes from silole-containing copolymers. Chem Mater 15:3496–3500
208. Mo Y, Tian R, Shi W, Cao Y (2005) Ultraviolet-emitting conjugated polymer poly(9,9′-alkyl-3,6-silafluorene) with a wide band gap of 4.0 eV. Chem Commun 4925–4926
209. Yang W, Hou Q, Liu C, Niu Y, Huang J, Yang R, Cao Y (2003) Improvement of color purity in blue-emitting polyfluorene by copolymerization with dibenzothiophene. J Mater Chem 13:1351–1355
210. Janietz S, Bradley DDC, Grell M, Giebeler C, Inbasekaran M, Woo EP (1998) Electrochemical determination of the ionization potential and electron affinity of poly(9,9-dioctylfluorene). Appl Phys Lett 73:2453
211. Chan KL, Watkins SE, Mak CSK, McKiernan MJ, Towns CR, Pascu SI, Holmes AB (2005) Poly(9,9-dialkyl-3,6-dibenzosilole): a high energy gap host for phosphorescent light emitting devices. Chem Commun 5766–5768
212. Scherf U, List EJW (2002) Semiconducting polyfluorenes: towards reliable structure-property relationships. Adv Mater 14:477–487
213. van Dijken A, Bastiaansen JJAM, Kiggen NMM, Langeveld BMW, Rothe C, Monkman A, Bach I, Stössel P, Brunner K (2004) Carbazole compounds as host materials for triplet emitters in organic light-emitting diodes: polymer hosts for high-efficiency light-emitting diodes. J Am Chem Soc 126:7718–7727
214. S Yamaguchi, T Endo, M Uchida, T Izumizawa, K Furukawa, K Tamao (2000) Toward new materials for organic electroluminescent devices: synthesis, structures, and properties of a series of 2, 5-diaryl-3,4-diphenylsiloles. Chem Eur J 6:1683–1692

215. Kawamura Y, Yanagida S, Forrest SR (2002) Energy transfer in polymer electrophosphorescent light emitting devices with single and multiple doped luminescent layers. J Appl Phys 92:87
216. Wang E, Li C, Mo Y, Zhang Y, Ma G, Shi W, Peng J, Yang W, Cao Y (2006) Poly(3,6-silafluorene-co-2,7-fluorene)-based high-efficiency and color-pure blue light-emitting polymers with extremely narrow band-width and high spectral stability. J Mater Chem 16:4133–4140, Doi: http://dx.doi.org/10.1039/b609250k
217. Wang E, Li C, Peng J, Cao Y (2007) High-efficiency blue light-emitting polymers based on 3,6-silafluorene and 2,7-silafluorene. J Polym Sci A Polym Chem 45:4941–4949
218. Wang F, Luo J, Yang K, Chen J, Huang F, Cao Y (2005) Conjugated fluorene and silole copolymers: synthesis, characterization, electronic transition, light emission, photovoltaic cell, and field effect hole mobility. Macromolecules 38:2253–2260
219. Wang Y, Hou L, Yang K, Chen J, Wang F, Cao Y (2005) Conjugated silole and carbazole copolymers: synthesis, characterization, single-layer light-emitting diode, and field effect carrier mobility. Macromol Chem Phys 206:2190–2198
220. Wang E, Li C, Zhuang W, Peng J, Cao Y (2008) High-efficiency red and green light-emitting polymers based on a novel wide bandgap poly(2,7-silafluorene). J Mater Chem 18:797–801
221. Horst S, Evans NR, Bronstein HA, Williams CK (2009) Synthesis of fluoro-substituted silole-containing conjugated materials. J Polym Sci A Polym Chem 47:5116–5125
222. Wang E, Wang L, Lan L, Luo C, Zhuang W, Peng J, Cao Y (2008) High-performance polymer heterojunction solar cells of a polysilafluorene derivative. Appl Phys Lett 92:033307
223. Liao L, Dai L, Smith A, Durstock M, Lu J, Ding J, Tao Y (2007) Photovoltaic-active dithienosilole-containing polymers. Macromolecules 40:9406–9412
224. Hou J, Chen HY, Zhang S, Li G, Yang Y (2008) Synthesis, characterization, and photovoltaic properties of a low band gap polymer based on silole-containing polythiophenes and 2,1,3-benzothiadiazole. J Am Chem Soc 130:16144–16145
225. Huo L, Chen HY, Hou J, Chen TL, Yang Y (2009) Low band gap dithieno[3,2-b:2′,3′-d]silole-containing polymers, synthesis, characterization and photovoltaic application. Chem Commun 5570–5572
226. Ohshita J, Nodono M, Watanabe T, Ueno Y, Kunai A, Harima Y, Yamashita K, Ishikawa M (1998) Synthesis and properties of dithienosiloles. J Organomet Chem 553:487–491
227. Zhu Z, Waller D, Gaudiana R, Morana M, Mühlbacher D, Scharber M, Brabec C (2007) Panchromatic conjugated polymers containing alternating donor/acceptor units for photovoltaic applications. Macromolecules 40:1981–1986
228. Beaujuge PM, Pisula W, Tsao HN, Ellinger S, Müllen K, Reynolds JR (2009) Tailoring structure-property relationships in dithienosilole-benzothiadiazole donor-acceptor copolymers. J Am Chem Soc 131:7514–7515

Polycarbosilanes Based on Silicon-Carbon Cyclic Monomers

E.Sh. Finkelshtein, N.V. Ushakov, and M.L. Gringolts

Abstract This review is devoted to analysis of the scientific data concerning polycarbosilanes and some of their functional derivatives, primarily, published in the last ten years. The scope is limited to highly molecular weight products of the above-mentioned type, prepared via polymerization of cyclic monomers as the most effective and flexible synthetic approach. The chapter consists of two main parts: heterochain and carbochain polycarbosilanes. It includes description of ring-opening polymerization (ROP) via rupture of endocyclic Si–C bonds in strained silacarbocycles, ring-opening metathesis polymerization (ROMP) via rupture of endocyclic C=C bond in silylcycloolefins, and vinyl type addition polymerization (AP) of silylnorbornenes. The review pays much attention to structure and physical chemical properties of the obtained polymers as well as possible ways for their applications. The mechanisms of some polymerization processes are also discussed.

Keywords Addition polymerization · ROMP · ROP · Silylnorbornenes · Strain silacarbocycles

Contents

1   Heterochain Polycarbosilanes 112
     1.1   Ring-Opening Polymerization of Saturated Silicon–Carbon Heterocycles 113
     1.2   Ring-Opening Polymerization of Unsaturated Silicon–Carbon Heterocycles 126

E.Sh. Finkelshtein (✉), N.V. Ushakov, and M.L. Gringolts
A.V. Topchiev Institute of Petrochemical Synthesis RAS,
Leninskii prospect 29, 119991 Moscow, Russia
e-mail: fin@ips.ac.ru

2   Carbochain Polycarbosilanes.. 133
    2.1   Ring-Opening Metathesis Polymerization of Silicon Substituted Cycloolefins.... 133
    2.2   Addition Polymerization of Silicon Substituted Norbornenes 142
3   Conclusions 147
References 148

Due to diverse applications of polysiloxane materials, the term "organosilicon polymers" is in common use in the chemical literature in relation solely to polysiloxanes. However, already by the late 1940s and early 1950s, papers devoted to the design of polycarbosilanes (PCS) or polysiliconhydrocarbons – polymers containing only Si, C, and H atoms – were published. In these polymers, Si atoms are incorporated either into the backbone (heterochain PCSs) or in side substituents (carbochain PCSs). Interest in these polymers is determined by the much higher chemical stability of a Si–C bond than that of a Si–O bond with respect to nucleophilic and electrophilic agents. In addition, polysiloxanes undergo various chain transformations and even depolymerize at elevated temperatures [1]. During the past half century, considerable progress has been achieved in the synthesis of various PCSs and study of their properties and application areas (e.g., such carbochain PCSs as polyvinylsilanes, polypropynesilanes, and Si-containing polynorbornenes show promise as materials for efficient membrane gas separation, adhesion, etc.).

1   Heterochain Polycarbosilanes

There are several approaches to the synthesis of heterochain PCSs, which can be divided into three large groups: chain ring opening and vinyl polymerizations, step-growth reactions (polymerization and polycondensation), and polymer-analogous transformations.

This part of the review is devoted to PCSs prepared from silicon-carbon cyclic compounds. In the first section, we will consider the polymerization of saturated and unsaturated silicon-carbon heterocycles (SCHs) via a Si–C bond rupture (ring-opening polymerization, ROP) and of unsaturated SCHs via a chain C=C bond rupture (ring-opening metathesis polymerization, ROMP). The polymerization of silicon homocycles is beyond the scope of this review since polymers containing only Si atoms in the backbone belong to a special branch of organosilicon chemistry.

The second section deals with the state-of-the-art in the polymerization of unsaturated cyclic monomers with silicon atoms occurring in substituents via both the ROMP scheme and additive (vinyl) polymerization (AP).

1.1 Ring-Opening Polymerization of Saturated Silicon–Carbon Heterocycles

The ROP of all known saturated SCHs proceeds via the rupture of endocyclic Si–C bonds. In the US patent issued in 1958, the first attempt was made to use a 4-membered SCH for the synthesis of heterochain PCSs [2]. From the late 1950s to the early 1980s, Vdovin and co-workers intensively investigated a new class of polymers – high-molecular-weight heterochain PCSs, prepared by the ROP of 4-membered SCHs – dimethyl-silacyclobutane (MSCB), tetramethyl-disilacyclobutane (DSCB) and their derivatives (see Sects. 1.1 and 1.2). The energy and structural characteristics of silacyclobutanes were analyzed in [3–10]. Though these publications are rather old, their results remain valid up to the present time.

MSCBs and DSCBs are highly strained 4-membered SCHs with strain energies (polymerization enthalpies) of 83.9, 90.3, and 71.5 kJ mol^{-1} for 1,1-dimethyl- and 1-phenyl-1-methyl-1-silacyclobutanes and 1,1,3,3-tetramethyl-1,3-disilacyclobutane, respectively [4].

Structural studies of MSCB and DSCB made it possible to estimate strain-induced deformations. Electron diffraction measurements (in the gas phase) showed that, in MSCBs, internal angles at a silicon atom are deformed to 80° or smaller; in the DSCB analog, the C–Si–C angle is 89° [5, 6]. The ^{29}Si, ^{13}C, and ^{1}H NMR data combined with quantum-chemical calculations performed for a number of substituted silacyclobutanes (SCBs) demonstrated the existence of a strong transannular interaction Si...β-C (π-overlap of atomic orbitals of silicon and β-carbon atom) in 4-membered SCHs. In these terms, a high reactivity of SCBs is associated with a decrease in the s-character of atomic orbitals of silicon in its endocyclic bonds with carbon atoms [7]. Further structural studies of crystalline α-naphthyl derivatives of 1-silacyclobutane testified that in fact strong intracyclic interaction occurs along with the feasible planar conformation of a silacyclobutane ring (in contrast to the case in the gas phase). The transannular distance Si...β-C, which is equal, for example, for dinaphthylsilacyclobutane, to 2.334 Å, is not only 0.47 Å shorter than the sum of single-bonded radii of silicon and carbon but is even smaller than the distance between valence nonbonded carbon atoms in α-positions (2.468 Å). Another reason for distortion of the tetrahedral configuration of bonds at the silicon atom that appears in the case of SCBs is a difference in the lengths of Si–C bonds (exo- and endocyclic). The dependences of the length of the Si–C endocyclic bonds on the size of cycles, constructed from the averaged lengths of these bonds available for 3-, 4-, 5-, 6-,7-, and 8-membered SCHs, suggest that, in the case of 4-membered cycles, the length of this bond is larger than that of other cycles [8–10]. Based on these data, it is suggested that a higher reactivity of the Si–C endocyclic bond in SCBs than that in larger silacarbocycles can be also explained by the fact that in SCBs this bond is the longest. The 4-membered SCHs are also characterized by an increased, compared to linear alkylsilanes and less strained heterocycles, tendency toward formation of associates in nonpolar solvents [11].

1.1.1 Ring-Opening Polymerization of Saturated Silicon–Carbon Heterocycles Carrying a Single Heteroatom

Ring-Opening Polymerization of MSCBs

The ROP of MSCBs proceeds via the scission of the Si–C endocyclic bond under the action of heating or various catalysts, which can be depicted by the following scheme:

$$\underset{\text{Si}}{\triangleleft}\!\!\!\overset{R^2}{\underset{R^1}{}} \xrightarrow{\Delta\,/\,[B^-]\,/\,[M]} \left[\!\!\begin{array}{c}R^2\\|\\\text{Si}\\|\\R^1\end{array}\!\!\!\sim\!\right]_x$$

Polymers prepared from silacyclobutanes by any methods (thermal or catalytic) are characterized by the same chemical structure, namely, their main chains are composed of repeating siltrimethylene units linked via the head-to-tail type, as evidenced by vibrational spectroscopy [12–14] and ^{1}H, ^{13}C, and ^{29}Si NMR studies (e.g., [3, 15, 16]).

The above structural features of MSCBs (the ratio between endo- and exocyclic angles CSiC and a large length of the Si–C endocyclic bond) [3] enable one easily to attach various substituents, including bulky groups, to the silicon atom via nucleophilic substitution of chlorine atoms (or alkoxy groups). This method allowed one to synthesize silacyclobutane monomers with various combinations of substituents – the same or different, hydrocarbon (saturated and containing multiple bonds or aromatic groups), and carbofunctional [17–23]. Such a variation in the nature of substituents in MSCBs coupled with a high activity of the polymerizable fragment (a silacyclobutane ring) makes it possible to prepare polymers with different properties. It should be noted that such syntheses for carbocyclic analogs are either impossible or very complex. The copolymerization of MSCBs with similar or different types of monomers widens the scope of means for tailoring the properties of the polymers.

Selection of polymerization procedures for every MSCB monomer depends on the nature of its substituents R^1 and R^2. MSCBs carrying purely hydrocarbon substituents, where R^1 and R^2 are alkyl, cycloalkyl, aryl, or alkylaryl radicals (free of polar groups), can be polymerized by all known methods available for this class of monomers. In the case of MSCBs containing double bonds in substituents, and Si–H-derivatives of MSCBs, the anionic low-temperature polymerization initiated by organolithium compounds in THF is applicable. If substituents carry polar groups of various natures, thermal polymerization (TROP) is sometimes possible. For example, a tertiary amino group or an ether bond (in the diphenyl oxide substituent) does not hamper TROP. Such halogens as chorine and bromine in the hydrocarbon substituent, including the aromatic ring, decrease the activity of monomers in TROP and the molecular mass of the polymers. If monomers contain three or a greater number of C–F bonds (F–Alk or F–Ar) in a substituent, they turn out to be incapable in TROP. In this case, polymerization may be catalyzed by the compounds of Pt or other transition metals [19].

Thermal Ring-Opening Polymerization of MSCBs

The TROP mechanisms were examined in detail, and the results of kinetic experiments were presented in [3]. With consideration of these data, the mechanism of TROP of silacyclobutanes was advanced. The TROP of silacyclobutanes proceeds through ring opening via the heterolytic scission of the endocyclic Si–C bond and occurs as a zwitterion process, in which the coordination of monomer molecules to active centers plays an important role. In this case, the process is limited by the activity of a positively charged part of the zwitterion.

If high-purity monomers were used, the molecular weight of the polymers showed an almost linear dependence on the polymerization temperature. This dependence followed the direct proportionality up to ~210°C and reflected the specific feature of SCBs. When the temperature was increased to above 210°C, this dependence deviated from linearity. This is probably associated with an increase in the contribution of inter- and intramolecular chain degradation processes (disproportionation reactions); as a consequence, the molecular weight of the polymer should decrease somewhat and its molecular weight distribution (MWD) should become wider.

When very high requirements are imposed on the dielectric characteristics of polymer materials (e.g., when they are used as recording layers in thermoplastic and photothermoplastic data recording), TROP is the most suitable polymerization procedure [18, 22, 23]; in the case of catalytic polymerization, there is a need to remove traces of metals from the polymers. The molecular weight of the polymers synthesized by TROP can be controlled with advantage due to the use of regulating polar additives, for example, n-hexyl alcohol in the polymerization of carbazolyl-containing monomers [22].

Elevated temperatures are necessary for the polymerization of MSCBs carrying bulky aromatic substituents (such as α-naphthyl). However, long impact of high temperatures (>240°C) may cause crosslinking of the polymers due to the homolytic scission of Si–Ar bonds. Attempts to prepare a higher molecular weight polymer from a monomer with the Si–ferrocenyl (Fc) bond (A) [3, 19] by increasing the temperature of polymerization also led to the crosslinked product apparently due to the homolysis of the Si-ferrocenyl bond.

Quite different causes are responsible for formation of the crosslinked polymers in the TROP of a mesogene-containing MSCB (B) [16]. The fact that, at a rather low temperature (120°C), spirocyclic trimethylene[1]silaferrocenophane containing silaferrocenophane and silacyclobutane groups (C) polymerized to afford insoluble and even not swelling in organic solvents crosslinked polymer suggested opening of both parts of the spiro system – the Si–C bond of silacyclobutane and the Si–Fc bond [24].

The TROP of MSCBs with multiple bonds in substituents, for example, such monomers as 1-vinyl-1-methyl-, 1-allyl-1-methyl-, 1,1-diallyl-MSCBs [20], and 4-silaspiro[3.5]non-6-en synthesized via metathesis [25], was also complicated by crosslinking [3].

In all these cases, the use of solvents decreased the amount of the crosslinked products, whereas the polymerization of bulky monomers may cause full crosslinking of the polymer.

The MSCBs having *tert*-butyl and phenyl substituents [26] (previously they were not involved in TROP) can be readily converted into soluble high-molecular-weight polymers [27]. The TROP of 1,1-bis(phenyldimethylsilylmethyl)-MSCB proceeded at a higher temperature with formation of a soluble macromolecular polymer as well [21].

*Thermal Copolymerization of MSCBs and MSCB
with Tetramethyl-Disilacyclobutane and Other Monomers Capable
of Thermal Ring-Opening Polymerization*

The thermal copolymerization (TROCP) of MSCB monomers was mostly used to impart desired properties to polymers necessary for solving various applied problems: for example, copolymers of di-α-naphthyl- and diphenyl-MSCBs, di-*p*-tolyl-MSCB, 1-*p*-dimethylaminophenyl-1-phenyl-MSCB, diphenyl-MSCB, and tetraphenyl-DSCB were used as materials for data recording; copolymers of 1,1-dimethyl- and 1-phenyl-1-methyl-1-silacyclobutanes were employed as rubbers; copolymers of 1,1-diphenyl-MSCB with 1,1-dimethyl- or 1,1,3-trimethyl-MSCB served as materials for resists; [28–30]; and the copolymers of 1,1-dimethyl-MSCB with 1,1,3,3-tetramethyl-1,3-DSCB were valuable materials for the separation of hydrocarbon gases [31, 32].

In the case of TROP of 1-methyl-1-norbornenylmethyl-MSCB performed at 220°C, the process was accompanied by a partial retrodiene reaction of the norbornene pendant fragments resulting in elimination of cyclopentadiene and formation of allyl side groups [33].

The reaction of spirocyclic [1] silaferrocenophane (in which the spiro system is composed of ferrocenophane and silacyclobutane cycles) with dimethyl [1] silaferrocenophane [24] may be classified as the TROCP of MSCBs. This reaction afforded crosslinked copolymers.

Anionic Polymerization of MSCBs in the Presence of AlkLi, ArLi, or Li and K

Due to a marked strain of the cycle and the existence of a dipole in the endocyclic bond $Si^+ - C^-$ in MSCBs, these monomers showed high activity in anionic polymerization. The above-mentioned structural features of MSCBs indicate the configuration of the Si-atom which is extremely favorable for formation of the pentacoordinated state (sp^3d) [34, 35].

Nucleophile attack under anionic impact and primary and subsequent ring opening via the Si–C bond (during chain propagation) proceed via formation of the pentacoordinated state. For the organolithium reaction the scheme looks as follow:

Lithium alkyls or aryls, metallic Li, Na, K, alcoholates, silanolates of alkali metals, and their hydroxides may be used as initiators of the AROP of MSCBs.

Polymerization initiated by AlkLi and metallic Li can be performed in nonpolar solvents at 20–80°C or in polar ones (THF, DME) at −80/−50°C [20]. The mechanism of the AROP of MSCBs initiated by AlkLi are surveyed in [3]. Specifically, it was shown that living chains can be formed sometimes, that is active centers can be easily transformed both in polar and nonpolar media [36, 37]. A decrease in the polymerization temperature in polar media (down to −70°C or below) allowed one to reduce significantly the rate of chain termination (elimination of LiH, etc.) and to synthesize higher molecular weight polymers [3].

In a polar solvent (THF) at low temperatures, the AROP of 1,1-dimethyl-MSCB showed a well-defined living mechanism since various side reactions of chain termination were suppressed [38]. In contrast, in the case of 1-methyl-1-phenyl-MSCB, the features of living polymerization were not so pronounced [38].

A wide use of polar solvents (THF, HMPA additives) at low temperatures allowed Weber et al. to prepare in the presence of BuLi siltrimethylene polymers from MSCBs containing vinyl-substituents at Si-atoms (1-methyl-1-vinyl-, 1-phenyl-1-vinyl-, and 1,1-divinyl-1-silacyclobutanes) [39]. Obviously, these syntheses were implemented because of much reduced rates of side reactions. Moreover, a decrease in temperature enabled one to suppress the easy replacement of hydrogen in hydrosilane derivatives of 1-silacyclobutane [40, 41], which occurred in highly polar media under the action of nucleophiles, and to prepare high-molecular-weight polymers [42]: for polyphenylhydrosiltrimethylene, polydihydrosiltrimethylene, and polymethylhydrosiltrimethylene ($M_w/M_n = 7,30,000/3,90,000, 84,400/23,900$, and $12,200/6,100$, respectively). Note that the molecular weight of the polymer prepared by the AROP of methylhydro-MSCB was not too high; the yield of the high-molecular-weight polymer based on phenylhydro-MSCB was low (∼15%), while the yield of the oligomer ($M_w/M_n = 900/800$) soluble in methanol was 40%; and the MWD of polydihydrosiltrimethylene was wider than that of the two other polymers.

In subsequent work, the scope of monocyclobutane monomers that can produce polymers with the regular unit structure via the low-temperature AROP became even wider. These are first of all monosilacyclobutanes carrying methylene groups in the third position which formally can be assigned to unsaturated SCBs but will be considered in this section since they do not contain the real endocyclic double bond even though they contain one sp^2-hybridized carbon atom in a cycle [43]. These 1,1-disubstituted monomers (having two methyl, *n*-propyl, or phenyl substituents at the silicon atom) with the exocyclic methylene group at the β-carbon atom were involved in anionic polymerization both at low temperatures in THF and at room temperature in nonpolar media. Low temperatures and polar solvents were not necessary for the block copolymerization of 1,1-dimethyl-3-methylene-1-silacyclobutane. This fact provided additional evidence that the AROP of MSCBs initiated by organolithium compounds in nonpolar medium shared common features with living polymerization [43].

The *n*-BuLi-initiated AROP of monomers carrying two methyl groups at the Si-atom and bulky substituents in the third position (1,1-dimethyl-3-α-naphthyl-, 1,1-dimethyl-3-α-naphthylmethyl-, and 1,1-dimethyl-3-diphenylmethyl-1-silacyclobutanes) also proceeded very efficiently and easily [44].

The authors of [45] compared the behavior of isomeric MSCBs with the phenyl substituents at α- and β-carbon atoms in the cycle – 1,1-dimethyl-2-phenyl-1-silacyclobutane and 1,1-dimethyl-3-phenyl-1-silacyclobutane – under the conditions of low-temperature AROP. The isomer with the phenyl group at the β-position obeyed all features of living polymerization, including growth of molecular weight upon addition of a fresh monomer portion (in this case, $M_w/M_n = 1.09$). However, if the phenyl substituent is moved to the α-position with respect to the silicon atom, the polymerization of such α-phenyl-substituted MSCB did not follow the living pattern. The same authors confirmed that the AROP of 1-methyl-1-phenyl-1-silacyclobutane may be complicated by nucleophilic attack via the Si–Ph bond; as a consequence, its polymerization was not living. At the same time, the presence of the methyl group in the α-position of the MSCB cycle did not hamper the living polymerization of this monomer [46]. It was also shown that the phenyllithium-initiated AROP of 1,1,2-trimethyl-1-silacyclobutane proceeded selectively – only scission of the endocyclic Si–CH$_2$ bond (1,4-insertion) took place; in the catalysis by Pt-compounds, both Si–CH$_2$ and Si–CH(Me) bonds broke (1,4- and 1,2-insertion, respectively), as was confirmed by experiments with a deuterated terminating agent [46].

To prepare amphiphilic siltrimethylene polymers, the authors of [47] synthesized two symmetrically substituted MSCB monomers carrying oxyethylene groups in substituents – 1,1-bis[4-(*tert*-butyldimethylsilyloxy-monooxyethylene)phenyl]-1-silacyclobutane and 1,1-bis[4-(ω-*tert*-butyldimethylsilyloxy-trioxyethylene)phenyl]-1-silacyclobutane. The polymerization of these monomers occurred at $-78°C$ or $-48°C$ in the presence of *n*-BuLi in THF without any HMPA additives. The use of a *tert*-butyldimethylsilyl group instead of the usually used trimethylsilyl group for silyl protection ensured stable protection under contact with *n*-BuLi. This made it possible to carry out the successful polymerization of these monomers and to prepare rather high-molecular-weight polymers stable in solution. After removal of silyl protection, amphiphilic polysiltrimethylenes bearing two side substituents in every unit, which contain oxyethylene fragments with end hydroxyl groups along with the oxyphenylene bridge, turned out to be soluble in polar solvents, such as ethanol and methanol.

Recently, 1-silacyclobutane with the planar configuration of the cycle, namely, racemic 1,1-(*rac*-1,1′-bi-2-naphthoxy)-1-silacyclobutane has been synthesized [48]. It was shown that the silicon-carbon cycle is planar, and the distance between the hetero atom and the β-carbon atom (Si....β-C) is 2.302 (5) Å, that is, close to the average value of this equilibrium distance in MSCBs (2.35 ± 0.02) [8, 17]. The authors of [48] used anionic polymerization since neither Karstedt's catalyst nor 1,5(cyclooctadiene)$_2$Pt(0) could initiate the polymerization of 1,1-(*rac*-1,1′-bi-2-naphthoxy)-1-silacyclobutane. In this case, 1,5(cyclooctadiene)$_2$Pt(0) eliminated cyclooctadiene and gave rise to 1,1′-bis-2-naphthol. Notably, the polymerization of this monomer was performed at room temperature in the absence of either HMPA or TMEDA.

The anionic copolymerization (AROCP) of different MSCBs with the monomer of another type capable of anionic polymerization in polar or nonpolar medium initiated by AlkLi mostly yielded random copolymers. The sequential polymerization (addition of the second monomer after polymerization of the first monomer was completed) enabled one to synthesize various block copolymers. As MSCBs, various symmetrically and unsymmetrically substituted derivatives were used. As monomers of other types, styrene, butadiene, isoprene, and 2,4-dimethylstyrene were tested [49].

Beginning from the 1990s, MSCBs have been primarily used for anionic copolymerization with monomers of other types, most frequently vinyl monomers, mainly styrene; in most cases, block copolymers were prepared at low temperatures in polar media (e.g., $-78°C$, THF). As an example, the scheme of the block copolymerization of 1,1-dimethyl-1-silacyclobutane with styrene under conditions more acceptable for formation of living chains ($-48°C$, THF: hexane $= 1:1$(v/v), without HMPA [38]) is presented below.

$$\text{SiMe}_2 \xrightarrow[\text{THF-Hexane}]{n\text{-BuLi}} \xrightarrow{\text{CH}_2=\text{CHPh}} \xrightarrow{\text{MeOH}} n\text{-Bu}\left[\begin{array}{c}\text{Me}\\|\\\text{Si}\\|\\\text{Me}\end{array}\right]_x \left[\begin{array}{c}\text{Ph}\\|\\\text{CH}_2-\text{CH}\end{array}\right]_y \text{H}$$

The molecular weight of such block copolymers was controlled by the monomer-to-initiator ratio. The block copolymer of 1,1-dimethyl-3-methylene-1-silacyclobutane with styrene can be prepared under common conditions (benzene, room temperature) [50].

To avoid the deactivation of active centers, which led to formation of allyl carbanions (and hence afforded homopolymers), a rapidly reacting 1,1-diphenylethylene was added immediately after interaction of the MSCB comonomer with RLi. This compound is known to produce a more stable carbanion but is not involved in polymerization. Since the activity of a carbanion derived from 1,1-diphenylethylene was sufficient for initiation of polymerization of other vinyl monomers, such as styrene or acrylates, then upon addition of the latter compounds, this carbanion provided initiation and further chain growth. As a result, block copolymers, in which siltrimethylene and polyvinyl blocks were separated by the 1,1-diphenylethylene unit, were formed. This approach made it possible to prepare amphiphilic diblock copolymers via the copolymerization of 1,1-dimethyl-, 1,1-diethyl-, and 1,1-di-*n*-butyl-1-silacyclobutanes with *tert*-butyldimethylsilyl methacrylate [51].

Anionic Ring-Opening Polymerization of MSCBs in the Presence of Alkalis and Alkali-Metal Alcoholates and Silanolates (Homo- and Copolymerization)

As was shown in the early 1970s, alkalis and their derivatives were able to catalyze the polymerization of MSCBs at 5–90°C, but the reaction proceeded at a low rate and led to polysiltrimethylenes with molecular weights much smaller than those of polymers prepared under thermal initiation. With the use of this method, the polymerization of MSCBs with various substituents (hydrocarbon and carbofunctional) [52] and the copolymerization of MSCBs [23] can be realized.

The study of the catalytic activity of mono- and disilanolates and alcoholates and hydroxides of alkali metals in the polymerization of 1,1-dimethyl-MSCB, and of the effect of the nature of solvent and initiator and other features of this reaction, showed that in the KOH catalysis, organopotassium compounds were responsible for the propagation of polymer chains [53]. For all three types of initiators (alcoholates, disilanolates, and hydroxides of alkali metals), catalytic activity showed a qualitative dependence on the nature of alkali metal. The activity decreased on passage from Cs- to K- and Na-containing initiators. The polarity of solvents in this process exerted the activating effect. Thus, the AROP of MSCBs in the presence of alkali derivatives can be described by the following scheme:

$$Me_3SiOK \xrightarrow{\triangle SiMe_2} Me_3SiOSi(Me_2)CH_2CH_2CH_2K \xrightarrow{\triangle SiMe_2} Me_3SiO[Si(Me_2)CH_2CH_2CH_2]_x K \xrightarrow{-LiH}$$
$$\longrightarrow Me_3SiO[Si(Me_2)CH_2CH_2CH_2]_{x-1} Si(Me_2)CH_2CH=CH_2$$

The above-described results laid the foundation for elaboration of catalytic systems based on MSCBs and alkali derivatives useful for the anionic polymerization of different vinyl monomers. As initiators, two-component systems (MSCB – silanolate or alcoholate) and single-component systems ("prepolymers" – products of interaction of disilanolates with MSCB) were tested.

The ability of MSCB to undergo ring opening with the subsequent formation of C–Li, C–Na, and C–K bonds was used to generate carbanions (i.e., SCB as a carbanion pump) via their reaction with 1,1-diphenylethylene with the aim of preparing various block copolymers [54–56]. This seems to be the most interesting polymerization application of alkali-metal catalysis of MSCBs. With the use of a fourfold excess of 1,1-diphenylethylene and a mixture of equal volumes of THF and hexane as a solvent, *tert*-butoxy anion can be converted into carbanion with a yield of ~97%. The rapid reaction with 1,1-diphenylethylene gave rise to a more stable carbanion which can initiate the polymerization of ethylene oxide and styrene.

Such a silacyclobutane "carbanion pump" system was used to synthesize various di- and triblock copolymers.

In [57], the efficiency of the "carbanion pump" system, which is the *tert*-butoxy anion – 1,1-dimethyl-1-silacyclobutane-1,1-diphenylethylene - was compared with that of the system in which α-methylstyrene was used instead of 1,1-diphenylethylene.

Ring-Opening Polymerization of MSCBs with Pt and Other Transition Metal Compounds: Homo- and Copolymerization of MSCBs, MSCBs with DSCB, and Monomers of Other Types

Compounds of Pt and other Pt-group metals served as the most versatile catalysts for the polymerization of SCBs: they catalyzed the polymerization of all types of SCB monomers. Frequently, catalysis by Pt-compounds appeared to be the only way suitable for preparing some siltrimethylene polymers.

With the use of Speier's catalyst, the first representative of mesogene-containing polysiltrimethylenes [16] and fluorine-containing polysiltrimethylenes [19] were synthesized.

Karstedt's catalyst, which is a complex of Pt with tetramethyldivinyldisiloxane (Pt$_2$[Vin(Me$_2$)SiOSi(Me$_2$Vin]$_3$), was found to be very active. With the use of this catalyst, soluble polymers were prepared from trimethylsiloxy- and triphenylsiloxy-derivatives of 1-silacyclobutane [58], mesogene-containing 1-silacyclobutanes carrying nitrile groups and siloxane bridges between a mesogenic group and an endocyclic silicon atom [59], and 1-methyl-1-phenylethynyl-1-silacyclobutane [60].

In the case of the latter monomer (with the triple bond in a substituent), the molecular weight of polysitrimethylenes produced by the Pt-catalyzed ROP of MSCBs was controlled by the dosed addition of Et$_3$SiH. The specific features of this process were examined by example of polymerization of 1,1-dimethyl-1-silacyclobutane carried out in the presence of Pt(1,5-cod)$_2$. This gave a chance to develop the general mechanism of the ROP of MSCBs catalyzed by Pt-compounds

[61]. In accordance with this mechanism, the catalyst containing the reduced platinum interacts with a monomer molecule to form a 5-membered intermediate, which in turn interacts with the next monomer molecule to give a 9-membered cyclic Pt-containing intermediate. The latter forms the stable dimer (as in the case of Pt-phosphine catalysis) via reductive elimination. When phosphine ligands are absent, the approach of the monomer to the Pt-center becomes apparently much easier; therefore, it can efficiently interact with the Pt-cyclic intermediate to produce the final polymer.

To implement the polymerization and copolymerization of various silacyclobutanes, we used catalysis by the propene complex of Pt, which was prepared by heating Speier's catalyst directly in the reaction ampoule before the experiment [21, 32]. In addition, the propene complex of Pt allowed one to achieve high yields of high-molecular-weight polymers of MSCBs carrying two phenyldimethylsilylmethyl substituents and carbazolyl and diphenyl oxide moieties in substituents [23]. At 7–15°C, the propene Pt-complex made it possible to prepare random soluble copolymers of diallylsilacyclobutane and tetramethyldisilacyclobutane of various compositions [21, 32]. In those works, gas transport parameters of copolymers of dimethylsila- and tetramethyldisilacyclobutanes were studied.

We have determined the parameters of hydrocarbon penetration through these copolymers and found that they retained very high selectivity and produced stable films. Selectivity (α) for the copolymers with $x/y = 1/1$ and $x/y = 3/1$ with respect to the butane and methane reached 19.4 and 20.7, respectively, together with high permeability. In this case, the permeability coefficients differed slightly from the corresponding parameters of the silmethylene homopolymer.

1.1.2 Ring-Opening Polymerization of 1,3-Disilacyclobutanes

Thermal Ring-Opening Polymerization of 1,3-Disilacyclobutanes

Due to the presence of the second silicon atom, 1,3-DSCBs differ appreciably from MSCBs in terms of their chemical behavior. This difference manifests itself (as was mentioned above) as a high tendency of MSCBs toward nucleophilic ring opening compared to DSCBs, and vice versa – in reactions with electrophiles MSCBs are less active than DSCBs. Nevertheless, the main feature of 4-membered SCBs in which they differ from their carbon analogs is their ability to undergo thermally initiated polymerization in the liquid phase. A comparison of methyl derivatives of MSCBs and DSCBs (1,1-dimethyl-MSCB and 1,1,3,3-tetramethyl-DSCB) showed that MSCBs were more active in the case of thermal initiation (its polymerization begins even at 85°C, whereas the thermal polymerization of DSCB proceeded at temperatures above 130°C). This is also evidenced by a lower activation energy of MSCB (11.6 kcal mol^{-1}) [3, 62] than that of DSCB (17.1 kcal mol^{-1}) [62, 63].

We would like to emphasize the distinctive feature of TROP of DSCB: the silicon-functional monomer with the Si–Cl function at one Si-atom can be easily polymerized. In general, electron-acceptor substituents showed a peculiar effect on the process [13]. Tetrachloro-DSCB was not involved in thermal polymerization (250°C) as well as 1-bis(trimethylsilylamino)-1,3,3-trimethyl-1,3-disilacyclobutane. However, soluble polymers were formed on the basis of 1,1,3-trimethyl-3-vinyl-1,3-disilacyclobutane [64] and trimethyl-DSCB [65]. Finally, as opposed to 1,1-dichloro-MSCB (which is incapable of thermal polymerization), 1,1-dichloro-3,3-dimethyl-1,3-disilacyclobutane was readily involved in thermal polymerization. Moreover, it turned out to be the most active in a series of chlorine-containing 1,3-DSCBs. In terms of decreasing activity, these monomers can be arranged as follows [35]:

Cl$_2$Si⟨⟩SiMe$_2$ > ClMeSi⟨⟩SiMe$_2$ > Me$_2$Si⟨⟩SiMe$_2$ > Cl$_2$Si⟨⟩SiMeCl >> Cl$_2$Si⟨⟩SiCl$_2$

A similar tendency followed from the qualitative comparison of activities of 1,3-DSCBs with such electron-acceptor substituents as γ-trifluoropropyl [35, 66]. This trend can be rationalized by different distributions of electron density in MSCB and DSCB molecules. In the case of the MSCB skeleton, the electron-accepting effect of a substituent causes a reduction in electron density on α-carbon atoms of the cycle, thereby making cyclic Si–C–bonds stronger. At the same time, in the case of DSCB, the electron-accepting effect decreases electron density on the endocyclic Si–C- bonds located near electron-acceptor substituent and the far-off Si–C bonds are weakened because of this effect.

It should be noted that the polymerization of 1,3-DSCBs with methyl substituents at the carbon atom of the cycle and two Cl atoms at the Si atom were readily involved in TROP [67]. TROP may be used to synthesize polysilmethylenes with alkyl and aryl substituents in the decoration. Molecular weights were also as high as in the polymerization of MSCB [68, 69]. In all cases, poly-DSCBs had regular structures in which repeating units were composed of alternating silicon and carbon atoms [12, 70].

The TROP of mono- and disilacyclobutanes are characterized by similar behavior, and the fact that copolymers can be produced from the combination of MSCBs and DSCBs indicates that the TROP of these monomers obeys a common mechanism [3, 62].

For the last two decades, interest in the polymerization of 1,3-disilacyclobutanes has shifted to the applied research due to high elastomeric, mechanical, and thermostable characteristics of polysilmethylenes and -trimethylenes [71, 72]. These polycarbosilanes are also of interest as precursors for production of photoresists and Si–C-ceramics [71, 72]. Therefore, many papers and patents have been devoted to the elaboration of effective methods of polymer syntheses. For example, traditional thermal initiation was used in the case of 1,3-dimethyl-1,3-diphenyl-1,3-disilacyclobutane (170°C) and 1,1,3,3-tetraphenyl-1,3-disilacyclobutane (up to 300°C) [70–72]. Sometimes, thermal initiation was coupled with other impacts, for example, laser irradiation, to embed heterogeneous catalysts, such as nanosized particles of Pt, Cu, and Ag [70]. In the case of thermal initiation the methodology of molecular weight control by addition of dosed amounts of trimethylchlorosilane to the reaction mixture was elaborated [72]. The TROP of 1,3-dimethyl-1,3-diphenyl-1,3-disilacyclobutane at 230°C in the absence of Me$_3$SiCl resulted in a polymer with $M_w = 2,640,000 (M_w/M_n = 4.1)$ and a yield of 75%. At Me$_3$SiCl: monomer $= 0.08, 0.37,$ and 3.9, the yields of the polymers with molecular weights 648,000 ($M_w/M_n = 4.1$), 269,000 ($M_w/M_n = 2.1$), and 90,000 ($M_w/M_n = 2.0$) were 79, 81, and 55%, respectively.

The thermal copolymerization of DSCBs can be illustrated by the reaction of tetramethyldisilacyclobutane with tetraphenyldisilacyclobutane [72] and the above-described reaction of tetraphenyl-DSCB with diphenyl-MSCB [28].

The thermal copolymerization of DSCBs was mostly carried out with such monomers as tetramethyldisilacyclobutane, tetraphenyldisilacyclobutane, and 1,3-dimethyl-1,3-diphenyl-1,3-disilacyclobutane [72].

Anionic Ring-Opening Polymerization of 1,3-Disilacyclobutanes

Although DSCBs were less prone to ring opening under the action of nucleophiles, such compounds as hydroxides (Na, K) and their derivatives (specifically, silanolates), catalyzed the AROP of 1,3-DSCBs at moderate temperatures (even at 20°C) but very slowly [73, 74]. The addition of cryptands activated this process, albeit its rate was also very small [75, 76]. Accordingly, the molecular weight of polysilmethylenes (as polysiltrimethylenes) was much smaller compared to that obtained by other catalytic polymerization procedures.

It was shown that the efficient AROP of tetramethyldisilacyclobutane proceeded on lithium alkyls in the presence of HMPA at a low temperature [77]. At monomer: initiator = 20, butyllithium initiation yielded a lower molecular weight polymer than PhLi and trimethylsilylmethyllithium did, although its M_n was not higher than 8,000 in both cases. This monomer did not polymerize in the absence of HMPA, as well as upon the addition of TMEDA. No polymerization of tetramethyldisilacyclobutane was observed, when bis(trimethylsilylmethyllithium) was used as a catalyst. This phenomenon can be expected since stabilization of the α-C–Li bond by two silicon atoms is too high.

Ring-Opening Polymerization of 1,3-Disilacyclobutanes with Compounds of Platinum and Other Transition Metals

At present, new ROP catalysts were designed on the basis of compounds of platinum group metals, and efficient polymerization procedures were developed. In the polymerization of DSCB and MSCBs, molecular weight was controlled by dosing dimethylphenylhydrosilane [71]. Pt-compounds which were inactive under ordinary conditions, for example, $Pt(acac)_2$, were activated with the help of UV radiation [78].

In the presence of Pt-based catalysts, DSCB monomers were readily involved in copolymerization. The copolymerization of tetramethyl-DSCB with some MSCBs initiated by Speier's catalyst has been already mentioned above in the section devoted to the copolymerization of MSCBs [21, 32].

Catalysis by compounds of Pt and other transition metals was used for the copolymerization of DSCBs with dimethyl[1]silaferrocenophane, in which one cyclopentadiene cycle was unsubstituted, while in the second cyclopentadiene cycle, all hydrogen atoms were replaced by methyl groups [79]. Since the polymerization of the ferrocenophane comonomer in the presence of $PtCl_2$ proceeded at a much higher rate than the polymerization of tetramethyl-DSCB, the polymerization of their equimolar mixture at 20°C predominantly yielded homopolymers. The random copolymer can be prepared at tetramethyl-DSCB: ferrocenophane = 15:1; however, the yield of the copolymer was also low ($\sim$21%).

Ring-Opening Polymerization of Tetramethyl-Disilacyclobutanes Under the Action of Metal (Ag, Cu, Bi, etc.) Halides

Heterogeneous metal halides (specifically, Ag, Cu, and other elements) [80] were found to be more accessible and suitable than homogeneous Pt-catalysts in polymerization of DSCBs. The indubitable advantage of this type of catalysts is that they allowed preparation of polymers on the basis of tetra-γ-trifluoropropyl-1,3-disilacyclobutane, which cannot polymerize according to the TROP mechanism. The polymerization of vinyltrimethyl-1,3-disilacyclobutane on metal halides led to fully soluble noncrosslinked products (as was shown above, TROP was complicated by crosslinking). The authors of [72] reported the successful polymerization of dimethyldiphenyl-1,3-disilacyclobutane in the presence of metal halides.

The kinetic study of the polymerization of tetramethyl-DSCB [81] showed that there was an inverse relation between the molecular weight of polydimethylsilmethylene and $CuCl_2$ amount, whereas molecular weight appeared to be independent of conversion. The experimental reaction order with respect to the monomer was found to be 1.5; with respect to $CuCl_2$, it was close to zero (~0.15). These data confirmed the heterogeneous character of the process. The activation energy was estimated as 23.0 kcal mol^{-1} in the temperature range 75–90°C.

1.2 Ring-Opening Polymerization of Unsaturated Silicon–Carbon Heterocycles

Silacarbocycles with sp^2-hybridized endocyclic carbon atoms showed much higher reactivity, including ROP, than silacycloalkanes. This is associated not only with the participation of multiple bonds in some reactions (as in the case of silacycloalkenes) but also with increased strain and manifestation of properties typical for alkenylsilanes. For example, at the allyl location of the sp^2- hybridized system in a cycle (with respect to the Si-atom), the endocyclic bond Si–C$_{All}$ (or Si–C$_{Benzyl}$) becomes weaker. As a result, not only 4-membered but also 5-membered silacarbocycles can undergo ROP with formation of high-molecular-weight products.

1.2.1 Ring-Opening Polymerization of 3,4-Benzo-1-silacyclobutenes

As was shown in [82], 1,1-dimethyl-2,3-benzo-1-silacyclobutene exhibited an extremely high activity in polymerization. This highly strained monomer was so active that its TROP proceeded even at 60°C.

The polymer synthesized at 80°C (with a yield of ~90%) was a high-melting (m.p. 190–193°C) product with a high molecular weight ($[\eta] = 2.2 \, \text{dl g}^{-1}$). Its X-ray analysis indicated a high degree of crystallinity. As was proposed in [83], the ROP of 1,1-dimethyl-2,3-benzo-1-silacyclobutene proceeded via the scission of the Si–C$_{Benzyl}$ bond. The symbate dependence of the polymer M_w on reaction temperature was observed in the TROP of this monomer as well. The nature of TROP of this 4-membered heterocycle condensed with the benzene ring was apparently the same as that of MSCB and DSCB; that is, zwitterions served as active centers. Salamone and Fitch also believed that the TROP of dimethyl-2,3-benzo-1-silacyclobutene obeyed the ionic mechanism [83].

It seems that the TROP of 1,1-dimethyl-1-silacyclobut-2-ene is the only example of the TROP of silacyclobutenes [84]. The polymer prepared at 140°C in xylene solution was a viscous resin with $M_w/M_n = 21,850/9,300$. An increase in temperature led to the crosslinked product.

In addition, the anionic polymerization of 1,1-dimethyl-2,3-benzo-1-silacyclo butene with the use of K-mirror in THF (or n-BuLi in a hydrocarbon solvent) was performed in [82]. As a result, powderlike polymers with lower molecular weight and melting point than those in the case of TROP were isolated.

It should be noted that relationships described in [85] did not evidence the living character of polymerization. This was presumably related to the existence of equilibrium between monomers and an alkyllithium reagent on the one hand, and pentacoordinated silicon intermediates on the other. The study of polymerization of 2,3-benzo-1-methyl-1-α-naphthyl-1-silacyclobut-2-ene [86] demonstrated that, under the low-temperature initiation by n-BuLi and even Ph$_2$(Me)SiLi in THF, the process followed the living chain mechanism. The molecular weight of the polymers increased with the monomer-to-initiator ratio and the addition of a fresh monomer portion after polymerization completion caused further polymerization and formation of a higher molecular weight polymer with a narrow MWD. It is of importance that the process was living both when a carbanion (in the case of n-BuLi initiation) was involved in chain propagation and a sila anion was located at the active chain end (in the case of Ph$_2$(Me)SiLi initiation). Preliminary experiments with a racemic α-naphthyl-containing monomer showed that the regioselectivity of ring opening with excess amounts of both n-BuLi and Ph$_2$(Me)SiLi was >99%. Polymers prepared from optically active (+)-monomers under initiation with BuLi and Ph$_2$(Me)SiLi were also optically active and had similar optical rotation but much lower stereoregularity.

The Pt-initiated polymerization of 2,3-benzo-1,1-diphenyl-1-silacyclobut-2-ene and 2,3-benzo-1,1dimethyl-1-silacyclobut-2-ene was described in [83, 87]. The comprehensive research into the polymerization of racemic 2,3-benzo-1-methyl-1-α-naphthyl-1-silacyclobut-2-ene and its optically pure isomer on Karstedt's catalyst (complex of tetramethyldivinyldisiloxane with Pt) was published in 2002 [88]. The happy choice of 2,3-benzo-1-methyl-1-α-naphthyl-1-silacylobut-2-ene as a subject for the study of polymerization catalyzed by compounds of Pt-group metals allowed one to gain knowledge about the main features of this process. Above all, it was found that the molecular weight of the polymers can be controlled by a change in the amount of the added telogen (triethylsilane). In addition, it was shown that the polymerization of a pure monomer (without regulating additives) gave a cyclic dimer formed as a minor product. Based on this evidence, it was supposed that the closure of polymer molecules to macrocycles took place. The mechanism of polymerization of 2,3-benzo-1-methyl-1-α-naphthyl-1-silacyclobut-2-ene catalyzed by compounds of Pt-group metals envisaged formation of just cyclic macromolecules and scission of the Si–C$_{Aryl}$ bond during ring opening.

Concluding the section devoted to the catalytic polymerization of 2,3-benzo-1-silacyclobutenes in the presence of Pt-containing compounds, we would like to note two examples of PtCl$_2$- or Pt(1,5 − cod)$_2$–catalyzed copolymerization of

2,3-benzo-1,1-dimethyl-1-silacyclobut-2-ene with symmetrical dimethyl[1] ferrocenophane [89] and nonsymmetrical dimethyl[1]ferrocenophane containing one unsubstituted cyclopentadiene ring and another ring having methyl substituents at four carbon atoms [79].

1.2.2 Ring-Opening Polymerization of Silaindanes and Silacyclopent-3-enes

Based on properties of silacarbocycles annulated with the benzene ring [90–93], conditions of preparing high-molecular-weight polymers from dimethylsilaindan were found.

The process occurred at low temperatures in the solid phase on the metallic K-mirror. Later on, low-temperature polymerization of silaindans with various substituents was catalyzed by n-BuLi in THF or n-BuLi/HMPA in THF. In addition to dimethylsilaindan, methyl- and phenylsilaindans and dihydrosilaindan were synthesized and polymerized in the presence of n-BuLi [94–96]. Since the molecular weight of the polymers was low ($M_w/M_n = 2,000/1,700$), ^{13}C and ^{29}Si NMR measurements made it possible to reveal that chain termination occurred via elimination of lithium hydride in the pentacoordinated state at the chain end [95]. These experiments extended anionic polymerization to organosilicon monomers with Si – H bonds of other types – the first example was the polymerization of p-dimethylsilylstyrene [97]. In [98, 99], the syntheses and AROP of silaindans with Si–H bonds in substituents of the aromatic part of a molecule were described.

There is no large strain in silaindan and silacyclopentene structures containing sp^2-carbon atom in benzyl or allyl position with respect to the silicon. Therefore, the driving force of the catalytic ROP of the above monomers is the combination of a small strain with the presence of chemically active endocyclic Si–C$_{Benzyl}$ or Si–C$_{Allyl}$ bonds. The existence of two active moieties in silacyclopentene molecules (the unsubstituted double bond and the Si–C$_{Allyl}$ bond) provides two ways for polymerization: via rupture of endocyclic Si–C bonds (cationic and anionic ROP) and via metathesis polymerization by cleavage of endocyclic C=C bonds (ROMP).

In experiments on metallation of silacyclopentenes by organolithium compounds, the ROP was observed [100]. Weber et al. [101–108] showed that polymerization can be performed by treating monomer with catalytic amounts of alkyllithium compounds complexed with polar compounds (HMPA, TMEDA) in THF at −78°C. In terms of phenomenology, symmetrically [101–103] and unsymmetrically [101, 103, 105] disubstituted derivatives of 1-silacyclopent-3-ene with methyl, phenyl, and vinyl radicals, hydro derivatives of 1-silacyclopent-3-ene [106, 107], 1,1,3-trimethyl- and 1,1,3,4-tetramethyl-1-silacyclopent-3-enes

[101, 108], three spiran compounds in which the spiro systems contain silacyclopentene [109], and silaindan [110] fragments were examined. Monomers and polymers carrying mesogenic groups in side-chain substituents [111], amphiphilic substrates having oxyethylene chains of various lengths [112], and germasilacyclopentene monomers and polymers [101, 113] were prepared.

1-Silacyclopent-3-enes lacking groups sensitive to RLi have been polymerized to give rise to polymers with a structure which was a result of ring opening via the Si–C$_{Allyl}$ endocyclic bond. The same is true for 1,1,3-trimethyl-1-silacyclopent-3-ene and 1,1-dimethyl-1-germanocyclopent-3-ene. In their polymers, double bonds occurred in *cis*-configuration; that is, the polymerization proceeded stereoselectively.

Despite certain factors complicating polymerization, in all the above-mentioned cases the resulting polymer chain consisted of identical units.

$R^3 = H$; $R^1 = R^2 = Me$, Ph, Vin
$R^3 = H$; $R^1 = Ph$; $R^2 = Me$, Vin, H
$R^3 = R^1 = R^2 = Me$
$R^3 = H$; $R^1 = Me$; $R^2 = H$

Spiro compounds containing two silacyclopentene rings with unsubstituted double bonds and one monosubstituted double bond afforded fully soluble polymers being brought in contact with the *n*-BuLi/HMPA or MeLi/HMPA complexes in THF [109], whereas the spiran monomer with one disubstituted double bond gave only a dimer as a single product [110].

To gain better understanding of the behavior of silaspirocyclic compounds containing silacyclopentene rings in low-temperature anionic polymerization in THF, 5-silaspiro[4,4]non-2-ene was synthesized and its interaction with *n*-BuLi in THF at −78°C was investigated [114]. It turned out to be that this spiran monomer also showed a tendency toward formation of the cyclic dimer and a low-molecular-weight oligomer (even without HMPA additives). Moreover, the oligomer was a set of predominantly cyclic molecules. The number of linear molecules in the oligomer with the same degree of polymerization was an order of magnitude smaller.

To prepare comb-shaped silbutylidene polymers carrying mesogenic groups, 1-methyl-1-silacyclopent-3-enes and 1-phenyl-1-silacyclopent-3-enes containing phenoxy-, biphenyloxy, and β-naphthoxy groups linked to the silicon atom via the trimethylene bridge, were synthesized and successfully polymerized [111]. Two methods were developed for the synthesis of 1-methyl-1-[ω-methoxyoligo(oxyethylene)propyl]-1-silacyclopent-3-enes, which are necessary for obtaining solvent-free electrolytes based on polymers having polysilacyclopent-3-ene main chains and pendant groups responsible for hydrophilic properties. The AROP of these monomers produced carbosilane polymers with the unsaturated silicon-carbon backbone (polysilbutylidenes), in which substituents with hydrophilic oligooxyethylene groups were attached to the silicon atom [112].

Furthermore, the synthesis of 1-methyl-1-(methylene-16'-crown-5')-1-silacyclopent-3-ene in combination with its isomer containing the double bond in the second position was described [112]. The desired polymer can be obtained using the initial mixture of isomers.

When two various derivatives of 1-silacyclopent-3-ene (1,1-dimethyl- and 1,1-diphenyl-1-silacyclopent-3-enes) were loaded simultaneously, random copolymers were obtained. The ratio of monomer units in them depended on the composition of the initial monomer mixture [115]. In this case, characteristics of the copolymers, such as the glass transition temperature and the melting point, changed accordingly. It was unexpectedly found that the copolymer of 1,1-dimethyl- and 1-methyl-1-phenyl-1-silacyclopent-3-enes was characterized by the block structure [116]. In the patent [117], the copolymerization of 1-phenyl-1-silacyclopent-3-ene with 1-methyl-1-phenyl-1-silacyclopent-3-ene was realized and various derivatives of 1-silacyclopent-3-ene were used as monomers and comonomers. The copolymerization of 1-methyl-1-phenyl-1-silacyclopent-3-ene with 1-phenyl-1-vinyl-1-silacyclopent-3-ene, carried out in THF in the presence of n-BuLi at $-78°C$, resulted in the random copolymer with a yield of 78% and $M_w/M_n = 13,000/8,000$ [118].

Silacyclopent-3-enes can form block copolymers with MSCBs (n-BuLi, $-78\,°C$, THF/TMEDA) [119], as was demonstrated by two examples, namely, the copolymerization of 1,1-dimethyl- and 1-methyl-1-phenyl-1-silacyclopent-3-enes with 1,1-dimethyl-1-silacyclobutane. Since both comonomers of the first pair formed living chains, block copolymers were produced if any of the comonomers was polymerized first. In the case of the second comonomer pair, a mixture of homo- and block copolymers was obtained.

1.2.3 Ring-Opening Metathesis Polymerization of 1-Silacyclopent-3-enes

The ROMP of silacyclopentenes led to polymers of the same structure as the AROP.

The ROMP of 1,1-dimethyl-1-silacyclopent-3-ene was performed in the presence of homogeneous catalytic system WCl_6–$Al(i\text{-}Bu)_3$–Na_2O_2 [120] and heterogeneous system Re_2O_7/Al_2O_3–Bu_4Sn (M_n 5,000) [121]. Unexpectedly, high molecular weight of poly(1,1-diphenyl-1-silacyclopent-3-ene) was attained, when ROMP was carried out in the presence of WCl$_6$–Ph$_4$Sn in combination with cyclopentene (or cyclohexene) as cocatalysts [116] (a yield of 66%; $M_w/M_n = 4,36,309/143,611$). Thermodynamics of the metathesis polymerization of 1,1-dimethyl-1-silacyclopent-3-ene in bulk was studied in [122]. It was demonstrated that the germanium analog – 1,1-dimethyl-1-germacyclopen-3-ene-also yielded the polymer through the ROMP mechanism in the presence of homogeneous system (CO)$_5$W:C(OMe)Ph-TiCl$_4$ [123]. In the case of this catalyst, the ROMP of 1,1-diphenyl-1-silacyclopent-3-ene yielded the polymer with the main chains consisted of solely *cis*-double bond units.

1.2.4 Ring-Opening Polymerization of Silaferroceno-, Silametalloceno-, and Silaarenophanes

[1]Silametalloareno- (or -ceno)phane systems are heteroannular derivatives of arene and cene structures with bridges between cycles containing one atom of Si, Ge, or Sn. These compounds, which were synthesized for the first time less than two decades ago, are of great interest as subjects for ROP and are useful for preparing promising polymer materials with a set of diverse characteristics. In all known structures of this type – [1]sila(germa-, stanna-)metalloareno- (or –ceno)phanes (ferroceno-, chromareno-, vanadareno-, etc.), valence angles between bonds of the bridge Si-atom and CPD or arene ligands are distorted.

M=Si (I), Ge (II), Sn (III)   (IV)   (V)   (VI)   (VII)   M=V;E=Si(VIII)   (XI)
M=Cr;E=Si(IX)
M=Cr;E=Ge(X)

This effect is accompanied with the nonparallel position of ligand planes and causes strain of the cyclophane system. For example, in dimethyl- and dihydro[1]silaferrocenophanes (I), angles between planes of CPD cycles are 20.8° [124, 125], and 19.1° [125], respectively; in diphenyl[1]silachromarhenophane, dimethyl [1]silachromarhenophane (IV), bis(methyl[1]silavanadarenophane) (V), and trimethylene[1]silavanadarenophane (VI), the angles between planes of arene ligands are 14.4° [126], 16.6° [127], 20.8° [128], and 19.9°, respectively [129].

A system of [1]silaferrocenophanes is so strained that all ROP methods typical for SCBs, including TROP, became possible. Although the class of polyferrocenylenesilanes had been discovered in the 1960s, the TROP of [1]silaferrocenophanes was not described until 1992 [130]. The polymers had molecular weights in the range from 500,000 to 2,000,000. Later on, the anionic polymerization of [1]silaferrocenophanes [131, 132] and its polymerization catalyzed by transition metal compounds [133, 134] was reported.

The TROP methods made it possible to synthesize both the homopolymer of dihydro[1]silaferrocenophane and its random copolymers with dimethyl[1]silaferrocenophane [125, 135]. High activity of spirocyclic[1]ferrocenophane (VII) in TROP has already been mentioned in 1.1.2 [24]. The polymerization of dimethyl[1]silaferrocenophane and of its germanium analog under the action of various complexes of Pt and Pd performed at room temperature and their random copolymerization were reported in [134]. Based on study of polymerization of [1]ferrocenophanes with the silicon-containing bridge, the mechanism of their ROP on Pt catalysts was developed [136]. Manners et al. investigated the AROP of dimethyl[1]silaferrocenophane catalyzed by butyllithium, phenyllithium, and ferrocenyllithium and the multiblock copolymerization of this monomer with styrene and hexamethylcyclotrisiloxane [137]. The TROP (homopolymerization) of dimethyl[1]silachromoarenophane at 180°C caused its decomposition and did not proceed at a lower temperature. However, at 140°C, this monomer can be involved in copolymerization with dimethyl[1]silaferrocenophane [138, 139]. The copolymer of the same structure was prepared with the AROP of these comonomers.

For 17 years since appearance of the first publication concerning the ROP of [1]silaferrocenophanes (1992 year), many papers have described the syntheses of various metalloceno and areno-cyclophane compounds capable of producing polymers under thermal, anionic, or transition-metal initiation [133]. Recently, the synthesis and polymerization of various new compounds of this type, for example, telechelic poly(ferrocenylsilanes) [140], and [1.1]silaferrocenophane containing pentacoordinated silicon moiety, have been reported [141]. A range of new 1-sila-3-metallacyclobutanes (η^5-C$_5$H$_4$Fe)(CO)$_2$CH$_2$SiR$_2$ was synthesized. The study of their ring opening resulted in preparation of a new class of organometallic polymers [142]. The synthesis and properties of polyferrocenylbutylmethylsilane with unsymmetrical silyl groups [143], [2]silatrovacenophanes (η^7–C$_7$H$_6$V(η^5–C$_5$H$_4$SiR$_2$), R = Me, Ph) [144], and dimethyl [1]sila-trochrocenophane (-chromocenophane) [Cr(η^5–C$_5$H$_4$) (η^7 C$_7$H$_6$SiMe$_2$] (IX) were described. The latter was polymerized on Karstedt's catalyst [145]. There were no attempts to apply this catalyst in the case of germatrochrocenophane (-chromocenophane) (X) [146], and the first synthesized manganese cene-arene [147] while it was successfully used for ROP of dimethyl derivatives of and silacyclobutane-bridged spirocyclic ansa-zirconocene complexes containing two Cl atoms at the Zr atom (XI) [148].

The photolytic anionic polymerization [149–151] and the spontaneous polymerization of [1]stannacenophane monomers leading to high-molecular-weight polymers and their cationic polymerization were also performed [152].

2 Carbochain Polycarbosilanes

2.1 Ring-Opening Metathesis Polymerization of Silicon Substituted Cycloolefins

Metathesis reaction of unsaturated compounds proceeds via opening of double bond and intra- or inter- molecular exchange of alkylidene fragments using catalyst systems containing Ru, Mo, W, Re, and some other transition metals [153, 154].

In the case of cycloolefins, this reaction is named ring-opening metathesis polymerization (ROMP).

Now ROMP is one of the most developed approaches to synthesis of polymers having different desirable side substituents and main chain architecture [153–157]. In principal, ROMP of strained cycloolefins has thermodynamic driving force because of a release of ring strain at opening monomer cyclic molecules. Table 1 demonstrates strain energy of some monocycloolefins and norbornene as well as Gibbs free energy. $\Delta G°$ for cyclohexene is negative in contrast to other cycloolefins because of low strain energy. That is why cyclohexene is not involved in ROMP.

Earlier it was shown that the presence of Si–C- and Si–H-fragments in unsaturated substrates did not prevent their metathesis including ROMP. Therefore, many original papers and several reviews devoted to metathesis of silicon-containing compounds have been published over the last 40 years [153, 154, 162–166]. However, the number of scientific results on ROMP of monocycloolefins bearing silyl-containing substituents is limited due to complication of monomer syntheses.

Below we demonstrate a few examples of ROMP of monocycloolefin Me_3Si-derivatives.

Katz et al. realized ROMP of 1-(trimethylsilyl)cyclobutane in the presence of Casey's carbene with formation of polymer with exclusively *cis*-double bonds and head-to-tail arrangement [167].

Table 1 Strain energy and $\Delta G°$ for liquid cycloolefin ROMP into solid polymer at 25°C [158–161]

Cycloolefin	C3	C4	C5(polymer)	C6	C8	NB
Cyclic strain energy, kJ mol^{-1}	232	119.1	18.9	0	Trans 70 Cis 16	100.3
$-\Delta G°$ ROMP, kJ mol^{-1}	–	105	0.3 (*cis*) 2.6 (*trans*)	–6.2 –7.3	13 (48% *trans*)	47

In general, cycloolefins having substituents connected directly with double bond are not active in metathesis because of higher stability of metallo-carbene intermediates [153]. However, probably in the case of the cyclobutene derivative, a high thermodynamic driving force became a predominate factor.

ROMP of trimethylsilylcyclooctatetraene was realized in the presence of the tungsten-based Schrock complex as a catalyst [168, 169]. It led to conjugated polymer as a result of isomerization of the structure with a predominantly *cis* configuration to a predominantly *trans* one accomplished either thermally or photochemically.

Some data about oligomerization of 1-(trichlorosilyl)cyclododeca-4,8-diene in the presence of typical ROMP catalyst $WCl_6 \cdot EtOH-Alk_2AlCl$ resulting in low molecular weight products of undefined structure ($M_w \approx 1,000$) were also published [170].

Strained norbornene and its derivatives are more popular substrates for ROMP. Norbornene was the first monomer effectively involved in metathesis polymerization [171].

Ever since, a lot of articles and patents devoted to ROMP of a great variety of norbornene derivatives have been published, including a number of special reviews and book chapters [153, 154, 156, 172, 173]. There are two reasons: (1) high thermodynamic driving force realized as a result of opening strained bicyclic norbornene skeleton and (2) accessibility of norbornene derivatives due to effective and flexible Diels-Alder condensation of cyclopentadiene (or dicyclopentadiene) with practically any olefinic compounds [174, 175].

It should be noted that today there are industrial processes based on ROMP of norbornenes such as manufacturing of polynorbornene rubber (trade mark "Norsorex") and polycyclopentadiene thermoset material (trade marks "Telene" and "Metton"). Both processes use effective "classical" Ru- and W- catalysts of the first generation [176].

Syntheses of norbornenes and norbornadienes having Si-containing substituents as a rule were carried out by the above-mentioned Diels-Alder condensation of cyclopentadiene (or dicyclopentadiene) and Si-containing ethenes or ethynes [177] commonly followed by modification of Si-containing groups. The presence of electron – withdrawing substituents such as – SiR_nCl_{3-n}, Si(OAlk), accelerated reaction.

The other approaches to norbornenes and norbornadienes preparation start from previously synthesized bicyclic structures:

1. Hydrosilylation of norbornadiene (nbd) by chloro- or alkyl-hydrosilanes in the presence of Pd- or Mo-catalysts [178, 179] led to the corresponding silylnorbornenes. Norbornenes with ethane spacer between bicyclic nuclear and silicon atom can be synthesized by hydrosilylation of 5-vinyl-2-norbornene [180].
2. Metallation of norbornadienes by Li or Na followed by interaction with chloroalkylsilanes [181] as well as interaction of Cl (or Br) Mg-methylnorbornene with organochlorosilanes also resulted in norbornenes bearing silicon-containing groups [182].

The presence of Si-containing substituents in norbornene is important for design of desirable monomer and polymer structures. Highly reactive Si–Hal, Si–H bonds allow introduction of such chemical groups which are difficult to incorporate using traditional methods of organic chemistry. Furthermore, Si-containing groups are responsible for useful polymer properties, such as adhesion, high gas permeability, transparency, and others [183–186].

2.1.1 Polynorbornenes with Alk₃Si-Substituents

Mono-substituted norbornenes as a rule contain a substituent in the 5-position and consist of *endo-* and *exo-*isomers. Since they are obtained by Diels-Alder condensation, the *endo*-isomer is dominant in reaction isomer mixtures. 5-Trimethylsilylnorbornene (NBSiMe₃), obtained by condensation of cyclopentadiene and vinyltrichlorosilane accompanied by methylation, contains predominantly *endo*-isomer (70–75%) [177]. The interaction of dicyclopentadiene with vinyltrimethylsilane at higher temperature (200–210°C) led to equimolar mixture of *endo-* and *exo-*5-trimethylsilylnorbornene [187]. It was shown that silyl-substituted norbornenes are extremely active in ROMP (Table 2), *exo-*isomers being more active than *endo-*forms [153, 188].

Different catalyst systems have been used for this ROMP: from so called "poorly characterized" catalysts (RuCl₃ · H₂O, RuCl₂(PPh₃)₃, WCl₆/1,1,3,3-tetramethyldisilacyclobutane-1,3 (TMSB), WCl₆/PhC≡CH) to "well-defined" Ru- and Mo- carbene Grubbs (Cl₂(PCy₃)₂Ru=CHPh – Grubbs 1) and Schrock complexes [182, 187–190].

Table 2 ROMP of mono substituted silicon containing norbornenes

R	Catalyst	Mol ratio, [M]/[cat]	Yield, %	M_n	I_p	Cis, %	References
SiMe$_3$	RuCl$_3$· 3H$_2$O	180	76	215,000	2.7	1–2	[187]
	WCl$_6$/TMSB	1,000	99	4.0 (dL g^{-1}) –	–	–	
	Grubbs 1	720	98	150,000	2.4	28	[188]
SiMe$_2$CH$_2$SiMe$_3$	Re$_2$O$_7$/Al$_2$O$_3$/SnBu$_4$	350	30	1.7 (dL g^{-1}) –	–	–	[187]
	WCl$_6$/PhC≡CH	750	81	2.0 (dL g^{-1}) –	–	–	
-SiMe$_2$(CH$_2$)$_3$Crbz[a]	RuCl$_2$(PPh$_3$)$_3$	92	58	25,000	2.4	Pref. trans	[189]
-SiMe(Crbz)$_2$	Grubbs 1	200	100	57,000	1.3	–	[190]
—CH$_2$–Si(Me)< (silacyclobutane)	RuCl$_3$· 3H$_2$O	125	99	0.2 (dL g^{-1}) –	–	trans	[32]
	WCl$_6$/PhC≡CH	500	99	0.1 (dL g^{-1}) –	–	–	

[a]*Crbz* carbazol

Polymer molecular weight can be controlled by adding a chain transfer agent (CTA) or by monomer/catalyst ratio (for Grubbs-carbene 1). The *cis:trans* bond ratio in polymer chains depended on catalyst type (NMR signals of allylic carbons and protons located around them in polymers are quite informative for determination of the stereo situation) [153]. The most selective catalyst is RuCl$_3$H$_2$O providing 90–98% of *trans* double bonds (Table 2). At the same time, Ru-carbene Grubbs-complex turned out to be more sensitive to monomer structure [154, 156]. So, in the case of silylnorbornene derivatives, greater amounts of *cis* forms in the corresponding polymers were observed. Polymers obtained in the presence of W- catalysts contained about equal amount of *cis* and *trans* double bonds.

Introduction of reactive substituents in norbornene gave a chance for cross-linking or polymer modifications. Thus, 1-methyl-1-norbornenylmethyl-1-silacyclobutane containing two reactive strained fragments is able to polymerize according to two routes with formation of either polycyclopentylenevinylene with silacyclobutane side groups or polysiltrimethylenes with norbornene side groups. Simple heating of the polycyclopentylenevinylene up to 180–200°C provided crosslinking by opening silacyclobutane rings. Similar thermal treatment of the polysiltrimethylenes resulted in retrodiene decomposition of norbornene side groups [33].

Polynorbornenes with highly reactive Si–Cl-containing substituents would be attractive for transformations into other functional groups. It should be noted that Si–Cl bond did not hinder ROMP of 5-(trichlorosilyl)norbornene in the presence of WCL$_6$-iBu$_2$AlCl [170, 182]. However, primarily soluble polymer underwent crosslinking when it was isolated. This could complicate polymer modifications with help of side Si–Cl-containing groups. On the other hand, the introduction of desirable substituents directly in monomer followed by ROMP is a much simpler and steady method than polymer-analog reactions.

Polymers with carbazole pendant groups are promising materials for optoelectronic devices [191–193]. Norbornenes having one or two carbazole fragments in silicon-containing substituents were polymerized on Ru-catalysts [189, 190] with high yields of polymers bearing carbazolyl-containing side groups.

Ru–carbene Grubbs I catalyst did not polymerize 5-trimethylsilylnorbornene in a living manner ($I_p \gg 1$, Table 2). At the same time, it was shown that the presence of functionalized substituents in monomer molecule is able to coordinate the catalyst, decrease to some extent its activity, and lead a living process [156]. Thus, block-copolymers of 5-(dicarbazolylmethylsilyl)norbornene and 5-(trimethylsilyloxymethyl)norbornene (NBCH$_2$OSiMe$_3$) were obtained by living polymerization in the presence of Grubbs 1 complex [190].

Also a "protection–deprotection" technique was used for final introduction of polar alcohol groups in target ROMP copolymers to avoid poisoning the catalyst. In this case, OH-functions were protected with trimethylsilyl group by interaction with trimethylchlorosilane. After copolymerization the OH-groups were regenerated by simple deprotection reaction.

A similar technique was applied in copolymerization of NBSiMe$_3$ with a series of norbornenes having acid sensitive and polar groups, including nitrile, carboxylic acid, hydroxyl, and anhydride functions in the presence of Grubbs 1 and Schrock

Mo- catalysts to achieve random copolymers with suitable properties to be applied as resist materials [194]. The original approach was employed to involve inactive symmetrical cyclopentendiol-3,5 in ROMP [195]. The protected strained cycloolefin monomer was polymerized followed by hydrogenation of the obtained polymeric silane and deprotection to give regioregular methylene-(vinyl alcohol) polymer.

It should be noted that copolymerization of NBSiMe$_3$ with unprotected NBCH$_2$OH (mol ratio 3:1) in the presence of Grubbs 1 catalyst (yield 73%) is also possible [196].

2.1.2 Polynorbornadienes with Alk3Si-Substituents

Norbornadienes with silicon-containing substituents at the 2-position were also effectively polymerized according to the ROMP way (Table 3). Substituted double bond is inactive in this reaction and therefore polymers obtained were soluble in contrast to polynorbornadiene itself. [153]. Norbornadienes did not polymerize in the presence of simple RuCl$_3$-catalysts, but W-containing and Grubbs systems were very effective.

Table 3 ROMP of mono substituted silicon containing norbornadienes

Monomer (M) R^1	Catalyst	Mol ratio [M]/[cat]	Yield, %	M_n	I_p	Cis, %	References
–SiMe$_3$	WCl$_6$/SnBu$_4$	50	46	25,903	3.3	50	[197–200]
	WCl$_6$/TMSB	250	70	180,000	3.1	42	
	Grubbs 1	500	88	53,000	1.9	–	
–SiMe$_2$(CH$_2$)$_3$SiMe$_3$	WCl$_6$/TMSB	92	40	106,000	3.0	43	[199]
–SiMe$_2$H	WCl$_6$/SnBu$_4$	43	53	11,700 $T_g=71°C$	1.7	–	[201]
SiMe$_2$CH$_2$–	WCl$_6$/SnMe$_4$	39	98	M_w332000. $T_g=42°C$	–	–	[197]

It is interesting that ROMP could proceed along with hydrosilylation of nbd. Thus, photochemically initiated hydrosilylation of nbd by triethylsilane in the presence of a catalytic amount (ca. 2 mol%) of the molybdenum complexes [Mo(CO)$_6$] or [Mo(CO)$_4$(η^4-nbd)] gave 5-triethylsilylnorbornene (99:1 mixture of *endo*- and *exo*-isomers).

R$_3$=Et$_3$, Cl$_3$, Et$_2$H, Ph$_2$H

The latter under these reaction conditions was next transformed to the ROMP polymer poly(5-triethylsilylnorbornene) (yield only 2%) [179]. This one-pot combined reaction proceeded more effectively in the case of nbd and diethylsilane interaction (80% yield of polymer and nbd conversion is 40%).

Polynorbornadienyl oligomer containing pentasilane bridge was prepared by ROMP of a 1,5-bis(7-norbornadienyl)-decaphenylpentasilane on Grubbs 1 catalyst [202], as a precursor for formation of electrically conductive films.

Me$_3$Si-containing norbornenes and norbornadienes with two substituents in different positions also demonstrated high activity as monomers in ROMP (Table 4).

All prepared polymers were used for searching correlations between their structures and gas permeability. Along with mono-silyl substituted polynorbornenes and polynorbornadienes, they showed the dramatic influence of SiMe-containing substituents on gas permeation properties (Table 5).

Table 5 demonstrates strong dependence between polymer gas permeability and number and location of Me$_3$Si-groups. The introduction of Me$_3$Si-substituent in monomer unit substantially enhanced gas permeability of ROMP polynorbornene (Table 5 rows 1 vs 2, 4 vs 5). The introduction of the second Me$_3$Si-group in monomer unit increased permeability coefficients approximately as much as the first one did (rows 2, 11, 12, 14, 16).

Two MeSi-groups in the same substituent reduced gas permeability (Table 5 rows 3, 7), probably due to self-plasticization. The polymers showed much lower T_g: thus, polymers with SiMe$_3$CH$_2$SiMe$_2$- and SiMe$_3$(CH$_2$)$_3$SiMe$_2$- side groups had T_g 24 and 5°C, correspondingly, that is, lower than T_g of unsubstituted polynorbornene (rows 1, 3, 15 in Table 5). Therefore, these low T_g mark the polymers as rubbers. Their gas permeability also reflects elastomeric properties: for example, the permeability coefficients of heavier hydrocarbons are higher than those of lighter

Table 4 ROMP of bis-substituted silicon containing norbornenes and norbornadienes

Monomer	Catalyst	Mol ratio [M]/[cat]	Yield, %	M_n	I_p	Cis, %	References
SiMe₃ / SiMe₃	RuCl₃	40–200	50–98	137,000	1.8	6	
	WCl₆/TMSB	50	93	100,000	2.9	54	[199]
	Grubbs 1	700	99	116,000	2.4	65	
SiMe₃ / SiMe₃	RuCl₃	50	50	62,000	1.8	1–2	[203]
	WCl₆/TMSB	100	98	500,000	3.2	60	
	Grubbs 2	200	32	33,400	2.2		[196]
SiMe₃ / SiMe₃	WCl₆/TMSB	100	98	570,000	2.8	71	[200]
	Grubbs 1	1000	98	183,000	1.2	11	
SiMe₃ / SiMe₃	RuCl₃	50	50	56,000	1.8	1–2	
	WCl₆/TMSB	100	98	625,000	1.6	47	[200]
	Grubbs 1	1500	98	444,000	1.8	22	
SiMe₃ / CN	WCl₆–Et₃Al	150	24	63,600	3.0	–	[204]
SiMe₃ / CN	WCl₆–iBu₃Al	177	17	43,700	2.4	–	[204]
SiMe₃ / COOMe	WCl₆–iBu₃Al	57	43[a]	233,000	1.9	–	[204]

Grubbs 1: Cl₂(PCy₃)₂Ru=CHPh; Grubbs 2: [structure with Mes–N, N–Mes, Cl, Cl–Ru=, PCy₃, Ph]

[a]Silyl substituents were cleaved in the purification procedure

ones [31, 207]. A double bond in cyclopentene ring in the polynorbornadiene main chain did not affect gas transport parameters of polymers (Table 5, rows 2 vs 14 and 11 vs 16). Polymers having two Me₃Si-substituents located at different carbon atoms in every monomer unit – polynorbornenes, polynorbornadienes, or polytricyclononenes (Table 5, rows 11, 13, 16) – demonstrated better permeability with respect to light gases, than in the case of geminal substituted polymers (run 12). CN-group reduced gas permeability of polynorbornene (Table 5, rows 1, 4). The replacement of Me-groups at the silicon atom for -OSiMe₃ groups (Table 5, rows 8–10) substantially improved gas permeability of the corresponding polymers. Thus, polynorbornene with (Me₃SiO)₃Si- substituent had the highest permeability coefficient among all known ROMP polynorbornenes (Table 5, row 10) [206].

Table 5 Gas-permeation properties of different ROMP-polynorbornenes and polynorbornadienes

Polymer	№	R^1	R^2	R^3	T_g, °C	$P(O_2)$, Barrer	$\alpha(O_2/N_2)$	References
	1	H	H	H	31	2.8	1.9	[205]
	2	H	SiMe$_3$	H	113	30	4.2	[205]
	3	H	SiMe$_2$CH$_2$SiMe$_3$	H	24	16	4.3	[187]
	4	H	H	CN	140	0.53	6.3	[204]
	5	H	SiMe$_3$	CN	211	7.6	4.6	[204]
$R^1\ R^2 R^3$	6	H	CH$_2$SiMe$_3$	CN	128	11	4.2	[204]
	7	H	SiMe$_2$SiMe$_3$	CN	120	4.4	4.6	[204]
	8	H	SiMe$_2$OSiMe$_3$	CN	88	15.3	4.4	[204]
	9	H	SiMe(OSiMe$_3$)$_2$	H	27	99	3.3	[206]
	10	H	Si(OSiMe$_3$)$_3$	H	101	290	3.1	[206]
	11	SiMe$_3$	SiMe$_3$	H	167	95	3.8	[199]
	12	H	SiMe$_3$	SiMe$_3$	203	54	3.9	[188]
$R^1\ R^2$	13	SiMe$_3$	SiMe$_3$	H	129	89	3.7	[188]
	14	H	SiMe$_3$	–	108	20	4.1	[199]
	15	H	SiMe$_2$(CH$_2$)$_3$SiMe$_3$	–	5	Rubbery	–	[199]
$R^1\ R^2$	16	SiMe$_3$	SiMe$_3$	–	183	71	3.6	[188]

Recently, ROMP of silicon-containing norbornenes was used for creation of inorganic-organic hybrid materials. Particularly, incorporation of the polyhedral oligomeric silsesquioxanes molecule (POSS) in norbornene and its ROMP copolymerization with norbornene or norbornene-dicarboxylic acid gave POSS-containing copolymers [208, 209].

R=cyclohehyl, cyclopentyl

They are attractive for various applications such as liquid crystals, nanocomposites, CVD coatings, and photoresists in lithographic technologies, based on their high temperature and oxidation resistance properties compared to non-POSS containing polymers [210].

In nanopattern fabrication, for the preparation of "smart" materials, surface grafting methods have been employed to attach covalently polymer films to various inorganic substrates. Among different approaches, surface-initiated ring-opening metathesis polymerization (SIROMP) is an effective method to attach polymer chains to substrates including gold [211], silicon wafers [212], silicon (111) [213], silica nanoparticles [214], and others. As a rule a surface was first treated with 5-trichlorosilylnorbornene [215], linear alkenylsilanes [212], or norbornenyltriethoxysilane [216, 217] that interacted with surface OH-groups. Then Grubbs Ru-catalyst was affixed to norbornene fragment on the surface and initiated ROMP of norbornene or its derivatives [214, 218].

The presence of a double bond in each monomer unit of ROMP-polymers imparts to them some disadvantages, in particular a rather high chemical activity and poor thermooxidative stability. These problems can be solved with hydrogenation of unsaturated polymers. [206]. However, it led to a decrease in permeability. This phenomenon may be connected with either lower rigidity of hydrogenated polymer main chain or its crystallinity or affinity to gases after hydrogenation [206].

2.2 Addition Polymerization of Silicon Substituted Norbornenes

Addition ("vinyl") polymerization (AP) of norbornene and its derivatives proceeds through opening of double bonds with formation of completely saturated polymers containing bicyclic units in backbone.

AP of norbornenes requires catalyst systems on the basis of compounds containing some transition metals such as Zr, Ni, and Pd, that is in principle different in comparison with ROMP catalysts. Polymerization and copolymerization performing according to the addition mechanism has recently attracted much attention of researchers working in polymer chemistry [219, 220]. Such a situation arose due to very useful properties of addition polynorbornenes. Among them, one can mention high decomposition temperatures, good transparency, high plasma etch and chemical resistance, low birefringence, and very small tendency to water absorption, all of which makes them very promising for use in the development of microelectronic and optoelectronic devices [221, 222].

Introduction of proper pendant groups in monomer unit of addition polynorbornene is the way to get polynorbornenes with desired properties. However, the appearance of substituents in norbornene molecules resulted in a decrease of their activity as monomers [220, 223]. In the case of ROMP, this effect is softened to some extent by substantial thermodynamic driving force in the process appearing as a result of opening highly strained bicyclic norbornene skeleton (see Table 1). AP is not such a thermodynamically favorable process. Therefore, in this case the introduction of side substituents, especially bulky or functional groups, led to a dramatic decrease in activity of norbornene derivatives [224–227]. Probably, this fact is one of the reasons for the limited number of publications devoted to AP of silyl substituted norbornenes.

Several works contain information of addition homo- and copolymerization of 5-(triethoxysilyl)norbornene (NBSi(OEt)$_3$). It underwent AP under influence of (1,5-cod)Pd(CH$_3$)(Cl)]/PPh$_3$/Na$^+$[3,5-(CF$_3$)$_2$C$_6$H$_3$]$_4$B$^-$ catalyst with 88% yield of saturated polymer characterized with low $M_w = 12,700$ and $I_p = 1,5$ [223].

Also, 5-(triethoxysilyl)norbornene took part in addition copolymerization with 5-ethylidene-2-norbornene [223], 5-n-butylnorbornene [228], and unsaturated norbornene [229] in the presence of Pd [223, 228] and Ni-catalysts [229]. Copolymer's thin films were crosslinked or hardened for use as polymer-based dielectric material in the interconnection of semiconductor devices [230].

Notably, copolymers obtained with various catalysts substantially differed in properties. For example, copolymer of NBSi(OEt)$_3$ with norbornene obtained on [(η^6-toluene)Ni(C$_6$F$_5$)$_2$] catalyst, exhibited suitable mechanical properties (a tensile modulus of 1.4 GPa, an elongation-to-break of 15%) while catalyst [(η^3-crotyl)Ni(1,4-COD)]PF$_6$ gave completely brittle polymers, despite the fact that they had equal molecular weights [231]. Ludovice et al. suggested that these differences could be connected with polymers microstructure: a more atactic polymer was brittle, and a more stereoregular one had better mechanical properties [232]. However, determination of microstructure of substituted polynorbornenes is a very intricate problem because they have unresolved NMR-spectra. Till now, addition polynorbornene has been depicted as 2,3- or 2,7-structures.

2,3-polynorbornene   2,7-polynorbornene

Only addition polynorbornene obtained with iPr(Ind)$_2$ZrCl$_2$/MAO (Ind = indenyl, MAO = methylalumoxane) catalyst was actually proved to have 2,3- and 2,7- units simultaneously [233].

Behavior of NBSiMe$_3$ was studied under the conditions of addition polymerization initiated by some Ni- and Pd-containing catalysts [234]. The known Pd-containing catalytic systems – [(η^3-allyl)Pd(SbF$_6$)] [235] and σ,π-bicyclic complex [NB(OMe)PdCl]$_2$ [236] – turned out to be practically inactive in polymerization of this monomer. In contrast, Ni-based complexes demonstrated a real activity. As a result, saturated cyclolinear polymers with yields up to 80% were formed according to the AP.

The absence of any unsaturation in these polymers was confirmed by both IR (no bands in the 1620–1680 cm^{-1} region) and ^{1}H NMR spectroscopy (no signals at 5–6 ppm). Among studied Ni-based catalyst systems, Ni(II)naphthenate–MAO and (π-C$_5$H$_9$NiCl)$_2$–MAO were the most active. Polymers obtained in the presence of the catalyst Ni (II)naphthenate–MAO had the highest molecular weights and demonstrated better film-forming properties. They didn't show any glass transition up to 340°C (DSC). GLC analysis of the final polymerization mixtures indicated that, in the course of the reaction, the *exo*-conformer was consumed much faster than the *endo*-form independently of the type of catalyst system employed. It should be noted that AP of NBSiMe$_3$ proceeded substantially slower than that of unsubstituted norbornene. Unlike ROMP, activity in AP depended to a great degree on the presence of *exo*-conformation in monomer. The presence of a bulky group in the *endo* position prevents the *endo* face coordination of monomers. As a result, activity of *endo* norbornene derivatives was substantially lower in comparison with that of *exo*-isomers [225, 234, 237], and in some cases *endo*-isomers were completely inactive [227]. According to [224] some steric hindrances arise even at more a suitable *exo* site.

Norbornenes and norbornadienes with two Me$_3$Si-substituents turned out to be practically inactive in the presence of Ni- and Pd-catalysts [234, 238]. At the same time their copolymerization with norbornene, 5-*n*-hexyl-2-norbornene, and NBSiMe$_3$ could be readily realized with catalyst system Ni(II) naphthenate–MAO.

1. $R^1=R^2=SiMe_3$; $R^3=H$
2. $R^1=R^3=SiMe_3$; $R^2=H$

In the case of the studied bis-Me₃Si-norbornenes, one of the substituents is always in the disadvantageous *endo*-conformation. Probably this is the reason for their inactivity in AP. But when the mixture of different conformers is used, a part of the *endo*-form could participate in copolymerization with the *exo*-form. This fact excludes the possibility of homopolymerization for these monomers but does not exclude the proceeding of their copolymerization with norbornene itself nor with its substituted *exo*-derivatives.

It is interesting that 2,3-bis(trimethylsilyl)norbornadiene-2 in AP conditions formed only *exo-trans-exo* [2+2]-cyclodimer [238]. No formation of any polymeric products in this reaction was observed.

The same inactivity of norbornene monomers having geminal electron-withdrawing ester and trifluoromethyl substituents in AP in the presence of common Ni and Pd- catalyst systems was found by Grubbs et al. [239].

In contrast to bis-SiMe₃-norbornene derivatives, disubstituted tricyclononenes turned out to be active monomers in AP [238, 240]. In tricyclononene molecule both Me₃Si-groups are moved by an additional one C–C bond away from the double bond and therefore from the reaction catalytic center. Synthesis of bis-Me₃Si-substituted tricyclononene was carried out from quadricyclane and *trans*-1,2-bis(trichlorosilyl)ethylene. This route of synthesis provided formation of norbornene-type monomers with 100% *exo*-configuration of cyclobutane fragment that reduced steric hindrances in AP. That is why this monomer was active in AP catalyzed with common Ni- and Pd-catalyst systems. As a result, the formation of highly molecular weight polymer (M_w up to 500,000 was observed [196].

Gas transport properties of unsubstituted addition type polynorbornene and some polyalkylnorbornenes have been under examination since the end of the last century [241–243]. Some results were reported for norbornene-ethylene copolymers of different composition [244]. Permeability coefficients of addition polynorbornenes bearing alkyl and Si-containing side groups are presented in Table 6. For comparison, some data for poly(1-trimethylsilyl-1-propyne) – the most permeable polymer – are also given. Table 6 shows that Me$_3$Si-polynorbornene is characterized by much

Table 6 Permeability coefficients P^a of addition polynorbornene with different substituents R

R	Cat	P(He)	P(H$_2$)	P(O$_2$)	P(N$_2$)	P(CO$_2$)	P(CH$_4$)	P(C$_4$H$_{10}$)
H [237]	Ni	29.4	41.5	6.9	1.5	33.6	2.6	–
CH$_3$ [242]	Pd	174.4	–	–	12.6	202.1	16.9	–
	Ni	88.8	–	–	4.3	81.1	5.6	–
n-C$_4$H$_9$ [243]	Pd	66.7	110.7	33.3	11.2	141.9	28.4	–
n-C$_{10}$H$_{21}$ [243]	Pd	38.9	62.4	25.3	8.7	111.1	28.1	–
Si(OSiMe$_3$)$_3$ (Copolymer with norbornene) [245]	Ni	–	–	239	–	–	–	–
SiMe$_3$ [237]	Ni	840	1,800	980	390	4,480	993	17,500

[a] P, Barrer

higher gas permeability coefficients than other addition polynorbornenes. Based on the value of gas permeability coefficients, it can be considered among the group of highly permeable polymers including poly(methylpentyne) (PMP) and other polyacetylenes [207, 246–248]. Recently measured data on gas permeability of addition poly{bis(trimethylsilyl)tricyclononene} gave evidence that they were much higher than that of poly(trimethylsilylnorbornene). For example, $P(O_2) = 1,800$ Barrer [196]. It can be also concluded that silyl-substituted addition polynorbornenes are much more permeable than their structural isomers, the ROMP polymers containing $Si(CH_3)_3$ groups [187, 249, 250].

3 Conclusions

The presented data confirm that organosilicon polymers are not only polysiloxanes, polysilazanes, or polysilanes having main chains consisting of just Si–O-, Si–N-, or Si–Si bonds. Now it is obvious that polycarbosilanes are a special type of organosilicon materials with their own distinctive range of promising physicochemical properties.

Depending on macromolecular structure, the polycarbosilanes can be glassy, elastomeric, or thermosetting materials having T_g values in the range $(-)100$-$(+)350°C$. They are of interest as thermostable adhesives, liquid crystals, photoresist components, electrosensitive materials, precursors for ceramics, highly permeable membranes for separation of light and hydrocarbon gases, etc. At the present time, polycarbosilanes attract the attention of many scientists in different countries and are the subjects of many publications and patents

At the same time, there are some difficulties preventing wide application of polycarbosilanes in modern technologies.

In the case of heterochain polymers, the main problem is the absence of technologically promising methods of monomer syntheses.

In the case of carbochain organosilicon polymers, addition polymerization of norbornene derivatives became problematic when they had more than one bulky Si-containing group. For example, norbornenes with $Si(OSiMe_3)_3$- or two $SiMe_3$-substituents gave only low molecular weight products or even didn't polymerize at all on simple Ni- and Pd-catalysts. According to detailed study of ROMP of polynorbornenes bearing the above-mentioned pendant groups, the corresponding highly molecular weight addition polymers are expected to have distinguished membrane and film forming properties.

From our point of view, of the problems which should be solved in the near future with respect to polycarbosilanes, the most important ones are elaboration of technologically suitable methods of monosilacyclobutane, disilacyclobutane, silaferroceno-, and silaarenophane preparation, and finding the effective catalysts for addition polymerization of norbornenes with bulky substituents.

Acknowledgements This work was supported by The Russian Foundation of Basic Research (Project 09–03–00342-a).

References

1. Bazant V, Chvalovsky V, Rathousky J (1960) Silicones. Goskhimizdat, Moscow
2. Knoth WH (1958) Carbosilane ring-opening polymerization. US Patent 2850514
3. Ushakov NV, Finkel'shtein ESh, Babich ED (1995) Polymerization of silacyclobutanes. Polym Sci SerA 37:470–492
4. Lebedev BV, Rabinovich IV, Lebedev NK, Ushakov NV (1978) Thermodinamics of polymerization of silacycloalkanes range. Dokl Acad Nauk SSSR 239:1140–1143
5. Vilkov LV, Kusakov MM, Nametkin NS, Oppengeim VD (1968) The electronographical study of molecular structure of 1,1,3,3-tetrachloro-1,3-disilacyclobutane. Dokl Acad Nauk SSSR 183:830–833
6. Alexanyan VT, Kuz'yanz GM, Vdovin VM et al. (1969) IR- spectra of some silacyclobutane derivatives and structure of silacyclobutane ring. Zhurn Strukt Khimii 10:481–484
7. Krapivin AM, Magi M, Svergun VI et al. (1980) The NMR study and CNDO/2 molecular orbital calculation of silacyclobutanes. J Organomet Chem 190:9–33
8. D'yachenko OA, Sokolova Yu A, Atovmyan LO, Ushakov NV (1982) The planar silicon–carbon heterocycle in dinaphthylsilacyclobutane. Izv Acad Nauk SSSR Ser Khim 9:2060–2065
9. D'yachenko OA, Sokolova Yu A, Atovmyan LO, Ushakov NV (1984) X-Ray study of 1,1-bi(1-α-naphthyl-1-silacyclobutyl) – first monosilacyclobutane derivative with Si–Si bond. Izv Acad Nauk SSSR Ser Khim 6:1314–1319
10. D'yachenko OA, Sokolova Yu A, Atovmyan LO, Ushakov NV (1985) The structure of 1,1,3,3-bis(trimethylene)-1,3-di-α-naphthyldisiloxane. Izv Acad Nauk SSSR Ser Khim 5:1030–1034
11. Babich ED, Pozdnyakova IV, Karelova II, Vdovin VM, Nametkin NS (1969) A molecular complexation of small silicon-carbon heterocycles and their analogues. Khim Geterotsikl Soedin 4:736–739
12. Nametkin NS, Oppengeim VD, Zav'yalov VI et al. (1965) IR-spectra of 1,1-substituted silacyclobutanes, silacyclopentanes and corresponding polymers. Izv Akad Nauk SSSR Ser Khim 9:1547–1553
13. Babich ED (1966). Multifuctional derivatives of 1-silacyclobutane. Ph.D. Thesis, Academy of Sciences of the SSSR, Topchiev Institute of Petrochemical Synthesis Moscow
14. Nametkin NS, Vdovin VM, Zav'alov VI, Grinberg PL (1969) The polymers with silicon–carbon chains and pendant silacycloalkane groups. Khim Geterotsikl Soedin 2:357–363
15. Finkel'shtein ESh, Ushakov NV, Pritula NA et al. (1992) Synthesis of polysil-trimethylenes with trimethylsilyl groups in the side-chain. Izv Acad Nauk SSSR Ser Khim 1:223–225
16. Ushakov NV, Yarysheva AYu, Tal'roze RV et al. (1992) The first representative of polysilmethylenes with mesogen in pendant group. Dokl Acad Sci Russia 325:964–966
17. Nametkin NS, Ushakov NV, Vdovin VM (1971) Polymers with silicon-carbon chains and functional groups containing carbon substituted at the silicon atom. Vysokomolek Soedin A13:29–37
18. Radugina Yu E, Sakharova IN, Avilov GV et al. (1973) The method for preparation of thermoplasticity material for an information recording. USSR Patent 225015 Bull of Inventions 10
19. Ushakov NV, Fedorova GK, Finkelshtein ESh (1995) Synthesis and polymerization of Si-substituted silacyclobutanes. Izv Acad Nauk Ser Khim 2:2475–2477
20. Nametkin NS, Vdovin VM, Zav'yalov VI (1965) Polymerization of 1,1-substituted silacyclobutanes. Izv Acad Nauk Ser Khim 8:1448–1453
21. Finkelshtein ESh, Ushakov NV, Krasheninnikov EG, Yampolskii Yu P (2004) New polysilalkylenes: synthesis and gas-separation properties. Russ Chem Bull Int Ed 53:2604–2610 transl from Izv Acad Nauk Ser Khim 11:2497–2503
22. Radugina Yu E, Ushakov NV, Malakhova IA, Pritula NA (1991) 9-Carbozolyl-containing polyorganosiltrimethylene as a photoconductor of electrical-photographic material. USSR Patent 1680714 Bull of Inventions 36

23. Ushakov NV, Pritula NA, Rebrov AI (1993) Synthesis and polymerization of 9-carbazolyl-containing 1-silacyclobutane derivatives. Russ Chem Bull 42:1372–1376 [translated from (1993) Izv Acad Nauk Ser Khim 8:1437–1441]
24. MacLachlan MJ, Lough AJ, Manners I (1996) Spirocyclic [1]ferrocenophanes: novel crosslinking agents for ring-opened poly(ferrocenes). Macromolecules 29:8562–8564
25. Ushakov NV, Portnykh EB, Pritula NA, Finkelshtein ESh (1989) Synthesis of silicon–carbon spiranes by metathesis reaction with alumina-rhenium catalyst. Izv Acad Nauk SSSR Ser Khim 12:2797–2803
26. Makarov IG, Kasakova VM, Zhil'tsov VV et al. (1983) The study of anion-radical structures of organosilicon compounds. VII. ESR spectra of alkylarylsilacyclobutane anion-radicals. Zhurnal Obshchei Khimii 53:1315–1320
27. Ushakov NV (2010) unpublished results
28. Radugina YuE, Sakharova IN, Avilov GV et al. (1973) The method of preparation of thermoplasticity material for an information recording. USSR Patent 228525 Bull of Inventions 5
29. Babich ED, Paraszcak J, Hatzakis M et al. (1985) A comparison of the electron beam sensitivities and relative oxygen plasma etch rates of various organosilicon polymers Microelectronic Eng 3:279–291
30. Babich ED, Paraszcak J, Hatzakis M et al. (1989) A comparison of the E-beam sensitivities and relative O_2-plasma stabilities of organosilicon polymers. Part III. Lithographic characteristics of poly-1,1,3-trimethyl-1-sila- and poly-1,1,3,3-tetramethyl-1,3-disilacyclobutenes and related silmethylene polymers. Microelectronic Eng 9:537–542
31. Gringolts M, Ushakov NV, Rogan Yu, Finkelshtein ESh (2007) ROMP, ROP and addition polymerization of silico-containing cyclic monomers a way to new membrane materials. NATO Science II. Math Phys Chem 243:395–411
32. Finkelshtein Sh, Ushakov NV, Yampolskii YuP (2003) The method of higher hydrocarbons removal from natural and petroleum gases. Patent Russian Federation 2218979 Bull of Inventions 15
33. Finkelshtein ESh, Ushakov NV, Portnykh EB et al. (1993) Ring-opening metathesis and thermoinitiated polymerization of 1-methyl-1-norbornenylmethyl-1-silacyclobutane. Vysokomolec Soedin A35:242–247
34. Sommer LH (1965) Stereochemistry mechanism and silicon. McGraw-Hill Book Company, USA
35. Vdovin VM (1968) Investigation in the field of compounds with silicon–carbon heterocycles. PhD Thesis Academy of Sciences of the USSR, Topchiev Institute of Petrochemical Synthesis, Moscow
36. Nametkin NS, Bespalova NB, Ushakov NV, Vdovin VM (1973) Polymerization of monosilacyclobutanes initiated with *n*-butyllitium. Dokl Acad Nauk SSSR 209:621–623
37. Ushakov NV, Vdovin VM, Pozdnyakova MV, Pritula NA (1983) Interaction of Li diphenylphosphide with monosilacyclobutanes. Izv Acad Nauk SSSR Ser Khim 9:2125–2129
38. Matsumoto K, Shimazu H, Deguchi M, Yamaoka H (1997) Polymerization of silacyclobutanes. J Polym Sci Part A Polym Chem 35:3207–3215
39. Liao Ch X, Weber WP (1992) Synthesis and characterization of poly(1-methyl-1-vinyl-1-silabutane), poly(1-phenyl-1-vinyl-1-silabutane),and poly(1,1-divinyl-1-silabutane). Macromolecules 25:1639–1641
40. Gilman H, Zeuch EA (1957) Some selective reactions of the silicon–hydrogen group with organometallic compouns. J Am Chem Soc 79:4560–4561
41. Ushakov NV, Pritula NA (1992) Synthesis of asymmetrically substituted 1-silacyclobutane and 1-sila-3-cyclopentene derivatives. Zhurnal Obshchei Khimii 62:1318–1324
42. Liao Ch X, Weber WP (1992) Synthesis and characterization of poly(1-methyl-1-silabutane), poly(1-phenyl-silabutane) and poly(1-silabutane). Polym Bull 28:281–286
43. Matsumoto K, Miyagawa K, Yamaoka H (1997) Anionic polymerization of 3-methylenesilacyclobutanes and reactivity of poly(3-methylenesilabutane)s. Macromolecules 30:2524–2526
44. Matsumoto K, Shinohata M, Yamaoka H (2000) Synthesis and anionic polymerization of silacyclobutanes bearing naphthyl or biphenyl groups at the 3-position. Polym J 32:1022–1029

45. Matsumoto K, Shinohata M, Yamaoka H (2000) Anionic ring-opening polymerization of phenylsilacyclobutanes. Polym J 32:354–360
46. Komuro K, Kawakami Y (1999) Polymerization of 1,1,2-trimethylsilacyclobutane. Polym J 31:138–142
47. Matsumoto K, Shimazu H, Yamaoka H (1998) Synthesis and characterization of polysilabutane having oligo(oxyethylene)phenyl groups on the silicon atom. J Polym Sci Part A Polym Chem 36:225–231
48. Jain R, Brunskill APJ, Sheridan JB, Lalancette RA (2005) A planar silacyclobutane, 1-(*rac*-1,1'-bi-2-naphthoxy)-1-silacyclobutane and its unusual reaction with bis(1,5-cyclooctadiene)platinum(0). J Organomet Chem 690:2272–2277
49. Kawakami Y, Park SY, Uenishi K et al. (2003) Controlled synthesis of silicon-containing polymers by metal catalysts. Polym Int 52:1619–1624
50. Matsumoto K, Miyagawa K, Matsuoka H, Yamaoka H (1999) Synthesis and solution behavior of the silicon-containing amphiphilic block copolymer, polystyrene-b-poly(3-hydroxymethyl-silacyclobutane. Polym J 31:609–613
51. Matsumoto K, Mizuno U, Matsuoka H, Yamaoka H (2002) Synthesis of novel silicon-containing amphiphilic diblock copolymers and their self-assembly formation in solution and at air–water interface. Macromolecules 35:555–565
52. Nametkin NS, Vdovin VM, Poletaev VA, Alekhin NN et al. (1973) New method of synthesis of organosilicon derivatives of alkali-metals. Izv Akad Nauk SSSR Ser Khim 6:1434
53. Alekhin NN (1975) Investigation of polymerization of 1,1-dimethyl-1-silacyclobutane in the presence of alkali-metal silanolates, alkoholates, and alkalis. Ph D Thesis Academy of Sciences of the SSSR Topchiev Institute of Petrochemical Synthesis, Moscow
54. Sheikh MRK, Tharanikkarasu K, Imae I, Kawakami YY (1999) Application of silacyclobutanes as "Carboanion Pump". Anionic polymerization of styrene using potassium-tert-butoxide and silacyclobutanes. 1st International Workshop on Silicon Chemistry-Polymers (May 29–31 1999, Tatsunokuchi, Japan) [Poster presentation]
55. Sheikh Md RK, Imae I, Tharanikkarasu K et al. (2000) Silacyclobutanes as "Carbanion pump" in anionic polymerization. I. Anionic polymerization of styrene by potassium t-butoxide in the prosense of silacyclobutanes. Polym J 32:527–530
56. Hyun J-Y, Kawakami Y (2004) Silacyclobutane as "Carbanion pump" in anionic polymerization. III. Synthesis of di- and tri-block copolymer by "diphenylsilacyclobutane-potassium tert-butoxide system." Polym J 36:856–865
57. Sheikh Md RK, Tharanikkarasu K, Imae I, Kawakami Y (2001) Silacyclobutanes as "Carbanion pump" in anionic polymerization. 2. Effective trapping of the initially formed carbanion by diphenylethylene. Macromolecules 34:4384–4389
58. Komuro K, Toyokawa S, Kawakami Y (1998) Synthesis and polymerization of 1-trimethylsiloxy- or 1-triphenylsiloxysilacyclobutanes. Polym Bull 40:715–720
59. Komuro K, Kawakami Y (1999) Synthesis and characterization of side-chain liquid crystalline polycarbosilanes with siloxane spacer. Polym Bull 42:669–674
60. Greenberg S, Clendenning SB, Liu K et al. (2005) Synthesis and lithographic patterning of polycarbosilanes with pendant cobalt carbonyl clusters. Macromolecules 38:2023–2026
61. Yamashita H, Tanaka M, Honda K (1995) Oxidative addition of the Si–C bonds of silacyclobutanes to Pt(PEt$_3$)$_3$ and highly selective platinum (0)-catalyzed di- or polymerization of 1,1-dimethyl-1-silacyclobutane. J Am Chem Soc 117:8873–8874
62. Nametkin NS, Vdovin VM (1974) Ring-opening reactions of silacyclobutanes. Izv Acad Nauk SSSR Ser Khim 23:1153–1169
63. Nametkin NS, Poletaev VA, Zav'yalov VI, Vdovin VM (1971) Kinetic investigation of thermoinitiated polymerization of 1,1,3,3-tetramethyl-1,3-disilacyclobutane. Dokl Acad Nauk SSSR 198:1096–1098
64. Nametkin NS, Vdovin VM, Zelenaya AV (1997) Preparation of silmethylene polymers. Patent SSSR 214809 Bull of Inventions 21
65. Nametkin NS, Vdovin VM, Zelenaya AV (1997) Preparation and the use of silmethylene polymers bearing alkenyl pendant groups. Patent SSSR 216269 Bull of Inventions 21

66. Nametkin NS, Zav'yalov VI, Zelenaya AV et al. (1997) Synthesis of polymers having silicon–carbon main chain and F-atom in pendant groups. Patent SSSR 269485 Bull of inventions 21
67. Nametkin NS, Babich ED, Karel'skii VN, Vdovin VM (1969) Synthesis of 1,3-disilacyclopentane derivatives. Izv Akad Nauk SSSR Ser Khim 6:1336–1342
68. Nametkin NS, Vdovin VM, Zav'yalov VI (1965) Silylmethylene elastomers. Vysokomolek Soedin 7:757
69. Nametkin NS, Vdovin VM, Zelenaya AV (1966) 1,3-Disilacyclobutanes and their polymers. Dokl Akad Nauk SSSR 170:1088–1091
70. Song G, Yamagushi M, Nishimura O, Suzuki M (2006) Investigation of metal nanoparticles produced by laser ablation and their catalytic activity. Appl Surf Sci 253:3093–3097
71. Ogawa T, Murakami M (1996) Synthesis, thermal and mechanical properties of poly(methylphenylsimethylene)s. Chem Mater 8:1260–1267
72. Ogawa T (1998) Polymerization and copolymerization behavior of phenyl-substituted 1,3-disilacyclobutanes. Polymer 39:2715–2723
73. Nametkin NS, Vdovin VM, Zav'yalov VI (1964) The catalysts of polymerization of silacyclobutanes. Izv Akad Nauk Ser Khim 1:203
74. Nametkin NS, Vdovin VM, Zav'yalov VI (1965) Polymerization of 1,1,3,3-tetraphenyl-1,3-disilikacyclobutane. Dokl Akad Nauk SSSR 162:824–826
75. Zundel T, Tronc F, Lestel L, Boileau S (1996) Polydimethylsilmethylene: synthesis and chemical modification. XI Internation Symposium on Organosilicon Chemistry, Montpellier, France, 1–6 September 1996. Abstr PA-106.
76. Zungel T, Baran J, Mazurek M et al. (1998) Climbing back up the nucleophilic reactivity scale. Use of cyclosila derivatives as reactivity boosters in anionic polymerization. Macromolecules 31:2724–2731
77. Matsumoto K, Nishimura M, Yamaoka H (2000) Organolithium-induced anionic polymerization of 1,1,3,3-tetramethyl-1,3-disilacyclobutane in the presence of hexamethylphosphoramide. Macromol Chem Phys 201:805–808
78. Wu X, Neckers DC (1999) Photocatalyzed ring-opening polymerization of 1,1,3,3-tetramethyl-1,3-disilacyclobutane. Macromolecules 32:6003–6007
79. Gomez-Elipe P, Resendes R, Macdonald PM, Manners I (1998) Transition metal catalyzed ring-opening polymerization (ROP) of silicon-bridged [1]ferrocenophanes: facile molecular weight control and the remarkably convenient synthesis of poly(ferrocenes) with regioregular, comb, star, and block architectures. J Am Chem Soc 120:8348–8356
80. Nametkin NS, Vdovin VM, Poletaev VA, Zav'yalov VI (1974) The compounds containing elements I-B group as catalysts of polymerization 4-membered silicon–carbon heterocycles. Patent SSSR 216270 Bull of Inventions 15
81. Poletaev VA, Vdovin VM, Nametkin NS (1973) Catalytic polymerization of silacyclobutane monomers in the presence of transition metal halides. Dokl Akad Nauk SSSR 208:1112–1115
82. Nametkin NS, Vdovin VM, Finkelshtein ESh et al. (1969) Synthesis of new high molecular weight heterochain polymer with phenylen rings in the main chain. Vysokomol Soedin Part B 11:207–209
83. Salamone JC, Fitch WL (1971) Ring-opening polymerization of 1,1-dimethyl-2,3-benzo-1-silacyclobutene. J Polym Sci PartA-1 Polym Chem 9:1741–1745
84. Ushakov NV, Finkelshtein ESh (1993) Thermoinitiated polymerization of dimethylsilacyclobutene. Xth International Symposium on Organosilicon Chemistry, Poznan, Poland, 15–20 August 1993. Abstracts P-159:279
85. Theurig M, Sargeant SJ, Manuel G, Weber WP (1992) Anionic ring-opening polymerization of 2,3-benzo-1-silacyclobutene. Characterization of poly(2,3-benzo-1-silabutenes). Macromolecules 25:3834–3837
86. Kakihana Y, Uenishi K, Imae I, Kawakami Y (2005) Anionic ring-opening polymerization of optically pure 1-methyl-1-(1-naphthyl)-2,3-benzosilacyclobut-2-ene. Macromolecules 38:6321–6326
87. Bamford WR, Lovie JC, Watt JAC et al. (1966) Preparation and properties of polysilmethylenes: use of various compounds of group VIII metals as catalysts. J Chem Soc C 13:1137–1140

88. Uenishi K, Imae I, Shirakawa E, Kawakami Y (2002) Synthesis of stereoregular and optically active poly[{methyl(1-naphthyl)silylene}(o-phenylene)methylene] by platinum-catalyzed ring-opening polymerization. Macromolecules 35:2455–2460
89. Sheridan JB, Gomez-Elipe P, Manners I (1996) Transition metal-catalysed ring-opening copolymerization of silicon-bridged [1]ferrocenophanes and sila- or disilacyclobutanes: synthesis of poly(ferrocenylsilane)-poly(carbosilane) random copolymers. Macromol Rapid Commun 17:319–324
90. Vdovin VM, Nametkin NS, Finkelshtein ESh (1964) The transformation of vinylben-zylderivatives of silicium in the presence of alkylation catalysts. Izv Akad Nauk SSSR Ser Khim 9:458–464
91. Nametkin NS, Vdovin VM, Finkelshtein ESh (1964) Synthesis of 3,4-benzosila-cyclobutenes. Dokl Akad Nauk SSSR 154:383–386
92. Nametkin NS, Vdovin VM, Finkelshtein ESh (1965) Polymerization of 3,4-benzo-1,1-dimethylsilacyclopentene. Dokl Akad Nauk SSSR 162:585–588
93. Finkelshtein ESh (1966) Ph.D. Thesis Academy of Sciences of the USSR Topchiev Institute of Petrochemical Synthesis, Moscow
94. Park Y-H, Zhou SQ, Weber WP (1989) Anionic ring opening polymerization of 3,4-benzo-1,1-dimethyl-1-silapentene). Polym Bull 26:349–353
95. Zhou SQ, Weber WP (1990) Anionic ring opening polymerization of 2-methyl-2-silaindan. Characterization of the polymer and mechanism of polymerization. Macromol Chem Rapid Commun 11:19–24
96. Ko Y-H, Weber WP (1991) Synthesis and characterization of poly(3,4-benzo-1-phenyl-1-silapentene) and poly(3,4-benzo-1-1-silapentene). Polym Bull 26:487–492
97. Hirao A, Hatayama T, Nakahama S (1987) Polymerization of monomers containing functional silyl groups. 3. Anionic living polymerization of (4-vinylphenyl)dimethylsilane. Macromolecules 20:1505–1509
98. Park Y-T, Kim SO (1997) Synthesis and properties of novel poly[3,4-(silylisopropyl)benzo-1-silapentene]. Bull Korean Chem Soc 18:232–235
99. Park Y-T, Park SU, Kim HC, Lee K (1998) Synthesis and properties of organosilicon polymers containing 3,4-benzo-1-silacyclopentene derivatives. Bull Korean Chem Soc 19:328–332
100. Horvath RF, Chan TH (1987) Metalation reactions of 1-silacyclo-3-pentenes. J Org Chem 52:4489–4494
101. Zhang X, Zhou SQ, Weber WP et al. (1988) Anionic ring-opening polymerization of sila- and germacyclopent-3-enes. Macromolecules 21:1563–1566
102. Weber WP, Park Y-T, Zhou SQ (1991) Mechanism of anionic ring-opening polymerization of silacyclopent-3-enes. Makromol Chem Macromol Symp 42/43:259–267
103. Sargeant SJ, Zhou SQ, Manuel G, Weber WP (1992) Anionic dimerization and ring opening polymerization of 1,1-divinyl-1-silacyclopent-3-ene. Macromolecules 25:2832–2836
104. Liao X, Weber WP (1991) Synthesis of poly(1-methyl-1-phenyl-1-silapentane) by chemical reduction of poly(1-methyl-1-phenyl-1-sila-*cis*-pent-3-ene with diimide. Polym Bull 25:621–624
105. Liao X, Leibfried RT, Weber WP (1991) Anionic ring opening polymerization of 1-phenyl-1-vinyl-1-silacyclopent-3-ene. Polym Bull 26:625–628
106. Liao X, Ko Y-H, Manuel G, Weber WP (1991) Synthesis and microstructure of poly(1-phenyl-1-sila-*cis*-pent-3-ene). Polym Bull 25:63–69
107. Zhou SQ, Weber WP (1990) Anionic polymerization of 1-methyl-1-silacyclopent-3-ene. Characterization of poly(1-methyl-1-sila-*cis* –pent-3-ene) by ^{1}H, ^{13}C, and ^{29}Si NMR spectroscopy and mechanism of polymerization. Macromolecules 23:1915–1917
108. Park YT, Manuel G, Weber WP (1990) Anionic ring-opening polymerization of 1,1,3-trimethyl-1-silacyclopent-3-ene. Effect of temperature on poly(1,1,3-trimethyl-1-sila-*cis*-pent-3-ene) microctructure. Macromolecules 23:1911–1915
109. Park YT, Zhou SQ, Manuel G, Weber WP (1991) Synthesis and polymerization of 5-silaspiro[4.4]nona-2,7-dienes. Macromolecules 24:3221–3226

110. Wang L, Ko Y-H, Weber WP (1992) Dimerization and polymerization of 2,3-benzo-5-silaspiro[4.4]nona-2,7-diene. Macromolecules 25:2828–2831
111. Sargeant SJ, Weber WP (1993) Synthesis of carbosilane monomers and polymers with mesogenic pendant groups. Preparation and characterization of aryloxy-subtituted poly(1-sila-cis-pent-3-enes). Macromolecules 26:2400–2407
112. Wang L, Weber WP (1993) Synthesis and properties of novel comb polymers: unsaturated carbosilane polymers with pendant oligo(oxyethylene) groups. Macromolecules 26:969–974
113. Liao X, Weber Mazerolles WPP et al. (1991) Synthesis and characterization of poly[(3,4,c)furano-1-germa-1,1-dimethylpentane]. Polym Bull 26:499–502
114. Ushakov NV, Finkelshtein ESh (2009) Synthesis, dimerization and oligomerization of 5-silaspiro[4,4]non-2-ene. Izv Acad Nauk Ser Khim 5:923–925
115. Zhou Q, Manuel G, Weber WP (1990) Copolymerization of 1,1-dimethyl-1-silacyclopent-3-ene and 1,1-diphenyl-1-silacyclopent-3-ene. Characterization of copolymer micro-structures by ^{1}H, ^{13}C and ^{29}Si NMR spectroscopy. Macromolecules 23:1583–1589
116. Sargeant SJ, Farahi JB, Weber WP (1993) Anionic ring-opening block copolymerization of 1,1-dimethyl-1-silacyclopent-3-ene and 1-methyl-1-phenyl-1-silacyclopent-3-ene. Macromolecules 26:4729–4731
117. Weber WP, Zhou SQ (1992) Crosslinkable poly(unsaturated carbosilane) copolymers and rization methods of making same. US Patent 5 171 810
118. Weber WP, Liao X (1992) Crosslinkable saturated and unsaturated carbosilane polymers and formulations. US Patent 5 171 792
119. Ushakov NV, Finkelshtein ESh (2000) Anionic copolymerization of 1-silacyclobutanes with 1,1-dimethyl-1-silacyclopent-3-ene. 33rd Symposium on Organosilicon Chemistry, Saginaw, Michigan, USA. Abstracts PB-24
120. Lammens H, Sartori G, Siffert J et al. (1971) Ring-opening polymerization of 1,1-dimethyl-1-silacyclopent-3-ene. J Polym Sci Part B Polym Lett 9:341–345
121. Finkelshtein ESh, Portnykh EB, Ushakov NV, Vdovin VM (1981) Metathesis of dialkenylsilanes and polymerization of 1,1-dimethyl-1-silacyclopent-3-ene in the presence alumina-rhenium catalysts. Izv Akad Nauk SSSR Ser Khim 3:641–645
122. Bykova TA, Lebedev BV, Ushakov NV, Finkelshtein ESh (2000) Thermodynamics of metathesis polymerization of 1,1-dimethyl-1-silacyclopent-3-ene in bulk and the thermodynamic properties of the resulting polymer in the 0–340 K range. Vysokomolek Soedin A42:1307–1315
123. Nefedov OM, Kolesnikov SP, Okhrimenko NI, Povarov SL (1979) W-carben complexes in reactions with 1,3-dienes and cycloolefines. I. Vsesoyuznaya konfer po metalloorg Khimii 1979 Moscow Tezisy dokl Chast' II:263
124. Fink W, Nang B-Z, Faucher DA (1993) The polymerization behavior of [1]- and [2]ferrocenophanes containing silicon atoms in the bridge: comparison of the molecular structure of the strained, polymerizable cyclic ferrocenylsilane Fe(.eta.-C_5H_4)$_2$(SiMe$_2$) with that of the cyclic ferrocenyldisilane Fe(.eta.-C_5H_4)$_2$(SiMe$_2$)$_2$. Organometallics 12:823–829
125. Pudelski JK, Rulkens R, Foucher DA et al. (1995) Synthesis and properties of poly(ferrocenyldihydrosilane) non random copolymers. Macromolecules 28:7301–7308
126. Elschenbroich C, Hurley J, Metz B (1990) Metal.π.-complexes of benzene derivatives. 34. Tetraphenylsilane as a chelating ligand: synthesis, structural characterization, and reactivity of the tilted bis(arene) metal complexes [(C_6H_5)$_2$Si(.eta.6- C_6H_5)$_2$]M (M – vanadium, chromium). Organometallics 9:889–897
127. Hultzsch KG, Nelson JM, Lough AJ, Manners I (1995) Synthesis, characterization, and homopolymerization and copolymerization behavior of the silicon-bridged [1]chromarenophane Cr(η–C_6H_5)$_2$SiMe$_2$. Organometallics 14:5496–5502
128. Elschenbroich C, Bzetschneider-Hurley A, Hurley J (1993) [μ(1,1,2,2-Tetrakis(η^6-phenyl)-1,2-diphenyldisilane]-divanadium: long distance exchange interaction mediated through a >SiPh-SiPh< Unit. Inorg Chem 32:5421–5424

129. Elschenbroich C, Bzetschneider-Hurley A, Hurley J et al. (1995) Metal complexes of benzene derivatives. 45. Dinuclear bis(.eta.6-arene)vanadium and chromium complexes containing 1,3-disilacyclobutane as a spacer. An EPR study of intermetallic communication. Inorg Chem 34:743–745

130. Foucher DA, Tang B-Z, Manners I (1992) Ring-opening polymerization of strained, ring-titled ferrocenophanes: a route to high molecular weight poly(ferrocenylsilanes). J Am Chem Soc 114:6246–6248

131. Rulkens R, Ni Y, Lough AJ, Manners I (1994) Anionic ring-opening oligomerization and polymerization of silicon-bridged [1]ferrocenophanes: characterization of short-chain models for poly(ferrocenylsilane) high polymers. J Am Chem Soc 116:797–798

132. Rulkens R, Ni Y, Lough AJ, Manners I (1994) Living anionic ring-opening polymerization of silicon-bridged [1]ferrocenophanes: synthesis and characterization of poly(ferrocenylsilane)-polysiloxane block copolymers. J Am Chem Soc 116:12121–12122

133. Ni Y, Rulkens R, Pudelski JK, Manners I (1995) Transition metal catalyzed ring-opening polymerization of silicon-bridged [1]ferrocenophanes at ambient temperature. Macromol Chem Rapid Commun 16:637–641

134. Reddy NP, Yamashita H, Tanaka M (1995) Platinum- or palladium-catalysed ring-opening homo- and co-polymerization of silicon- and germanium-bridged [1]ferrocenophanes. J Chem Soc Chem Commun 1995:2263–2264

135. Foucher DA, Ziembinski R, Nang B-Z et al. (1993) Synthesis, characterization, glass transition behavior, and the electronic structure of high-molecular-weight, symmetrically substituted poly(ferrocenylsilanes) with alkyl or aryl side groups. Macromolecules 26:2878–2884

136. Temple K, Jakle F, Sheridan JB, Manners I (2001) The nature of the active catalyst in late transition metal-mediated ring-opening polymerization (ROP) reactions: mechanistic studies of the platinum-catalysed ROP of silicon-bridged [1]ferrocenophanes. J Am Chem Soc 123:1355–1364

137. Ni Y, Rulkens R, Manners I (1996) Transition metal-based polymers with controlled architectures: well-defined poly(ferrocenylsilane) homopolymers and multiblock copolymerization of silicon-bridged [1]ferrocenophanes. J Am Chem Soc 118:4102–4114

138. Hultzsch KG, Nelson JM, Lough AJ, Manners I (1995) Synthesis, characterization, and homopolymerization and copolymerization behavior of the silicon-bridged [1]chromarenophane $Cr(.ete.-C_6H_5)_2SiMe$. Organometallics 14:5496–5502

139. Pudelski JM, Manners I (1995) A heterolytic cyclopentadienyl carbon–silicon bond cleavage mechanism for the thermal ring-opening polymerization of silicon-bridged [1]ferrocenophanes. J Am Chem Soc 117:7265–7266

140. Bartle-Scott A, Resendes R, Manners I (2003) Transition metal-catalyzed ring-opening polymerization of silicon-bridged [1]ferrocenophanes in the presence of functional silanes: molecular weight control and synthesis of telechelic poly(ferrocenylsilanes). Macromol Chem Phys 204:1259–1268

141. Bao M, Hatanaka Y, Shimada S (2004) The first synthesis and X-ray structure of [1.1]silaferrocenophane containing pentacoordinate silicon moieties. Chem Lett 33:520

142. Sharma HK, Cervantes-Lee F, Pannell KH (2004) Isolation and ring-opening of new 1-silametallacyclobutanes $(\eta^5-C_5H_4Fe)(CO)_2CH_2SiR_2$ leading to a new class of organometallic polymer. J Am Chem Soc 126:1326–1327

143. Wang L, Chen T, Wang X-J et al. (2006) Study on synthesis and properties of poly(N-ferrocenyl^{n}buthylmethylsilane) with unsymmetrical silicon substitution groups. Eur Pol J 42:843–848

144. Elschenbroich C, Paganelli F, Nowotny M et al. (2004) Trovacene Chemistry. 10[1] The[1] – and [2]silatrovacenophanes $(\eta^7-C_7H_6)V(5-C_5H_4SiR_2)$ and $(\eta^7-C_7H_6)V(5-C_5H_4SiR_2SiR_2)$ (R= Me,Ph):Synthesis, structure, and ring opening. Zeitschrift anorg Und allgemeine Chemie 630:1599–1609

145. Bartole-Scott A, Braunschweig H, Kupfer T et al. (2006) Synthesis of ansa-[n]silacyclopentadienyl-cycloheptatrienyl-chromium complexes ($n = 1, 2$): novel precursors for polymers bearing chromium in the backbone. Chem Eur J 12:1266–1273

146. Braunschweig H, Kupfer T, Lutz M, Radacki K (2007) Ansa[1]trochrocenophanes and their related unstrained 1,1'-disubstituted counterparts: synthesis and electronic structure. J Am Chem Soc 129:8893–8906
147. Braunschweig H, Kupfer T, Radacki K (2007) Selective dimetalation of $Mn(C_5H_5)(C_6H_6)$: crystal structure and conversion to strained [n]metalloarenophanes (n = 1, 2). Angew Chem 119:1655–1658
148. Peckman TJ, Nguyen P, Bourke SC et al. (2001) Ring-opening polymerization behavior of ansa- and spirocyclic ansa-Zirconocene complexes. Organometallics 20:3035–3043
149. Wang Z, Masson G, Peiris FC et al. (2007) Living photolytic ring-opening polymerization of amino-functionalized [1]ferrocenophanes: synthesis and layer-by-layer self-assembly of well-defined water-soluble polyferrocenylsilane polyelectrolytes. Chem -A Eur J 13:9372–9383
150. Tanabe M, Manners I (2004) Photolitic living anionic ring-opening polymerization (ROP) of silicon-bridged [1]ferrocenophanes via an iron-cyclopentadienyl bond cleavage mechanism. J Am Chem Soc 126:11434–11435
151. Tanabe M, Vandermeulen GWM, Chan WY et al. (2006) Photocontrolled living polymerizations. Nat Mater 5:467–470
152. Baumgartner T, Jakle F, Rulkens R et al. (2002) Nucleophilically assisted and cationic ring-opening polymerization of tin-bridged [1]ferrocenophanes. J Am Chem Soc 124:10062–10070
153. Ivin KJ, Mol JC (1997) Olefin metathesis and metathesis polymerization. Academic Press, San Diego, London
154. Grubbs RH (ed) (2003) Handbook of metathesis. Wiley-VCH, Weinheim
155. Buchmeizer MR (2000) Homogenous metathesis polymerization by well-defined group VI and group VIII transition-metal alkylidenes: fundamentals and applications in the preparation of advanced materials. Chem Rev 100:1565–1604
156. Bielawski CW, Grubbs RH (2007) Living ring-opening metathesis polymerization. Prog Polym Sci 32:1–29
157. Buchmeiser MR (2009) Ring-opening metathesis polymerization. In: Dubois Ph, Coulembier O, Raquez J-M (eds) Handbook of ring-opening polymerization, 1st edn. Wiley-VCH, Weinheim
158. Schleyer RVR, Williams JE, Blanchard KR (1970) Evaluation of strain in hydrocarbons. The strain in adamantane and its origin. J Am Chem Soc 92:2377–2386
159. Lebedev BV, Lityagov VY (1977) Thermodynamics of the reactions of polypentenamer synthesis. Vysokomol Soedin B19:558–560
160. Lebedev B, Smirnova N, Kiparisova Y, Makovetssky K (1992) Thermodynamics of norbornene, of its polymerization process and of polynorbornene from 0 to 400 K at standard pressure. Makromol Chem 193:1399–1411
161. North M (2002) ROMP of norbornene derivatives of amino-ester and amino-acids. In: Khosravi E, Szymanska-Buzar T (eds) Ring opening metathesis polymerization and related chemistry. NATO Sci Ser II, vol 56. Kluwer, Dordrecht, pp 157–166
162. Finkelshtein ESh, Marciniec B (1995) Preparation of organosilicon oligomers and polymers via the reaction of olefin metathesis. In: Marciniec B, Choinowski J (eds) Progress in organosilicon chemistry. Gordon and Breach Publication, Amsterdam, pp 445–465
163. Finkelshtein ESh (1995) Metathesis polymerization of unsaturated organosilicon compounds. Polym Sci B 37:185–202
164. Finkelshtein ESh (1998) Olefin metathesis in organosilicon chemistry In: Imamoglu Y (ed) Metathesis polymerization of olefins and polymerization of alkynes, NATO ASI Series C, vol 506. Kluwer, Dordrecht, pp 201–224
165. Matloka PP, Wagener KB (2006) The acyclic diene metathesis (ADMET) polymerization approach to silicon containing materials. J Mol Cat A Chem 257:89–98 and references herein
166. Dragutan V, Dragutan I, Fischer H (2008) Synthesis of metal-containing polymers via ring opening metathesis polymerization (ROMP). Part I. Polymers containing main group metals. J Inorg Organomet Polym 18:18–31

167. Katz TJ, Lee SJ, Shippey MA (1980) Preparations of polymers using metal-carbenes. J Mol Catal 8:219–226
168. Ginsburg EJ, Gorman ChB, Marder SR, Grubbs RH (1989) Poly (trimethylsilylcyclo-octatetraene): a soluble conjugated polyacetylene via olefin metathesis. J Am Chem Soc 11:7621–7622
169. Gorman C, Ginsburg E, Grubbs R (1993) Soluble, highly conjugated derivatives of polyacetylene from the ring-opening metathesis polymerization of monosubstituted cyclooctatetraenes: synthesis and the relationship between polymer structure and physical properties. J Am Chem Soc 115:1397–1409
170. Streck R (1982) Some applications of the olefin metathesis reaction to polymer synthesis. J Mol Catal 15:3–19
171. Anderson AW, Merckling NG (1955) Polymeric bicyclo[2.2.1]-2-heptene. US Patent 2721189. Chem Abstr(1956) 50:3008
172. Schrock RR (1990) Living ring opening metathesis polymerization catalyzed by well-characterized transition-metal alkylidene complexes. Acc Chem Res 23:158–165
173. Khosravi E, Szymanska-Buzar T (eds) (2000) Ring opening metathesis polymerization and related chemistry. NATO Science Ser.II: mathematics, physics and chemistry, vol 56. Kluwer Academic Publishers, Dordrecht
174. Nicolaou KC, Snyder SA, Montagnon T, Vassilikogiannakis G (2002) The Diels–Alder reaction in total synthesis. Angew Chem Int Ed 41:1668–1698
175. Onischenko AS (1963) Diene synthesis. Akad Sci SSSR Press, Moscow
176. Mol JC (2004) Olefin metathesis: early days. J Mol Cat 213:39–45
177. Cunico RF (1971) The diels-alder reaction of a,β-unsaturated trihalosilanes with cyclopentadiene. J Org Chem 36:929–932
178. Kuivila HG, Warner CR (1964) Trimethylsilyl-substituted norbornenes, norbornanes, and nortricyclene. J Org Chem 29:2845–2851
179. Stosur M, Szymanska-Buzar T (2008) Facile hydrosilylation of norbornadiene by silanes R_3SiH and R_2SiH_2 with molybdenum catalysts. J Mol Cat A Chem 286:98–105
180. Park JH, Ryu BG, Kim JY et al. (2008) Norbornene-based silsesquioxane copolymers, norbornene-based silane used for preparation of the same and method of preparing low dielectric insulating film comprising the same. IPN WO 2008/082128 A1, PCT/KR2007/006794
181. Kenndoff J, Polborn K, Szeimies G (1990) Generation and trapping of 1,5-dehydroquadricyclane. J Am Chem Soc 112:6117–6118
182. Makovetsky KL, Finkelshtein ESh, Ostrovskaya IYa, Portnykh EB et al. (1992) Ring-opening metathesis polymerization of substituted norbornenes. J Mol Cat 76:107–121
183. Finkelshtein ESh, Portnykh EB, Makovetskii KL et al. (1998) Synthesis of membrane materials by ROMP of norbornenes. In: Imamoglu Y (ed) Metathesis polymerization of olefins and polymerization of alkynes. NATO ASI Series C, vol 506. Kluwer, Dordrecht, pp 189–199
184. Jones RG, Ando W, Chojnowski J (eds) (2000) Silicon containing polymers: the science and technology of their synthesis and applications. Kluwer Academic Publishers, Dordrecht
185. Yamada S, Cho S, Lee JH et al. (2004) Design and study of silicone-based materials for bilayer resist application. J Photopolym Sci Technol 17:511–518
186. Ganachaud F, Boileau S, Boury B (eds) (2008) Silicon based polymers: advances in synthesis and supramolecular organization. Springer, The Netherlands
187. Finkelshtein ESh, Makovetskii KL, Yampolskii YuP, Portnykh EB et al. (1991) Ring-opening metathesis polymerization of norbornenes with organosilicon substituents. Gas permeability of polymers obtained. Makromol Chem 192:1–9
188. Gringolts ML, Bermeshev MV, Starannikova LE et al. (2009) Synthesis and gas separation properties of metathesis polynorbornenes with one and two groups $SiMe_3$ in monomer unit. Polym Sci Ser A 51:1233–1240
189. Finkelshtein ESh, Portnykh EB, Ushakov NV, Greengolts ML et al. (1994) Synthesis of polymers containing carbazolyl groups via ring-opening metathesis polymerization (ROMP) of a carbazolylsubstituted norbornene derivative. Macromol Rapid Commun 15:155–159
190. Gratt J, Cohen RE (1997) Synthesis of block copolymers containing pendant carbazole groups via living ring-opening metathesis polymerization. Macromolecules 30:3137–3140

191. Pearson JM, Stolka M (1981) Poly(N-vinylcarbazole). Gordon and Breach Science Publishers, New York
192. Gill WD (1972) Drift mobilities in amorphous charge-transfer complexes of trinitrofluorenone and poly-*n*-vinylcarbazole. J Appl Phys 43:5033–5040
193. Zhang C, von Seggern H, Pakbaz K, Kraabel B et al. (1994) Blue emission from polymer light-emitting diodes using non-conjugated polymer blends with air-stable electrodes. Synth Met 62:35–40
194. Kwark UJ, Bravo-Vasquez JP, Ober ChK (2003) Novel silicon containing polymers as photoresist materials for extreme UV lithography. In: Fedynyshyn TH (ed) Advances in resist technology and processing XX, Proceedings of SPIE, vol 5039, pp 1204–1211
195. Scherman OA, Kim HM, Grubbs RH (2002) Synthesis of well-defined poly((vinyl alcohol)2-*alt*-methylene) via ring-opening metathesis polymerization. Macromolecules 35:5366–5371
196. Gringolts ML, Bermeshev MV, Finkelshtein ESh (2010) unpublished results
197. Aoki T, Ohshima M, Shinohara K et al. (1997) Enantioselective permeation of racemates through a solid (+)-poly{2-[dimethyl(10-pinanyl)silyl]norbornadiene}membrane. Polymer 38:235–238
198. Stonish DA, Weber WP (1991) Synthesis and characterization of poly[(2-trimethylsilyl-2-cyclopentene-1,4-diyl)vinylene]. Polym Bull 26:493–497
199. Finkelshtein ESh, Gringolts ML, Ushakov NV et al. (2003) Synthesis and gas permeation properties of new ROMP polymers from silyl substituted norbornadienes and norbornenes. Polymer 44:2843–2851
200. Gringolts ML, Bermeshev MV, Neliubina YuV, Finkel'shtein ESh (2009) Catalytic transformations of mono- and bis-silylsubstituted norbornadienes. Petrol Chem 49:369–376
201. Stonish DA, Weber WP (1991) Synthesis and characterization of poly[(2-dimethylsilyl-2-cyclopentene-1,4-diyl)vinylene]. Polym Bull 27:243–249
202. Mera G (2005) Contributions to the synthesis of silicon-rich oligocarbosilanes and their use as precursors for electrically conductive films. Diss Ruhr-Universität Bochum Chair of Inorganic Chemistry I
203. Bermeshev MV, Gringolts ML, Lakhtin VG, Finkel'shtein ESh (2008) Synthesis and metathesis polymerization of 5,5-bis(trimethylsilyl)norbornene-2. Petrol Chem 48:302–308
204. Kawakami Y, Toda H, Higashino M, Yamashita Y (1988) Polynorbornenes with oligodimethylsiloxanyl substituents for selectively oxygen permeable membrane material. Polymer J 20:285–292. doi:10.1295/polymj.20.285
205. Bondar V, Kukharskii Yu, Yampolskii Yu, Finkelshtein E et al. (1993) Permeation and sorption in polynorbornenes with organosilicon substituents. J Polym Sci Part B Polym Phys 31:1273–1283
206. Katsumata T, Shiotsuki M, Sanda F, Masuda T (2009) Synthesis and properties of polynorbornenes bearing oligomeric siloxane pendant groups. Polymer 50:1389–1394
207. Yampolskii Yu, Freeman BD, Pinnau I (eds) (2006) Materials science of membranes for gas and vapor separation. Wiley, Chichester
208. Xu W, Chung Ch, Kwon Y (2007) Synthesis of novel block copolymers containing polyhedral oligomeric silsesquioxane (POSS) pendent groups via ring-opening metathesis polymerization (ROMP). Polymer 48:6286–6293
209. Mather PT, Jeon HG, Romo-Uribe A (1999) Mechanical relaxation and microstructure of poly(norbornyl-POSS) copolymers. Macromol 32:1194–1203
210. Pan G (2007) Polyhedral oligomeric silsesquioxane (POSS). In: Physical properties of polymers handbook. Springer, New York
211. Rutenberg IM, Scherman OA, Grubbs RH, Jiang W et al. (2004) Synthesis of polymer dielectric layers for organic thin film transistors via surface-initiated ring-opening metathesis polymerization. J Am Chem Soc 126:4062–4063
212. Harada Y, Girolami GS, Nuzzo RG (2003) Catalytic amplification of patterning via surface-confined ring-opening metathesis polymerization on mixed primer layers formed by contact printing. Langmuir 19:5104–5114

213. Juang A, Scherman OA, Grubbs RH, Lewis NS (2001) Formation of covalently attached polymer overlayers on Si(111) surfaces using ring-opening metathesis polymerization methods. Langmuir 17:1321–1323
214. Jordi MA, Seery TAP (2005) Quantitative determination of the chemical composition of silica-poly(norbornene) nanocomposites. J Am Chem Soc 127:4416–4422
215. Kim NY, Jeon NL, Choi IS, Takami S et al. (2000) Quantitative determination of the chemical composition of silica-poly(norbornene) nanocomposites. Macromol 33:2793–2795
216. Weck M, Jackiw JJ, Rossi RR et al. (1999) Ring-opening metathesis polymerization from surfaces. J Am Chem Soc 121:4088–4089
217. Feng J, Stoddart SS, Kanchana A, Weerakoon Chen KAW (2007) An efficient approach to surface-initiated ring-opening metathesis polymerization of cyclooctadiene. Langmuir 23:1004–1006
218. Jeon NL, Choi IS, Whitesides GM, Kim NY et al. (1999) Patterned polymer growth on silicon surfaces using microcontact printing and surface-initiated polymerization. Appl Phys Lett 75:4201–4203
219. Makovetskii KL (2008) Catalytic addition polymerization of norbornene and its derivatives and copolymerization of norbornene with olefins. Polym Sci Ser C 50:22–38
220. Blank F, Janiak Ch (2009) Metal catalysts for the vinyl/addition polymerization of norbornene. Coord Chem Rev 253:827–861
221. Janiak C, Lassahn PG (2001) Metal catalysts for the vinyl polymerization of norbornene. J Mol Catal A Chem 166:193–209
222. BL (2003) Cycloaliphatic polymers via late transition metal catalysis. In: Rieger BL, Saunders B, Kacker S, Striegler S (eds) Late transition metal polymerization catalysis. Wiley-VCH, Weinheim
223. Hennis AD, Polley JD, Long GS, Sen A et al. (2001) Novel, efficient, palladium-based system for the polymerization of norbornene derivatives: scope and mechanism. Organometallics 20:2802–2812
224. Funk JK, Andes CE, Sen A (2004) Addition polymerization of functionalized norbornenes: the effect of size, stereochemistry, and coordinating ability of the substituent. Organometallics 23:1680–1683
225. Myagmarsuren G, Lee Ki-Soo, Jeong O-Yong, Ihm Son-Ki (2005) Homopolymerization of 5-alkyl-2-norbornenes and their copolymerization with norbornene over novel Pd(acac)$_2$/PPh$_3$/BF$_3$OEt$_2$ catalyst system. Polymer 46:3685–3692
226. Kaita Sh, Matsushita K, Tobita M, Maruyama Y, Wakatsuki Y (2006) Cyclopentadienyl nickel and palladium complexes/activator system for the vinyl-type copolymerization of norbornene with norbornene carboxylic acid esters: control of polymer solubility and glass transition temperature. Macromol Rapid Commun 27:1752–1756
227. Jung IG, Seo J, Chung YK, Shin DM et al. (2007) Polymerization of carboxylic ester functionalized norbornenes catalyzed by (η^3-allyl)palladium complexes bearing N-heterocyclic carbene ligands. J Polym Sci Part A Polym Chem 45:3042–3052
228. Lipian J, Mimna RA, Fondran JC, Yandulov D et al. (2002) Addition polymerization of norbornene-type monomers. High activity cationic allyl palladium catalysts. Macromol 35:8969–8977
229. Barnes DA, Benedikt GM, Goodall BL, Huang SS et al. (2003) Addition polymerization of norbornene-type monomers using neutral nickel complexes containing fluorinated aryl ligands. Macromol 36:2623–2632
230. Park S, Krotine J, Allen SAB, Kohl PA (2006) Electron-beam hardening of thin films of functionalized polynorbornene copolymer. J Electronic Mat 35:1112–1121
231. Grove N, Kohl P, Allen S, Jayaraman S et al. (1999) Functionalized polynorbornene dielectric polymers: adhesion and mechanical properties. J Polym Sci Polym Phys 37:3003–3010
232. Ahmed S, Bidstrup-Allen S, Kohl P, Ludovice P (1998) Prediction of stereoregular-poly(norbornene) structure using a long-range RIS model. Macromol Symp 133:1–10
233. Karafilidis Ch, Hermann H, Rufinska A, Gabor B et al. (2004) Metallocene-catalyzed C7-linkage in the hydrooligomerization of norbornene by s-bond metathesis: insight into the microstructure of polynorbornene. Angew Chem Int Ed 43:2444–2446

234. Finkelshtein ESh, Makovetskii KL, Gringolts ML, Rogan YV et al. (2006) Addition polymerization of silyl-containing norbornenes in the presence of Ni-based catalysts J Mol Catal A Chem 257:9–13
235. Mathew JP, Reinmuth A, Melia J, Swords N, Risse W (1996) (η3-Allyl)palladium(II) and palladium(II) nitrile catalysts for the addition polymerization of norbornene derivatives with functional groups. Macromolecules 29:2755–2763
236. Green M, Hancock R (1967) The stereochemistry of the reaction of bicyclo [2.2.1] heptadiene-palladium chloride with methoxide anions. J Chem Soc A 12:2054–2057
237. Finkelshtein E, Makovetskii K, Gringolts M, Rogan Yu et al. (2006) Addition-type polynorbornenes with Si(CH$_3$)$_3$ side groups: synthesis, gas permeability and free volume. Macromol 39:7022–7029
238. Gringolts ML, Bermeshev MV, Makovetsky KL, Finkelshtein ESh (2009) Effect of substituents on addition polymerization of norbornene derivatives with two Me$_3$Si-groups using Ni(II)/MAO catalyst. Eur Polym J 45:2142–2149
239. Sanders DP, Connor EF, Grubbs RH, Hung RJ et al. (2003) Metal-catalyzed addition polymers for 157 nm resist applications. Synthesis and polymerization of partially fluorinated, ester-functionalized tricyclo[4.2.1.02,5]non-7-enes. Macromolecules 36:1534–1542
240. Gringolts ML, Bermeshev MV, Kaz'min AG, Finkelshtein ESh (2009) New quadricyclane-based cyclic polycarbosilanes. *Dokl Chem 424:49–51*
241. Zhao Ch, Ribeiro MR, de Pinho MN, Subrahmanyam VS et al. (2001) Structural characteristics and gas permeation properties of polynorbornenes with retained bicyclic structure. Polymer 42:2455–2462
242. Wilks BR, Chung WJ, Ludovice PJ, Rezac MR et al. (2003) Impact of average free-volume element size on transport in stereoisomers of polynorbornene. I. Properties at 35°C. J Polym Sci: Part B Polym Phys 41:2185–2199
243. Dorkenoo KD, Pfromm PH, Rezac ME (1998) Gas transport properties of a series of high T$_g$ polynorbornenes with aliphatic pendant groups. J. Polym Sci Part B Polym Phys 36:797–803
244. Poulsen L, Zebger I, Klinger M, Eldrup M et al. (2003) Oxygen diffusion in copolymers of ethylene and norbornene. Macromol 36:7189–7198
245. Tetsuka H, Isobe K, Hagiwara M (2009) Synthesis and properties of addition-type poly(norbornene)s with siloxane substituents. Polym J. doi:10.1295/polymj.PJ2009010
246. Alentiev AYu, Yampolskii YuP, Shantarovich VP et al. (1997) High transport parameters and free volume of perfluorodioxole copolymers. J Membr Sci 126:123–132
247. Morisato A, Pinnau I (1996) Synthesis and gas permeation properties of poly(4-methyl-2-pentyne). J Membr Sci 121:243–250
248. Nagai K, Masuda T, Nakagawa T, Freeman BD, Pinnau I (2001) Poly[1-(trimethylsilyl)-1-propyne] and related polymers: synthesis, properties and functions. Prog Polym Sci 26:721–798
249. Yampolskii YuP, Finkelshtein ESh, Makovetskii KL, Bondar VI et al. (1996) Effects of *cis-trans*-configurations of the main chains of poly(trimethylsilyl norbornene) on its transport and sorption properties as well as free volume. J Appl Polym Sci 62:349–357
250. Makovetsky KL (1999) Addition polymerization of cycloolefins: new polymeric materials for advanced technologies. Polymer Sci Ser B 41:269–285

New Synthetic Strategies for Structured Silicones Using $B(C_6F_5)_3$

Michael A. Brook, John B. Grande, and François Ganachaud

Abstract The dehydrocarbonative condensation of alkoxysilanes + hydrosilanes in the presence of the Lewis acid catalyst $B(C_6F_5)_3$ ($R_3SiOR' + HSiR''_3 \rightarrow R_3SiOSiR''_3 + R'H$) – described throughout this review as the Piers-Rubinsztajn reaction – provides a new, mild strategy for the controlled synthesis of silicones. In this review we examine the mechanistic parameters that control the reaction, and outline the types of accessible small molecules, linear, branched, and cross-linked materials (resins and elastomers) that can be prepared using this and related reactions.

Keywords $B(C_6F_5)_3$ · Controlled 3D silicone structures · Dehydrocarbonative condensation · Piers-Rubinsztajn reaction · Silicone synthesis

Contents

1	Introduction	162
	1.1 Traditional Methods for Silicone Chemistry Rely on Ionic Chemistry	162
	1.2 New Chemistry for Silicone Synthesis	163
2	Mechanistic Considerations	164
	2.1 From the Organic World	164
	2.2 Additional Mechanistic Subtleties: The Silicone World	166

M.A. Brook (✉) and J.B. Grande
Department of Chemistry and Chemical Biology, McMaster University, 1280 Main Street West, Hamilton, Ontario, Canada L8S 4M1
e-mail: mabrook@mcmaster.ca

F. Ganachaud
Institut Charles Gerhardt UMR5253 CNRS, Equipe «Ingénierie et Architectures Macromoléculaires», Ecole Nationale Supérieure de Chimie de Montpellier,
8 Rue de l'Ecole Normale, Montpellier, Cedex 34296, France
e-mail: francois.ganachaud@enscm.fr

	2.3	Metathesis	167
	2.4	Steric Effects and Thermal Control..	169
	2.5	Complexation of Methylsilicones with $B(C_6F_5)_3$ Does Not Catalyze Redistribution...	170
3	Silicone Synthesis		170
	3.1	Resins	171
	3.2	Silicones with Controlled 3D Structures	172
4	Analogous Reactions with Silanols: Silicone Copolymers		174
5	The Special Case of Water as a (Co)Solvent		176
	5.1	$B(C_6F_5)_3$: A Unique, Water-Tolerant Lewis Acid..................................	176
	5.2	Polymerization in Moist Conditions...	177
	5.3	Silicone Synthesis in Aqueous Medium..	178
6	Other $B(C_6F_5)_3$-Catalyzed Reactions Relevant to Silicones		179
	6.1	Hydrosilylation	179
	6.2	Coordinated Anionic Polymerization...	179
	6.3	Competing Reactions: Epoxide Ring Opening......................................	180
7	Conclusion		181
References			181

1 Introduction

1.1 Traditional Methods for Silicone Chemistry Rely on Ionic Chemistry

Silicones, polymers based on an $\sim R_2SiO\sim$ repeat unit, constitute an important class of materials [1, 2]. Their additional cost, when compared to organic polymers, is justified by the exceptionally useful properties that the materials possess in terms of thermal, electrical, biological, surface, and other behaviors.

These materials have been commercially produced since the 1940s, but little has changed in their manufacture in the intervening time. Silicone synthesis starts with nucleophilic substitution of chlorosilanes by water, leading to low molecular weight linear and cyclic products (Fig. 1). These materials are finished into high molecular weight and functional materials, also using nucleophilic substitution processes that are normally catalyzed by acid or base [2]. Unlike many polymers that are formed under kinetic control, the 'equilibration' process used for silicone manipulation is completely reversible. While high molecular weight silicones are readily formed under acidic or basic conditions, they also undergo degradation under these conditions

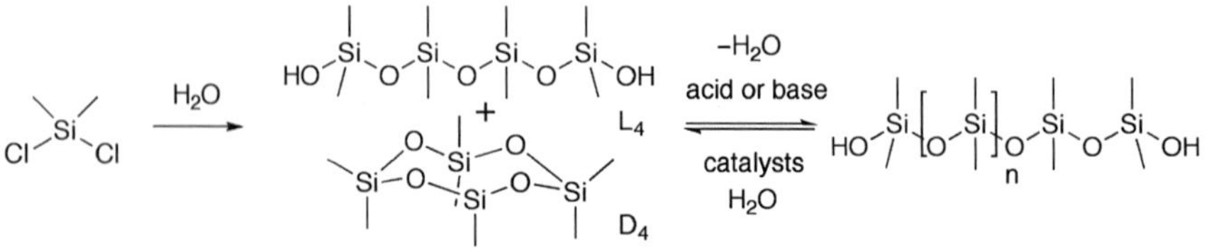

Fig. 1 Conventional industrial silicone synthesis

if water is present, or if low molecular weight cyclics (e.g., D_4, $(Me_2SiO)_4$) are able to leave the system, typically by evaporation.

One consequence of the reactivity of silicones towards acidic or basic conditions is that it is exceptionally difficult to reliably assemble complex silicone structures. Almost inevitably, the reaction conditions used to create a structure also lead to structural fragmentation. Even linear polymers of narrow molecular weight are difficult to prepare on a large scale and in good yield. The very few reports of controlled silicone synthesis, for example, of a silicone dendrimer [3], utilize difficult-to-reproduce and non-generic reaction conditions.

There is a need to create silicones with controlled 3D structures for a variety of purposes. As one example, the class of silicone polymers known as MQ resins [4, 5] exhibit very interesting surface activity and are utilized in high value formulations including liquid crystals [6], antifoaming compositions [7] and pressure sensitive adhesives [8]. There is significant interest for these applications in generating explicit MQ resins and other structures rather than mixtures of silicones, which is the current state of the art. Traditional chemistry does not fulfill these needs in most cases.

1.2 New Chemistry for Silicone Synthesis

In a series of seminal papers, summarized beautifully in an authoritative review [9], Piers demonstrated that the reduction by hydrosilanes of carbonyls (Fig. 2a), thiocarbonyls, imines, and other functional groups could be catalyzed by catalytic (but not insignificant 5–10 mol%) quantities of the hydrophobic Lewis acid $B(C_6F_5)_3$. As part of his careful synthetic and mechanistic studies, Piers noted that over-reduction of the intermediate alkoxysilane could lead to complete reduction to the alkane and the formation of a siloxane by-product (Fig. 2b) [10].

Rubinsztajn and Cella looked at this outcome from the perspective of the silicon chemist. What may be by-products to organic chemists could be the synthetic target for a polymer chemist [11–13]. Thus, the Piers-Rubinsztajn reaction refers to the reaction between a hydrosilane and alkoxysilane that leads, with loss of an alkane, to a siloxane (Fig. 2b). The process can be performed on small molecules or can lead to large, complex structures. In the following sections, we will provide the mechanistic background for the reaction from both the organic and silicone perspectives, and

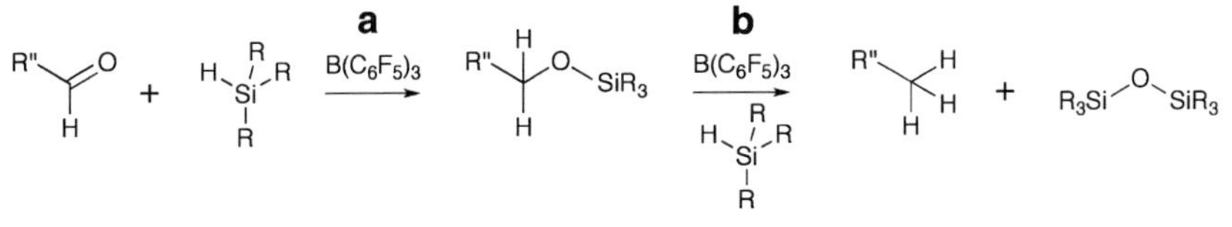

Fig. 2 Origins of the Piers-Rubinsztajn reaction. (**a**) reduction of a carbonyl, is followed by, (**b**) reduction at carbon of the alkoxysilane

then demonstrate the flexibility of the process in the synthesis of silicones and silicone copolymers. In addition, comments will be made about the analogous coupling of hydrosilanes with alcohols and silanols that lead to siloxane bonds. While we have tried to provide an overview of the entire field, we have done so selectively to provide representative examples rather than providing exhaustive referencing. Both the primary and patent literature are presented.

2 Mechanistic Considerations

2.1 *From the Organic World*

$B(C_6F_5)_3$ owes its Lewis acidity to the absence of necessary valence electrons to complete an octet and to the inductive effects provided by the perfluorosubstitution of the three aromatic rings. Most undergraduates when presented with such a highly electrophilic borane **1** would predict that its interactions would be dominated by strong Lewis bases such as amines, alcohols, ethers, etc. (Fig. 3a,c). Parenthetically, we have been to many conferences where experts have espoused this conventional view. However, in addition to the traditional type of Lewis acidity, $B(C_6F_5)_3$ effectively forms complexes with hydrosilanes (Fig. 3b) **2**, which is a much less easily predicted interaction based on Lewis acid theory, given that the silane possesses no Lewis basic electron pairs.

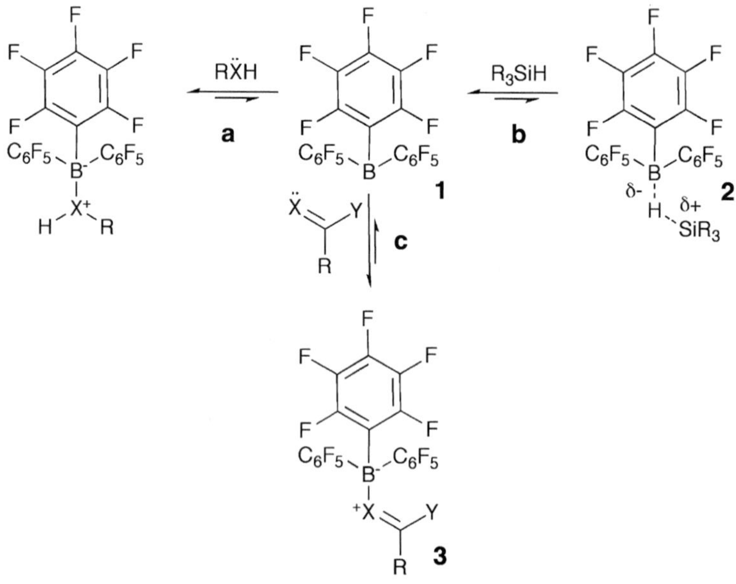

Fig. 3 Lewis acid/base complexes of $B(C_6F_5)_3$ (**a**) with active hydrogen compounds, (**b**) with hydrosilanes, and (**c**) with carbonyl groups and imines

A comparison of $B(C_6F_5)_3$ with other Lewis acids using either the Childs strategy (complexation with crotonaldehyde) [14, 15] or Gutman's acceptor number (AN) [16] – both NMR-based methods – suggests that the Lewis acidity of $B(C_6F_5)_3$ falls between $TiCl_4$ and BF_3. It is thus a very strong Lewis acid. When combined with an appropriate Lewis base, including phosphines or 2,6-lutidine, $B(C_6F_5)_3$ can even split H_2 [17, 18]. In addition to electronic effects, the complexation behavior of $B(C_6F_5)_3$ is particularly sensitive to steric factors. When assessing its reactivity, perhaps the most important factor to consider is the very fast exchange in solution of the Lewis acid/Lewis base complexes of $B(C_6F_5)_3$ when compared to the NMR time scale [9].

Piers examined the equilibria between B and B-carbonyl complexes in a study of C=O hydrosilylation (Fig. 3c, X = O, Y = H, Me, OEt). Equilibrium constants were found to favor the carbonyl complex **3** by about 10^2 depending on the nature of Y [19]. However, the rates of the hydrosilylation reaction were inversely proportional to the carbonyl concentration, suggesting that complexes such as **3** *inhibit* the reaction. These observations led to a mechanistic interpretation involving the reversible formation of a borohydride complex **2**, which is the active species in the subsequent reductive silylation process via **4** or **5**, shown for carbonyl reduction (Fig. 4). Analogous processes have been invoked for the conversion of aldehydes/ketones to alkanes [20], alcohols to alkoxysilanes [21], and then to alkanes [22], and the hydrosilylation of C=C bonds [23].

Further support for this hypothesis comes from studies of the hydrosilylation of imines [24]. Less basic imines were found to undergo hydrosilylation more readily, consistent with the necessity for free $B(C_6F_5)_3$ in solution to complex with the hydrosilane giving **2**. ^{11}B and ^{19}F NMR showed that less basic imines do not form complexes with the boron Lewis acid. Of greater significance, the hydridoborane complex $HB^-(C_6F_5)_3$ of ketimine derivatives could be detected directly using ^{11}B and ^{19}F NMR. The ^{29}Si NMR data of the same mixture showed the presence of a nitrogen/silicon complex analogous to **4** or **5**.

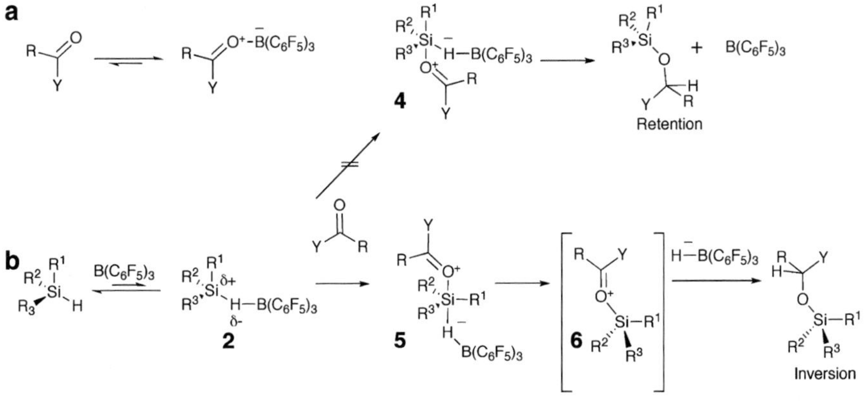

Fig. 4 Proposed mechanism for C=O hydrosilylation catalyzed by $B(C_6F_5)_3$ (**a**) equilibrium with heteroatoms, (**b**) competing equilibrium with hydrosilanes

These data convincingly show that borohydride complexes of $B(C_6F_5)_3$ are key intermediates in the reduction of a variety of oxygen (and nitrogen) based organic functional groups. Piers [10] and others [25] have observed the formation of disiloxanes when large excesses of $HSiR_3$ are present in the reaction mixture. This observation, which is of less interest to synthetic organic chemists, provides a new strategy to synthesize silicones.

The use of enantiomerically pure silanes has permitted a more detailed view of this mechanistic picture. Rendler and Oestreich found that the hydride transfer from the Lewis acid-complexed silane takes place enantioselectively, favoring inversion at silicon [26]. Normally, substitution reactions with hydrosilanes occur with retention (Fig. 4a) [27, 28]. The process of retention has been rationalized to occur through a trigonal bipyramidal complex (either intermediate or transition state) in which the incoming nucleophile is apical, but the only-slightly-polarized hydrogen remains equatorial, such as in 4 [29]. Displacement of hydride by heteroatom nucleophiles thus occurs from the same face of the molecule, giving retention of stereochemistry. In the case of $B(C_6F_5)_3$ catalysis, however, the exceptional Lewis acidity of the highly polarized boron-H complex should favor placement of hydride in the apical position, anti to the incoming carbonyl, see 5 (Fig. 4b). The substitution reaction is thus reminiscent of, or may be, an S_N2 process with inversion of stereochemistry.

Reduction ultimately involves transfer of a hydride to a silyloxonium ion. It remains unclear whether this involves direct reaction of a second borohydride complex 2 with 5, with a tetrahedral oxonium ion 6 or some other process. Irrespective, the data does not support the intermediacy of a free silylium ion intermediate [24].

2.2 Additional Mechanistic Subtleties: The Silicone World

Rubinsztajn and Cella, looking at the results with alkoxysilanes described above through the lens of a silicone chemist, saw not a strategy for reduction of various functional groups including silyl ethers, but rather an opportunity to convert alkoxysilanes (= silyl ethers [30]) into silicones [11]. The breadth of silicone structures that are synthetically available through the use of this reaction will be discussed below. Here, distinctions are made between the previously described organic mechanistic information and the chemistry that occurs at silicon centers.

Chojnowski and coworkers demonstrated that the formation of small siloxanes catalyzed by $B(C_6F_5)_3$ is a second order process. The complexation of the boron to hydrosilane is a rapid equilibrium step that precedes the rate determining step [31]. They proposed a mechanism in which the boron-hydrosilane complex is attacked by the alkoxysilane nucleophile leading to a zwitterionic oxonium ion (Fig. 5a).

The group of Kawakami observed inversion of stereochemistry at the hydrosilane in this process, and retention at the alkoxysilane partner during formation of a disiloxane (Fig. 5b) [32]. While retention at the oxygen bearing silicon is expected, inversion of the hydrosilane is somewhat unexpected as noted above. These processes can be understood by comparing Piers-Rubinsztajn disiloxane formation

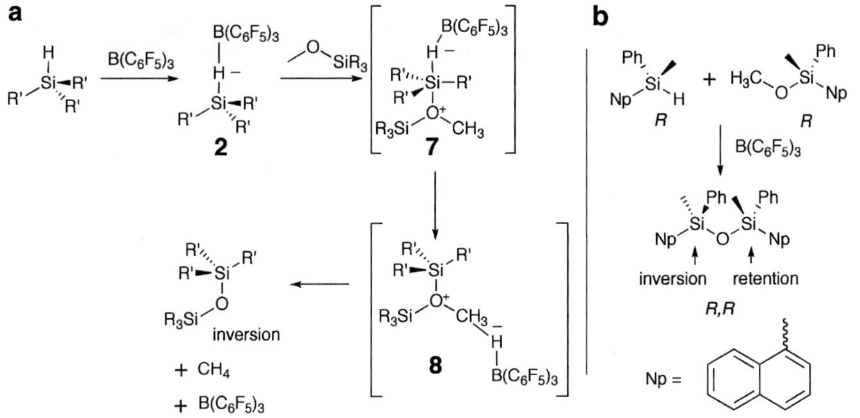

New Synthetic Strategies for Structured Silicones Using $B(C_6F_5)_3$

Fig. 5 (a) Proposed mechanism for dehydrocarbonative condensation. (b) Stereoselective siloxane formation

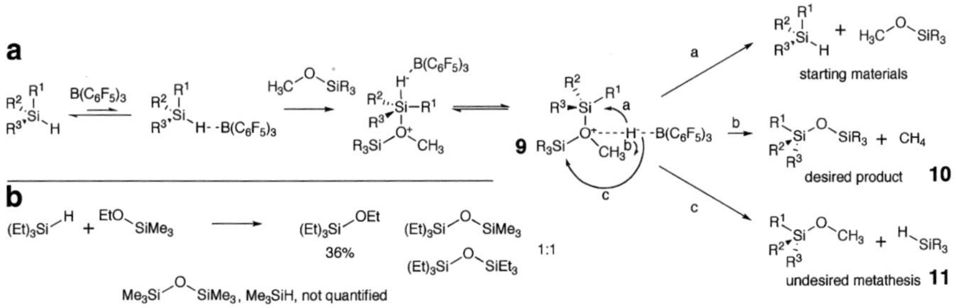

Fig. 6 Metathesis in the Piers-Rubinsztajn process. (a) Mechanistic hypothesis. (b) Example of metathesis (the only reported yield was for Et_3SiOEt; the ratio of disiloxane was determined but yields were not reported)

(Fig. 5a) with the analogous carbonyl hydrosilylation shown in Fig. 4. Formation of a pentacoordinate silyl-H oxonium complex **7** from **2** by addition of the alkoxysilane is followed by substitution at the SiH partner with inversion. Rearrangement to an alkane complex **8** is followed by decomposition to product.

2.3 Metathesis

Unfortunately, the reaction is not quite as simple as implied above. When more than one type of silicon compound is present, metathesis (Fig. 6a–c) can occur in addition to direct disiloxane formation (Fig. 6a,b) [31, 33]. That is, in addition to conversion of hydrosilane to silicone, it can also be converted to alkoxysilane. Chojnowski and coworkers rationalize both processes using a common intermediate. The borohydride-oxonium intermediate **9** can decompose in three distinct ways:

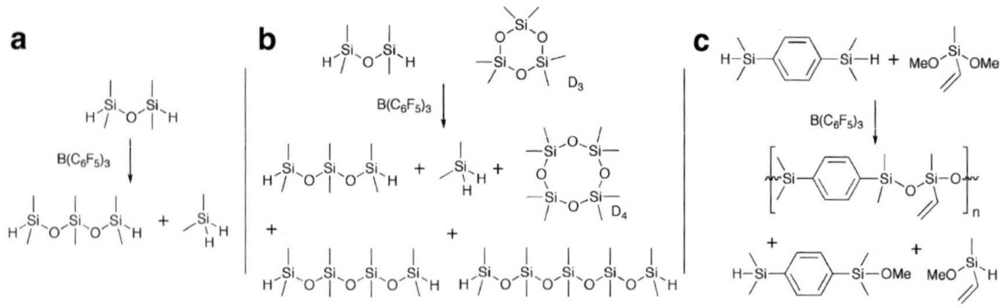

Fig. 7 (a, b, c) Structural manipulation through metathesis of various siloxanes

hydride migration back to starting materials (Fig. 6a); the desired hydride transfer to give the alkane and siloxane **10**; or migration of hydride to the other silyl group leading to metathesis **11**. The metathesis process is clearly disadvantageous as the formation of by-products removes a level of synthetic control from the reaction (Fig. 6b) [31]. As is noted in the silicone synthesis section below, complex resins can be a consequence of metathesis when the Piers-Rubinsztajn process is used with polyfunctional precursors [34].

Metathesis is particularly problematic with the hydrosilane reacting partners. For example, Cella and Rubinsztajn described linear, homologous polymer growth with concomitant loss of Me_2SiH_2 when starting from the simple, small silicone tetramethyldisiloxane (Fig. 7a) [35]: in certain circumstances such a process could be beneficial, but generally will lead to complex mixtures of products [36, 37]. The efficiency with which various silanes participate in such reactions has been determined by Chojnowski and coworkers. An examination of the efficiency of insertion of various silanes into the ring-strained [38] silicone monomer D_3 showed the compounds to exhibit the following order of reactivity: $PhMeSiH_2 > HMe_2SiOSiMe_2H > PhMe_2SiH > HSiMe_2OSiMe_3$ [33]. During the course of the reaction, larger linear and cyclic structures were formed, which increased in molecular weight over time (Fig. 7b).

The efficiency of a metathesis at a given silicon nucleus is affected by the nature of the organic groups found there. Arylsilanes undergo metathesis less efficiently than methylsilanes. Vinylsilanes also undergo metathesis provided that low concentrations ($\sim$1 mol%) of $B(C_6F_5)_3$ are used (Fig. 7c). At higher concentrations of catalyst the desired Piers-Rubinsztajn reactions compete with hydrosilylation of the vinyl group (see below) [39]. Fortunately, metathesis is not always observed.[1]

---

[1] Rubinsztajn, Chojnowski, and I have discussed why, in their hands, metathesis is frequently encountered, while we typically do not observe it. One possible explanation is the order of addition of reagents. Typically, we add catalyst to a mixture of the hydrosilane and alkoxysilane, while Chojnowski and coworkers use one of two alternate strategies (see Sect. 4.1). It is clear, nevertheless, that metathesis does not occur under all reaction conditions.

2.4 Steric Effects and Thermal Control

The Piers-Rubinsztajn reaction is normally initiated by adding $B(C_6F_5)_3$ to a mixture of alkoxysilane and hydrosilane (see, however, Sect. 3.1). There is normally an induction time during which, presumably, the equilibria described above become established. Thereafter, *there is a rapid evolution of flammable (alkane) gas that clearly has safety implications (hydrogen can form from the analogous reactions with silanols or alcohols under these conditions – see below)*. Increased steric bulk at either the hydrosilane or alkoxysilane centers retards the rate of the reaction, but this can be overcome at higher temperatures in most cases.

Steric control in the Piers-Rubinsztajn reaction was demonstrated by the reaction of tetraalkoxysilanes with small model silicones [40]. The unhindered methoxysilane **12** reacted completely with the slightly hindered $HSiMe_2$-terminated silicone **13** at room temperature to give the tetrasubstituted star **14**, as did ethoxy and propoxysilanes (Fig. 8a). The lower yield observed with $Si(OMe)_4$ is a consequence of the high volatility of the starting material and challenges of isolation. In all three cases, pure compounds were isolated simply by removal of residual starting materials at low pressure (and the boron catalyst was removed by adsorption on neutral alumina). $Si(OMe)_4$ also reacted cleanly with the branched hydrosiloxane **15** to give the tetrasubstituted star **16** (Fig. 8b). However, the bulkier $Si(OEt)_4$ did not react at all at room temperature, and could only be forced to the trisubstituted compound **17** at elevated temperatures. Attempts to stop substitution at the disubstituted compound, or to push the reaction thermally to the tetrasubstituted product **16**, were unsuccessful. These data are completely consistent with the intermediacy of oxonium ions like **9**. Steric pressure exerted either by the alkoxy group or the groups adjacent to the SiH group will decrease the efficacy of the reaction. Chojnowski and coworkers note that more bulky silanes, including isopropoxy-derivatives, are yet more selective in reactions and require higher temperatures to initiate reaction [31].

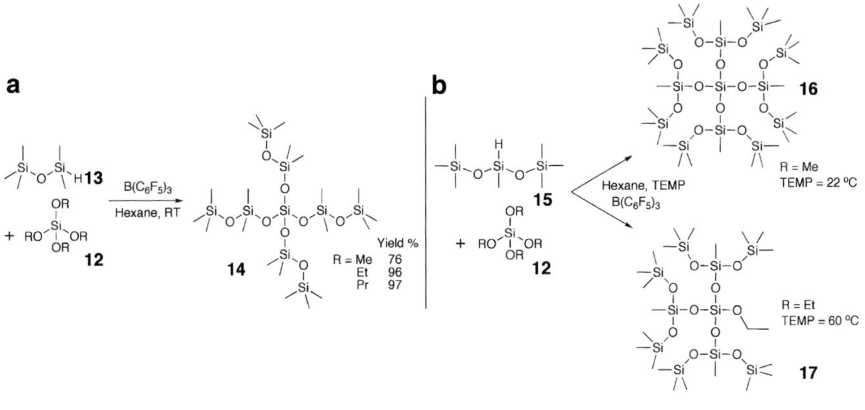

Fig. 8 Steric control over siloxane formation. (**a**) facile reaction with sterically undemanding silicones. (**b**) reduced reactivity when both alkoxysilane and hydrosilicone are encumbered

2.5 Complexation of Methylsilicones with $B(C_6F_5)_3$ Does Not Catalyze Redistribution

Brønsted acids and bases, the most common way to create or manipulate silicone structures, also facilitate degradation of silicones through redistribution, depolymerization processes, or hydrolysis. The presence of Lewis acids can also facilitate depolymerization [41, 42]. However, it is generally understood that a proton is the active agent in these cases [43]. As a consequence, high molecular weight materials can undergo chain scission and 'unzipping,' leading to lower molecular weight materials and well-defined materials that can lose their structure in the presence of acids or bases. Avoiding such acid/base reactions when performing traditional silicone synthesis with chlorosilanes or related materials that generate acid or base is extremely challenging [3].

$B(C_6F_5)_3$, like other Lewis acids, can complex with water, leading to the formation of Brønsted acids [9, 44]. An examination of the structures and reactivities of such structures is examined in detail in Sect. 6.1. Here we note that, while such Brønsted acids could in principle be deleterious to silicone structures, it is our experience and that of others [33] that redistribution of alkylsilicones during $B(C_6F_5)_3$-catalyzed reactions is not problematic. It is, for example, not necessary to dry scrupulously solvents and glassware. Some reactions leading to silicones and other polymers may even be run in water (see below) [45]. Similarly, when considering the synthesis of silicones, while hydrosilanes are more susceptible to Brønsted acid-catalyzed reactions than the alkoxysilane reaction partner, selective reactions can generally be realized by the use of an excess of the alkoxysilane in the reaction mixture.

3 Silicone Synthesis

Rubinsztajin and co-workers reported in 2004 that Piers' observations that alkoxysilanes and organohydrosilanes will react in the presence of $B(C_6F_5)_3$ could be more generally applied to silicone synthesis (Fig. 9) [11, 12, 46]. Very low levels of catalyst were required, typically less than 1 mol%. Under these conditions the condensation process proceeds to produce nearly perfectly alternating siloxane polymers (i.e., no

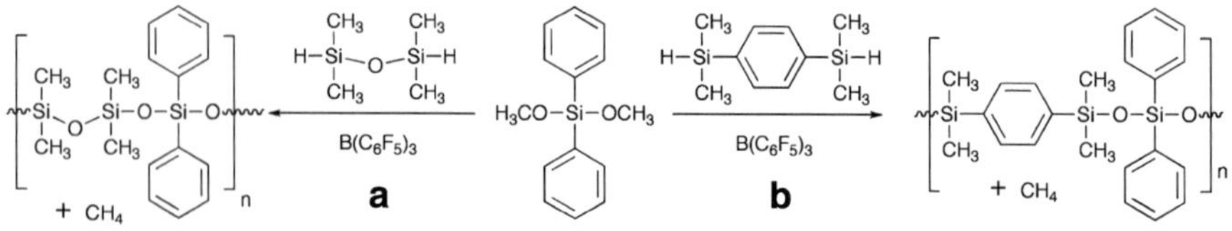

Fig. 9 (a, b) Two examples of polycondensation between diphenyldimethoxysilane and dihydrosilanes catalyzed by $B(C_6F_5)_3$

metathesis) containing phenyl moieties (58–95%), with M_w values of 10–50,000. By contrast, the use of dimethoxydimethylsilane was accompanied by "scrambling" (metathesis) of the silyl groups, which gave rise to more disordered structures. Metathesis was shown to be, among other things, dependant on the nature of the groups on the alkoxysilane reacting partners (see also Sects. 2.3 and 2.4).

3.1 Resins

The Piers-Rubinsztajn reaction has also proven useful as a simple and mild method for the synthesis of highly branched alkoxy substituted polysiloxane resins consisting of a combination of M, D, T, and Q units [5], as noted by Chojnowski and co-workers [34]. Initial experiments with low levels of catalyst had long induction times, followed by very fast, uncontrollable reaction. Two distinct methods were developed to minimize the effect of impurities such as water, to control catalyst concentration, and to manage the highly exothermic nature of the reaction. Both methods used higher concentrations of catalyst to reduce the induction time necessary for reaction. Variant I involves the slow introduction of the hydrosilane to a mixture of alkoxysilane and $B(C_6F_5)_3$ in toluene. Variant II, which involves the addition of both hydrosilane and alkoxysilane to a solution of $B(C_6F_5)_3$ uses yet higher catalyst concentrations, but is performed at lower temperatures, to minimize induction time (Fig. 10a). The polydispersity of the products can be controlled through either manipulation of the silane:alkoxysilane ratio, or through temperature manipulation. For example, use of $-25°C$ as reaction temperature was accompanied both by higher tetraalkoxysilane conversion and a lower PDI of the resin.

More recently, Chojnowski's group has expanded this strategy to the preparation of more highly crosslinked TQ resins. Phenylsilane ($PhSiH_3$):TMOS ($Si(OMe)_4$) were mixed in a molar ratio of less than 0.9 to yield alkoxy-rich products [47]. As with the DQ resins, the TQ resins are free of unwanted hydrophilic silanol groups, which leads to an increase in polymer hydrophobicity, solubility in organic solvents, and long-term stability (Fig. 10b).

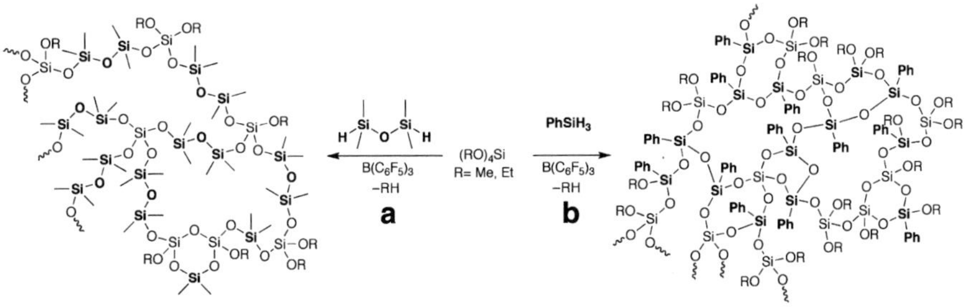

Fig. 10 Generic scheme depicting the synthesis of DQ and TQ resins from (**a**) tetramethyldisiloxane and, (**b**) phenylsilane

During the growth of both of these resins, significant evidence was provided that an undesired, reversible metathetic process was occurring leading to functional group exchange. Mechanistic studies, described in more detail above, convincingly suggest the reaction proceeds through a disilyl-oxonium ion intermediate (similar to Fig. 6) [31]. At low conversion, evidence for exchange, resulting in a decrease in control associated with the condensation process, was provided by ^{29}Si NMR spectroscopy and GC analysis: new reactants, bearing two Si–H groups, two Si–OR groups, or a mixture of the two, were formed. This leads, for example in the case of DQ [5] resins, to unequal distribution of D units in the product polysiloxanes, rather than perfectly alternating copolymers.

3.2 Silicones with Controlled 3D Structures

With appropriate control of the reaction conditions, and in particular by exploiting the rules of steric control described in Sect. 2.4, the Piers-Rubinsztajn reaction can be employed to synthesize well-defined complex 3D siloxane architectures. Small, readily available alkoxysilanes and silicones can be rapidly assembled into MDTQ [5] silicones (Fig. 8): in our work, metathesis was not observed [40]. These structures would be exceptionally difficult to synthesize using traditional means because of the susceptibility of siloxane bonds to strong electrophiles and nucleophiles, especially in the presence of water, and the ability of silicones to undergo redistribution under acidic or basic conditions.

A key intermediate for the synthesis of large structures was **17**. Although it is a Q [5] unit, it possesses a single residual ligand that will react in the presence of $B(C_6F_5)_3$. Only non-branched, terminal hydrosilanes are small enough to react with the ethoxysilane. Thus, linear or highly branched silicones are readily available by the rapid assembly of hydrosilane building blocks (Fig. 11).

Many interesting applications of silicones exploit their unusual surface activity. Mixtures of silicones and silica are used as defoaming agents [48]; MQ [5]

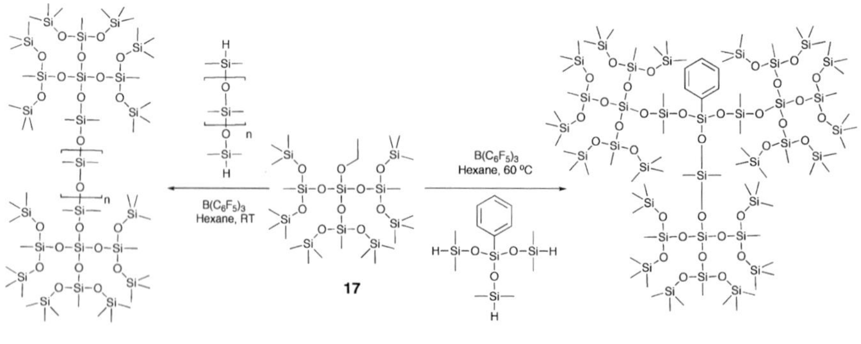

Fig. 11 Synthesis of nearly perfect 3D silicone structures

resins can destabilize foams as well [7]. More interesting, in terms of surface activity, are functional silicones including copolymers with polyethers, liquid crystals, and others, which rely for their properties, in part, on organic groups or polymers chains. Piers' studies on the mechanism of B(C$_6$F$_5$)$_3$-catalyzed hydrosilylation provide clues about the types of functional groups that will be tolerated during the synthesis of functional silicones [10, 24]. Very basic ligands will shut the process down by irreversible complexation with boron [9]. In addition, hydrosilylation of alkenes has been reported at high [23], but not low [39, 49], boron catalyst concentrations. Otherwise, little was known about the functional group tolerance of the reaction.

As was expected from Piers' results [21], even primary amines suppressed the reactivity of the Piers-Rubinsztajn reaction (Fig. 12a) [50]. Softer and less basic functional groups such as thiols participated in the reaction, but not exclusively (Fig. 12b): hydrosilanes show approximately equal reactivity towards alkoxysilanes and thiols in the presence of B(C$_6$F$_5$)$_3$ (see also silanol and alcohol coupling with hydrosilanes in Sect. 5). Attempts to direct the reaction exclusively to one or other functional groups have not yet been successful. A broad series of alkyl halides, including chlorides and iodides, and of allyl- and vinylsilanes (Fig. 12c,d), reacts cleanly and exclusively at the alkoxysilane. Normally, B(C$_6$F$_5$)$_3$ loadings of up to 5 mol% are required to facilitate hydrosilylation [23]. However, α,β-unsaturated carbonyl compounds behaved quite differently (Fig. 12e). Even at low catalyst concentrations (<0.5%), C=C hydrosilylation dominates the reaction process: both regioisomers were isolated, but the alkoxysilane was untouched. Also of interest is the fact that silylation of the carbonyl group (either 1,2- or 1,4-hydrosilylation), which often occurs in the presence of transition metal catalysts [51], was not observed. These observations broaden the applicability of the Piers-Rubinsztajn reaction, as they demonstrate it is compatible with a variety of organic transformations that can be used to functionalize silicones or convert them to copolymers.

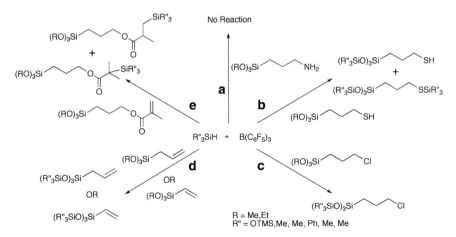

Fig. 12 The functional tolerance of the Piers-Rubinsztajn reaction. Reactions with (**a**) amines, (**b**) thiols, (**c**) alkyl halides, (**d**) alkenes, (**e**) α,β-unsaturated carbonyl compounds

4 Analogous Reactions with Silanols: Silicone Copolymers

A series of papers has reported that $B(C_6F_5)_3$ will also catalyze the related reactions of silanols [52] or alcohols with hydrosilanes to give alkoxysilanes (Fig. 13b) [53, 54]. Traditionally, such reactions are promoted by transition metals including tin- [55] and platinum-based catalysts (Fig. 13a), although such catalysts are not always necessary [56]. When transition metals are present, platinum and rhodium in particular, hydrogenation of any alkenes (or alkynes) present may occur using the hydrogen that spontaneously evolves during the process. The alternative boron-based catalyst system thus offers an advantage because, except in the absence of complementary Lewis bases [57, 58], hydrogenation of alkenes will not take place provided the catalyst concentration is kept low.

Tradition and experience in the silicone industry has led to an understanding that polymeric structures with embedded SiOC linkages are inherently unstable and, therefore, 'unsatisfactory.' However, there is now a large literature demonstrating that hydrolytic instability of alkoxysilanes can be overcome by use of appropriate steric congestion at silicon [59, 60]. In addition, certain SiOC linkages are intrinsically kinetically stable, including phenoxysilanes. Moreover, in some cases, the ultimate use of the material will occur in anhydrous environments, such as battery electrolyte carriers, in which case concerns about hydrolysis are avoided [53]. Finally, silicone recycling might be aided by the addition of appropriate catalysts to facilitate hydrolysis of SiOC linkages.

As a consequence of the potential utility of materials that contain an SiOC linkage, it is not surprising that several developments in the preparation of surface active, silicone-grafted polyethers have been reported in a series of publications and patents. The processes start with poly(ethylene glycol) or copolymers with poly(propylene oxide) that are typically capped on one end with a methyl group [61–63]. The residual alcohol can react with silicone polymers bearing terminal, or pendant SiH groups: the silicones and polyethers necessary for these reactions are commercially available in many molecular weights and SiH contents (Figs. 13b and 14a,b [53, 54, 64]).

The enhanced thermal stability that in-chain aromatic groups bring to polymer chains has been exploited in many contexts in silicone chemistry. The ability of the aromatic ring to self-assemble can lead to hard blocks, while the silicones provide exceptionally soft blocks. In an elegant study, Cella and Rubinsztajn have described the preparation of a series of such siloxane-modified aryl polymers starting from phenols, based on both 'soft' tetramethyldisiloxane and longer silicones, and 'hard' bis-silylaryl compounds, using catalysis by $B(C_6F_5)_3$. Polymers

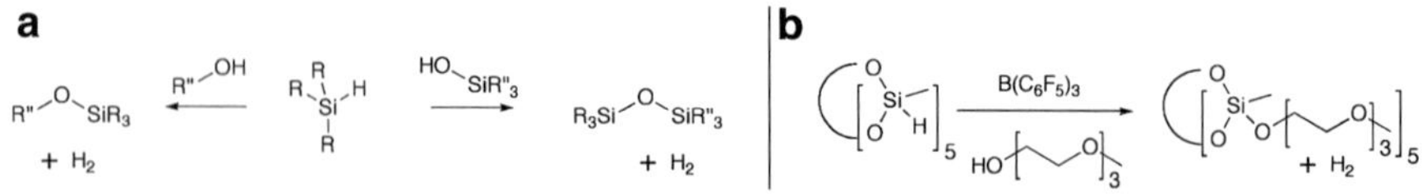

Fig. 13 (a, b) $B(C_6F_5)_3$-catalyzed condensation of silanols or alcohols with hydrosilanes

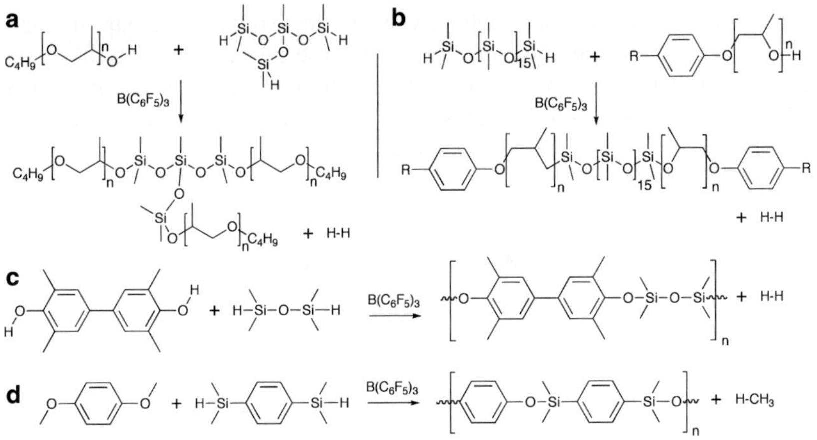

New Synthetic Strategies for Structured Silicones Using B(C$_6$F$_5$)$_3$ 175

Fig. 14 Examples of SiOC-containing polymers prepared using B(C$_6$F$_5$)$_3$. (**a**) poly(propylene oxide) derivatives, (**b**) poly(propylene oxide) derivatives with aryl end groups, (**c**) biaryl, and, (**d**) hydroquinone derivatives

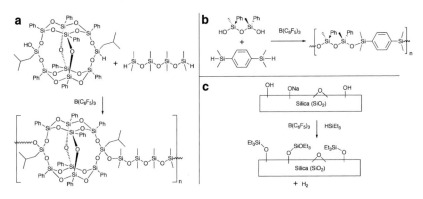

Fig. 15 Copolymer or surface capping using B(C$_6$F$_5$)$_3$-catalyzed condensation of silanols or alcohols with hydrosilanes. (**a**) modification of double-decker siloxanes, (**b**) bis-silylphenyl derivatives, (**c**) silica surfaces

with molecular weights from ca. 14,000 to 105,000 with T$_g$s ranging from 44 to 98 °C were prepared (Fig. 14c) [36]. Related processes were described starting from vinyl-containing alkoxysilanes, which may involve relatively simple [39] or much more complicated copolymers [49]. Another interesting outcome was the discovery that, analogous to the alkoxysilane reactions described above, methoxyphenol compounds will undergo reductive demethylation leading to aryloxysilanes (Fig. 14d), a process that can be exploited in polymer synthesis [36].

The combination of silanes and silanols is not limited to simple methylsilicones or arylsilicones but can be extended to much more complicated systems. Kawakami and coworkers described the synthesis of complex 'double decker' siloxanes and their alternating copolymerization with a dimethylsilicone chain (Fig. 15a) [65]. They have also shown that asymmetrical copolymers can be prepared enantioselectively using this strategy (Fig. 15b) [66].

Silica surfaces are often hydrophobized with Me_3Si or related groups to improve reinforcement in silicone polymers. Typically, base-catalyzed (which may include the surface itself) reaction with chlorosilanes completes this conversion. A related SiOH + HSi coupling process catalyzed by $B(C_6F_5)_3$ has been utilized to create hydrophobic silicone surfaces (Fig. 15c) [67].

5 The Special Case of Water as a (Co)Solvent

5.1 $B(C_6F_5)_3$: A Unique, Water-Tolerant Lewis Acid

The interaction of $B(C_6F_5)_3$ with water has been studied in the bulk and in different solvents (toluene d_8, acetonitrile) principally by ^1H and ^{19}F NMR. The generation of boron catalyst:water complexes with 1, 2, or 3 molecules of water inside the sphere of the molecule could be observed by simple titration [68]. Two of these three water molecules are found in the inner-sphere, whereas only one is B-bound, the two others being H-bound. The most abundant neutral complex in conventional non-dried conditions, $H_2O·B(C_6F_5)_3·2H_2O$, undergoes deprotonation in the presence of strong bases (e.g., KOH/crown ether [69] or a bulky tertiary amine [44]), to give a salt, $[HO·B(C_6F_5)_3]^-K^+·(crown)$, which in the bulk naturally crystallizes together with the adduct $H_2O·B(C_6F_5)_3$. The Brønsted acidity of this intermediate is rather weak, with a pKa value close to that of HCl in acetonitrile (pKa = 8.6). This fact explains why redistribution of silicones is not typically observed in reactions catalyzed by $B(C_6F_5)_3$ when adventitious water is present. It also explains the relative exchange rate of B-bound water molecule against acetonitrile molecules, 300:1, showing that the $B(C_6F_5)_3$:water complex is highly labile in the presence of other Lewis bases.

The mechanistic complexity in $B(C_6F_5)_3$-catalyzed reactions increases significantly when water is present. In wet solution, that is, in the presence of excess water compared to catalyst, or in aqueous media, both the Lewis acidity (of the free catalyst) and the Brønsted acidity (of the water bound complex) must be taken into account. In addition, the Si–H moiety is very sensitive to water in the presence of the catalyst and is converted into Si–OH with concomitant release of hydrogen gas (Fig. 16). Alkoxy groups can also hydrolyze in the presence of $B(C_6F_5)_3$ to generate silanol groups, but this reaction is less exothermic and occurs at a considerably slower rate than the reaction of Si–H groups. The silanol function can then react with the HSi group through a Piers-Rubinsztajn related process to generate a new siloxane bond (this is sometimes used as a powerful technique to dry a vessel, for instance using a hydrosilane similar to the one quoted in Fig. 15c). The silanol-silanol condensation does not occur in solution, but slow condensation reactions can be observed in aqueous media (see below).

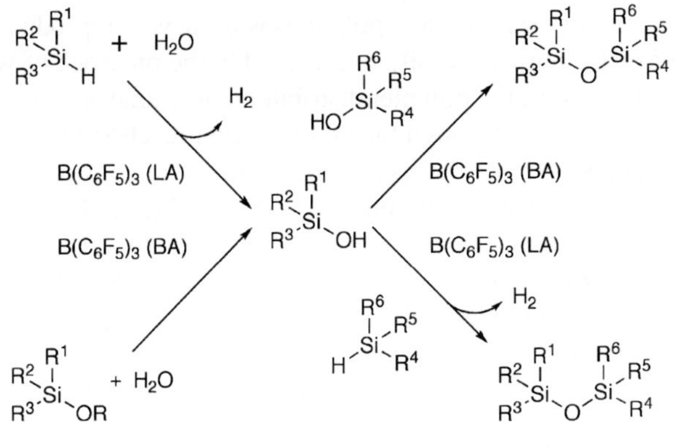

New Synthetic Strategies for Structured Silicones Using B(C$_6$F$_5$)$_3$

Fig. 16 Piers-Rubinsztajn-associated side reactions in the presence of water. LA: Lewis acid, BA: Brønsted acid

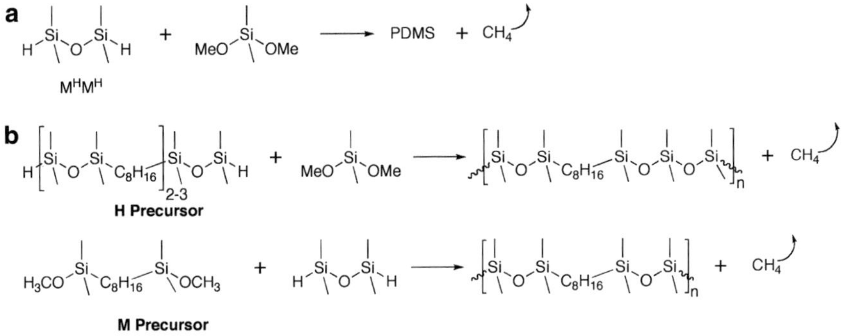

Fig. 17 Piers-Rubinsztajn reaction starting from various precursors. (**a**) Synthesis of PDMS from M^{H}M^H and dimethoxydimethylsilane. (**b**) Synthesis of hydrogenated hybrid silicones by two complementary pathways

5.2 Polymerization in Moist Conditions

Using conditions complementary to those of Chojnowski, Ganachaud and coworkers carried out model studies of the Piers-Rubinsztajn reaction in non-dried solvent conditions using simple bifunctional alkoxysilane and hydrosilane molecules to generate poly(dimethylsiloxane) (PDMS) chains (Fig. 17a) [70]. M^{H}M^H [5] could be added either in one shot or by several aliquots to the solvent/Me$_2$Si(OMe)$_2$ mixture that also contained a low concentration of B(C$_6$F$_5$)$_3$ (0.1 mol% compared to monomers). The main observation from these preliminary experiments was that a large content of cyclic molecules were initially formed in these conditions, and that the D$_4$–D$_x$ [5] content remained constant with time. Small linear polymer chains were also prepared under these conditions, although the process with respect to product ratio and molar mass of linear polymers was not very reproducible. Also, the fact that some Si–H groups are easily hydrolyzed in the presence of water may explain the important shift of the polymer distribution over time on a long run. Indeed, molecular weight increases through the condensation reaction between silanol terminated polymers and $M^H M^H$ chains (analogous to the Piers-Rubinsztajn reaction) were already apparent in blank experiments. NMR and MALDI mass spectrometric techniques both showed that chains bearing Si–OCH$_3$ and Si–OH moieties coexisted in the mixture. Note that, in these wet conditions, silanol-silanol condensation does not occur, demonstrating the low Brønsted acidity of the $B(C_6F_5)_3$:water complex when a low concentration of water is present in the medium.

We also prepared hybrid alkane-co-silicone polymers from precursor blocks containing Si–H or Si–OMe moieties, respectively, and a simple disiloxane or disilane, again in wet conditions (Fig. 17b) [70]. The best results were obtained using the precursor **M**, since the $M^H M^H$ co-reactant crossreacted readily with the methoxy precursor without undergoing self-condensation. Triple detection size exclusion chromatography revealed that, in toluene, large macrocycles of alternating alkyl and silicone moieties resulted from dominant end-biting reactions. This result was confirmed by the total absence of chain-end peaks in the ^{29}Si NMR spectra. The best strategy to generate longer linear chains would be to condense together both **M** and **H** precursors (Fig. 17b), work that remains to be done.

5.3 Silicone Synthesis in Aqueous Medium

Analogous dihydrosiloxane-dimethoxysilane reactions were carried out in an aqueous medium [45]. To that end, $M^H M^H$ (1 mol equiv) was added drop by drop to an aqueous solution of diethoxysilane (1 equiv) and $B(C_6F_5)_3$ (0.01 equiv) [31]. Using a stoichiometric equivalent of each, stable suspensions of long PDMS polymer chains with low contents of cyclics (D$_3$, D$_4$, D$_5$ [5], etc.) were generated. Initially, significant quantities of D$_3$ formed under these conditions, a compound that was not seen in the solution process. These strained cycles were nevertheless reopened by $B(C_6F_5)_3$, leading to the synthesis of higher molecular weight polymers. Condensation reactions between SiOH and SiH or SiOH groups occurred together under these conditions to generate linear silanol-terminated polymer chains of large molar mass. It was noted that the consumption of polar silanol groups, mainly located at the droplet interface, was accompanied by a decrease of the colloidal stability of the emulsions, in agreement with recent results published by Vincent and coworkers [71].

Of most interest is the specificity of the process. When conventional, strong Brønsted acids (e.g., HCl) were used under the same conditions, hydrosilane + ethoxysilane did not generate polymers. Each individual reactant gave reactions with $B(C_6F_5)_3$ that were too fast (Si–H molecules) or too slow (ethoxy molecules) to be used on their own to create PDMS polymers. Thus, cross-condensation is the preferred and controlled reaction, and leads in high yield to silicone polymers.

6 Other B(C₆F₅)₃-Catalyzed Reactions Relevant to Silicones

We have focused in this review on the ways in which B(C₆F₅)₃ can facilitate silicone and silicone co-polymer formation through the synthesis of siloxane bonds. It should be noted that the same catalyst can promote other reactions relevant to the field of silicones. Three examples are summarized briefly here.

6.1 Hydrosilylation

It was noted above that synthesis of silicones, including those possessing vinyl-containing silane groups, readily occurs from SiH and SiOR in the presence of B(C₆F₅)₃ (Fig. 7c). Note that the concentration of catalyst required for the Piers-Rubinsztajn reaction is typically less than 1 mol%, a concentration at which hydrosilylation is not observed. However, at higher concentrations of the catalyst it is possible to perform hydrosilylation [23]. This could be quite useful, as it may be possible first to assemble the silicone without crosslinking, and then to crosslink the elastomer simply by adding more B(C₆F₅)₃ (Fig. 18a). Regiocontrol strongly favors the introduction of the silyl group on the least hindered carbon.

6.2 Coordinated Anionic Polymerization

Allylic or vinyl molecules bearing strong electron-withdrawing (EWD) substituents such as sulfones or phosphonated groups behave differently from other olefin monomers in the presence of SiH/B(C₆F₅)₃ complexes. Indeed, the generation of an intermediate carbanion from the transfer of the hydride to the activated monomer gives rise to a fast chain oligomerization through an 'ate process,' until irreversible end-capping occurs (Fig. 18b) [72]. Such reactions only proceed in the total absence

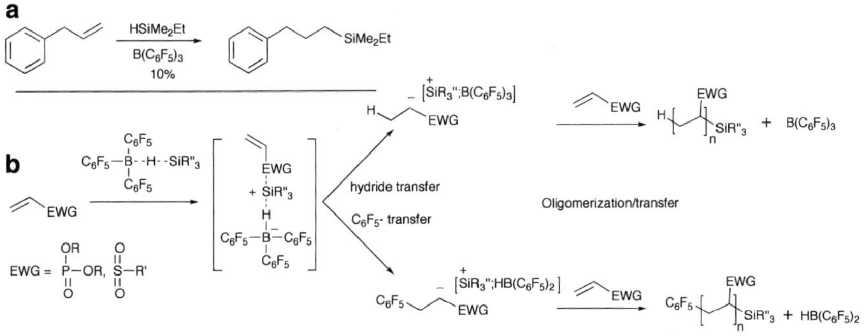

Fig. 18 (a) Example of aromatic allyl hydrosilylation and (b) ate-type oligomerization of EWD monomers catalyzed by B(C₆F₅)₃

of water, although water is frequently found in solutions of these polar monomers. The removal of water can be accomplished simply by adding a hydrosilane to the system, which converts the water to H_2 and a disiloxane (triisobutylaluminum can also be used as a drying agent, but with *caution*).

The fact that the polar monomers polymerize in a Markovnikov fashion may be ascribed to the strong withdrawal of electrons from the double bond, rendering the α carbon more electronegative than the β one. This method allows one to prepare polar block or grafted silicone copolymers of unique properties, e.g., thermoplastic elastomers [73].

6.3 Competing Reactions: Epoxide Ring Opening

An entire class of polymers is based on epoxide ring opening. Epoxy groups are readily introduced onto silicones, frequently through the hydrosilylation of allyl glycidyl ether with hydrosilanes or hydrosilicones. Epoxy groups undergo acid- or base-catalyzed ring opening: oligoamine curing agents are normally used for this process.

Epoxy-modified silicones were prepared by selective monohydrosilylation of H_2MeSi-silphenylenes. The remaining hydrosilane participated in $B(C_6F_5)_3$-catalyzed coupling with silanol-terminated silphenylenes to give compounds such as **18** (Fig. 19a) [55]. These materials underwent normal crosslinking processes with diamines. Notable is the absence of epoxy ring-opening in the presence of $B(C_6F_5)_3$. By contrast, Ganachaud and coworkers demonstrated that in water $B(C_6F_5)_3$ is sufficiently acidic to catalyze epoxide ring opening, which was used to create grafted poly(vinyl alcohol)-silicone copolymers (Fig. 19b) [74].

This example demonstrates the catalytic subtlety of the hydrophobic Lewis acid $B(C_6F_5)_3$. Careful control of reaction partners SiOH, SiOR, and R'_3SiH (including steric bulk of R and R') and reaction conditions, including catalyst concentration and the presence of water, allow for very high levels of control in assembling pure silicones or, alternatively, silicone/organic materials.

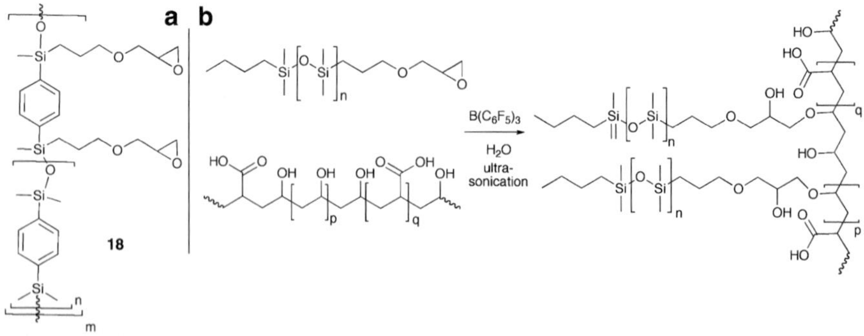

Fig. 19 (a) Epoxy-functional silicones and (b) their reaction with poly(vinyl alcohol)

7 Conclusion

The Piers-Rubinsztajn reaction provides a convenient and practical alternative to traditional nucleophilic substitution reactions catalyzed by acids and bases. The $B(C_6F_5)_3$-catalyzed coupling of alkoxy- and hydrosilanes, or the analogous coupling reactions of silanols and hydrosilanes, provide new, selective methods to create complex siloxane structures that can optionally bear synthetic organic entry points. A broad range of linear, highly branched, and explicit 3D structures are available from inexpensive readily available starting materials. It is likely that these processes will increasingly be exploited, including in commercial settings. In addition, one may expect many similarly useful reactions based on silicon chemistry and the unusual properties of $B(C_6F_5)_3$ to be discovered.

Acknowledgements We gratefully acknowledge the financial support of the Natural Sciences and Engineering Research Council of Canada (NSERC), Silcotech Canada, Siltech Canada, and Centre National de la Recherche Scientifique (CNRS). We also thank Prof. Alan Bassindale (Open University, UK), Prof. Warren Piers (Calgary), and Prof. Martin Oestreich (Münster) for helpful discussions.

References

1. Clarson SJ, Semlyen JA (1993) Siloxane polymers. PTR Prentice Hall, Englewood Cliffs, NJ
2. Noll WJ (1968) Chemistry and technology of silicones. Academic Press, New York
3. Uchida H, Kabe Y, Yoshino K, Kawamata A, Tsumuraya T, Masamune S (1990) J Am Chem Soc 112:7077–7079
4. Brook MA (2000) Organosilanes: where to find them, what to call them, how to detect them. In: Silicon in organic, organometallic and polymer chemistry. Wiley, New York, pp 1–26
5. General Electric silicone nomenclature: $M = Me_3SiO_{1/2}$, $D = Me_2SiO_{2/2}$, $T = MeSiO_{3/2}$, and $Q = SiO_{4/2}$. The subscript nomenclature is used to denote, for example with $SiO_{4/2}$, that there are four single bonds to oxygen from silicon, and that each oxygen bonds to another silicon through a single bond, i.e., $Si(OSi)_4$ rather than SiO_2, which might imply Si=O double bonds.
6. Ganicz T, Pakula T, Stanczyk WA (2006) J Organomet Chem 691:5052–5055
7. Araud C (1992) Polydimethylsiloxane resin antifoaming compositions. US 5,082,590, Rhone-Poulenc Chimie
8. Ulman KL, Thomas X (1995) Silicone pressure sensitive adhesives for healthcare applications. In: Satas D (ed) Advances in pressure sensitive adhesive technology, vol 2. Satas, Warwick RI, pp 133–157
9. Piers WE (2005) The chemistry of perfluoroaryl boranes. In: Advances in organometallic chemistry, vol 52. Elsevier Academic Press, San Diego, pp 1–76
10. Parks DJ, Blackwell JM, Piers WE (2000) J Org Chem 65:3090–3098
11. Rubinsztajn S, Cella J (2004) Polymer Prepr 45(1):635–636
12. Rubinsztajn S, Cella JA (2005) Silicone condensation reaction. European Patent Application, WO2005118682, General Electric
13. Rubinsztajn S, Cella JA (2006) Silicone condensation reaction. US 7064173, General Electric
14. Childs RF, Mulholland DL, Nixon A (1982) Can J Chem 60:801–808
15. Childs RF, Mulholland DL, Nixon A (1982) Can J Chem 60:809–812
16. Beckett MA, Brassington DS, Coles SJ, Hursthouse MB (2000) Inorg Chem Commun 3:530–533

17. Stephan DW (2009) Dalton Trans 3129–3136
18. Stephan DW (2008) Org Biomol Chem 6:1535–1539
19. Parks DJ, Piers WE (1996) J Am Chem Soc 118:9440–9441
20. Chandrasekhar S, Reddy CR, Babu BN (2002) J Org Chem 67:9080–9082
21. Blackwell JM, Foster KL, Beck VH, Piers WE (1999) J Org Chem 64:4887–4892
22. Gevorgyan V, Rubin M, Benson S, Liu JX, Yamamoto Y (2000) J Org Chem 65:6179–6186
23. Rubin M, Schwier T, Gevorgyan V (2002) J Org Chem 67:1936–1940
24. Blackwell JM, Sonmor ER, Scoccitti T, Piers WE (2000) Org Lett 2:3921–3923
25. Gevorgyan V, Rubin M, Liu JX, Yamamoto Y (2001) J Org Chem 66:1672–1675
26. Rendler S, Oestreich M (2008) Angew Chem Int Ed 47:5997–6000
27. Bassindale AR, Taylor PG (1989) Reaction mechanisms of nucleophilic attack at silicon. In: Patai S, Rappoport Z (eds) The chemistry of organic silicon compounds, vol 1. Wiley, Chichester, UK, p 839
28. Bassindale AR, Glyne SJ, Taylor PG (1998) Reaction mechanisms of nucleophilic attack at silicon. In: Rappoport Z, Apeloig Y (eds) The chemistry of organic silicon compounds, vol 2. Wiley, Chichester, UK, p 495
29. Corriu RJP, Guérin C, Moreau JJE (1984) Stereochemistry at silicon. In: Eliel EL, Wilen SH, Allinger NL (eds) Topics in stereochemistry, vol 15. Wiley, New York, pp 43–198
30. Brook MA (2000) Replacing H with Si: silicon-based reagents. In: Silicon in organic, organometallic and polymer chemistry. Wiley, New York, pp 189–255
31. Chojnowski J, Rubinsztajn S, Cella JA, Fortuniak W, Cypryk M, Kurjata J, Kazmierski K (2005) Organometallics 24:6077–6084
32. Shinke S, Tsuchimoto T, Kawakami Y (2005) Silicon Chem 3:243–249
33. Chojnowski J, Rubinsztajn S, Fortuniak W, Kurjata J (2007) J Inorg Org Polym Mater 17:173–187
34. Chojnowski J, Rubinsztajn S, Fortuniak W, Kurjata J (2008) Macromolecules 41:7352–7358
35. Chojnowski J, Fortuniak W, Kurjata J, Rubinsztajn S, Cella JA (2006) Macromolecules 39:3802–3807
36. Cella J, Rubinsztajn S (2008) Macromolecules 41:6965–6971
37. Rubinsztajn S, Cella JA, Chojnowski J, Fortuniak W, Kurjata J (2006) Process for synthesis of diorganosilanes by disproportionation of hydridosiloxanes. US 7148370, General Electric
38. Brook MA (2000) Silicones. In: Silicon in organic, organometallic and polymer chemistry. Wiley, New York, pp 256–308
39. Xunjun C, Yingde C, Guoqiang Y, Liewen L (2007) J Appl Polym Sci 106:1007–1013
40. Thompson DB, Brook MA (2008) J Am Chem Soc 130:32–33
41. Sigwalt P (1987) Polym J 19:567–580
42. Sigwalt P, Nicol P, Masure M (1989) Makromol chem. Supp 15:15–30
43. Jordan E, Lestel L, Boileau S, Cheradame H, Gandini A (1989) Makromol Chem Phys 190:267–276
44. Bergquist C, Bridgewater BM, Harlan CJ, Norton JR, Friesner RA, Parkin G (2000) J Am Chem Soc 122:10581–10590
45. Longuet C, Joly-Duhamel C, Ganachaud F (2007) Macromol Chem Phys 208:1883–1892
46. Rubinsztajn S, Cella JA (2005) Macromolecules 38:1061–1063
47. Kurjata J, Fortuniak W, Rubinsztajn S, Chojnowski J (2009) Eur Polym J 45:3372–3379
48. Owen MJ (1990) Siloxane surface activity. In: Zeigler JM, Fearon FWG (eds) Silicon-based polymer science: a comprehensive resource. American Chemical Society, Washington, D.C., pp 705–739
49. Chen X, Cui Y, Yin G, Jia Z, Liu Z (2008) Huagong Xuebao (Chinese edn), vol 59, pp 1143–1149
50. Grande JB, Thompson DB, Gonzaga F, Brook MA Controlled geometry functional silicones (in press)
51. Brook MA (2000) Formation of Si–C bonds: the synthesis of functional organosilanes. In: Silicon in organic, organometallic and polymer chemistry. Wiley, New York, pp 381–458
52. Li YN, Kawakami Y (1999) Macromolecules 32:6871–6873

53. Zhang Z, Lyons LJ, Jin JJ, Amine K, West R (2005) Chem Mater 17:5646–5650
54. Neumann T, Herrwerth S, Reibold T, Krohm H-G (2006) Solvent-free reaction of hydrosilyl-containing branched polyorganosiloxanes with alcohols. German Patent Application DE 102005004676 Goldschmidt AG
55. Xue L, Kawakami Y (2007) Polym J 39:379–388
56. Brook MA (2000) Hydrosilanes as reducing agents. In: Silicon in organic, organometallic and polymer chemistry, Wiley, New York, pp 171–188
57. Chase PA, Welch GC, Jurca T, Stephan DW (2007) Angew Chem Int Edit 46:9136
58. Geier SJ, Stephan DW (2009) J Am Chem Soc 131:3476–3477
59. Wuts PGM, Greene TW (2006) Greene's protective groups in organic synthesis, 4th edn. Wiley-Interscience, New Jersey
60. Thompson DB, Gonzaga F, Fawcett AS, Brook MA (2008) Silicon Chem 3:327–334
61. Knott W, Droese J, Klein K-D, Landers R, Windbiel D (2008) Method for manufacturing SiOC-linked, linear polydimethyl siloxane polyoxyalkyl block copolymers and their application. EP 1935922, Evonik Goldschmidt Gmbh
62. Neumann T, Knott W (2008) Method for converting polyorganosiloxanes and their application. EP 1935920, Evonik Goldschmidt
63. Henning F, Knott W, Dudzik H (2009) Method for producing branched SiH functional polysiloxanes and the use thereof for producting SiC- and SiOC-linked, branched organomodified polysiloxanes. DE 102007055485, Evonik Goldschmidt
64. Oestreich S, Scheiba M, Stadtmueller S, Weimann M (2006) Use of organo-modified siloxanes for improving the surface properties of thermoplastic elastomers. EP 1640418, Goldschmidt GMBH
65. Hoque MA, Kakihana Y, Shinke S, Kawakami Y (2009) Macromolecules 42:3309–3315
66. Zhou DQ, Kawakami Y (2005) Macromolecules 38:6902–6908
67. Casty GL, Rodriguez G (2009) Preparation of supported silyl-capped silica-bound anion activators and associated catalysts. US Patent Application 20090018290, ExxonMobil Chemical Company
68. Di Saverio A, Focante F, Camurati I, Resconi L, Beringhelli T, D'Alfonso G, Donghi D, Maggioni D, Mercandelli P, Sironi A (2005) Inorg Chem 44:5030–5041
69. Danopoulos AA, Galsworthy JR, Green MLH, Cafferkey S, Doerrer LH, Hursthouse MB (1998) Chem Commun 2529–2530
70. Longuet C, Ganachaud F (2008) Copolycondensation of functional silanes and siloxanes in solution using tris(pentafluorophenyl) borane as a catalyst in a view to generate hybrid silicones. In: Ganachaud F, Boileau S, Boury B (eds) Silicon based polymers. Springer, Netherlands, pp 119–134
71. Neumann B, Vincent B, Krustev R, Muller HJ (2004) Langmuir 20:4336–4344
72. Pouget E, Holgado-Garcia E, Vasilenko IV, Kostjuk SV, Campagne JM, Ganachaud F (2009) Macromol Rapid Comm 30:1128–1132
73. Pouget E, Ganachaud F, Boutevin B, Loubat C (2008) Silicone elastomer made by grafting hydrogen-polyorganosiloxane with alkenyl sulfone using optionally halogenated triphenylborane as Lewis acid. FR 2912410A1, Specific Polymers
74. Pouget E, Garcia EH, Ganachaud F (2008) Macromol Rapid Comm 29:425–430

Polyhedral Oligomeric Silsesquioxanes with Controlled Structure: Formation and Application in New Si-Based Polymer Systems

Yusuke Kawakami, Yuriko Kakihana, Akio Miyazato, Seiji Tateyama, and Md. Asadul Hoque

Abstract Features of the formation of cage oligomeric silsesquioxanes, including several new compounds, are described and possible reaction mechanism is proposed. Synthesis of phenyl oligomeric silsesquioxanes selectively functionalized at the 4-position is reported. Formation and utilization of incompletely condensed oligomeric silsesquioxanes are also described.

Keywords Cage oligomeric silsesquioxane · Cage scrambling · Completely and incompletely condensed oligomeric silsesquioxane · Functionalization · Higher order oligomeric silsesquioxane · Hydrolysis · Polymer · Trifunctional silane

Contents

1 Introduction 188
2 Cyclic Tetrasiloxanetetrol (R_4T_4-tetrol) 190
3 Octahedral Octasilsesquioxane (R_8T_8) 194
   3.1 Formation of R_8T_8 from R_4T_4-tetrol 194
   3.2 R_8T_8 from Condensate 197
   3.3 Scrambling of the Components in the Formation and Reaction of R_8T_8 198
4 Nitration of Cage 4-Trimethylsilyl-Substituted Phenyloligosilsesquioxanes 201
5 Incompletely Condensed Oligomeric Silsesquioxanes 202
   5.1 Heptaphenyltricyclo[7.3.3.1[5,11]]heptasilsesquioxane-3,7,14-triol (heptaphenylheptasilsesquioxanetriol: Ph_7T_7-triol) and Octaphenyltetracyclo[7.3.3.3[3,7]]octasilsesquioxane-5,11,14,17-tetrol (octaphenyloctasilsesquioxanetetrol: Ph_8T_8-tetrol) 202

Y. Kawakami (✉), Y. Kakihana, A. Miyazato, S. Tateyama, and Md. A. Hoque
Japan Advanced Institute of Science and Technology (JAIST), Asahidai 1-1, Nomi, 9231292 Japan
e-mail: kawakami@jaist.ac.jp

5.2	Formation of Ph₈T₈-tetrol	203
5.3	Reactivity of Four Silanol Groups of Ph₈T₈-tetrol	204
5.4	Scrambling of the Components in the Formation of Ph₈T₈-tetrol	206
6	Polymers and Higher Order Structures with Cage Oligomeric Silsesquioxane Core	208
6.1	Siloxane Bond Formation in the Presence of Tris(pentafluorophenyl)borane	209
6.2	Synthesis of Silane-, and Silanol-Functionalized Cage Oligomeric Silsesquioxanes	210
6.3	Reactivity of Silane and Silanol Functions in Cage Oligomeric Silsesquioxanes	213
6.4	Polymers with Cage Oligomeric Silsesquioxane in the Main Chain	215
7	Higher Order Polysilsesquioxanes with Oligomeric Silsesquioxane as Constitutional Units	220
7.1	Dehydrogenative Coupling Reactions for Synthesis of the Higher Order Silsesquioxane Structures	221
8	Conclusion	225
References		225

Abbreviations

AP	Azidopropyl
4-BrPh	4-Bromo-substituted phenyl
BzTMAH	Benzyltrimethylammonium hydroxide
CHCl₃	Chloroform
dba	Dibenzylideneacetone
DMF	*N,N*-Dimethylformamide
DMS	Dimethylsilyl
dvs	1,3-Divinyl-1,1,3,3-tetramethyldisiloxane
Et	Ethyl
EtOH	Ethanol
GPC	Gel permeation chromatography
i-Bu	2-Methylpropyl
i-Bu₇T₈Cl	1-Chloro-3,5,7,9,11,13,15-hepta(2-methylpropyl)-pentacyclo[9.5.1.1^{3,9}.1^{5,15}.1^{7,13}]octasiloxane
i-Bu₇T₈H	1-Hydro-3,5,7,9,11,13,15-hepta(2-methylpropyl)-pentacyclo[9.5.1.1^{3,9}.1^{5,15}.1^{7,13}]octasiloxane
i-Bu₇T₈OH	1-Hydroxy-3,5,7,9,11,13,15-hepta(2-methylpropyl)-pentacyclo[9.5.1.1^{3,9}.1^{5,15}.1^{7,13}]octasiloxane
i-Bu₇T₇-triol	1,3,5,7,9,11,14-Hepta(2-methylpropyl)-tricyclo[7.3.3.1^{5,11}]heptasiloxane-3,7,14-triol
i-Bu₇T₈ODMS	1-[Di(methylsilyl)oxy]-3,5,7,9,11,13,15-hepta(2-methylpropyl)pentacyclo[9.5.1.1^{3,9}.1^{5,15}.1^{7,13}]-octasiloxane
i-Bu,H;*i*-Bu,H-B₁₀	5,15-Di(2-methylpropyl)-5,15-dihydro-1,3,7,9,11,13,17,19-octaphenylpentacylo[11.7.1.1^{3,11}.1^{7,19}.1^{9,17}]decasiloxane

i-Bu,OH;*i*-Bu,OH-B$_{10}$	5,15-Di(2-methylpropyl)-5,15-dihydroxy-1,3,7,9,11,13,17,19-octaphenylpentacyclo[11.7.1.1^{3,11}.1^{7,19}.1^{9,17}]decasiloxane
i-Pr	1-Methylethyl
LAH	Lithium aluminum hydride
M	Methyl
M$_2$B$_9$-diol	3,3-Dimethyl-1,5,7,9,11,13,15,17-octaphenyl-tetracyclo[11.5.1.1^{5,11}.1^{7,17}]nonasiloxane-9,15-diol
M$_2$;BS-B$_{10}$	5,5-Dimethyl-15,15-propylene-1,3,7,9,11,13,17,19-octaphenylpentacyclo[11.7.1.1^{3,11}.1^{7,19}.1^{9,17}]-decasiloxane
M,EP;M,EP-B$_{10}$	5,15-Di[(2-triethoxysilyl)ethyl]-5,15-dimethyl-1,3,7,9,11,13,17,19-octaphenyl-pentacyclo[11.7.1.1^{3,11}.1^{7,19}.1^{9,17}]decasiloxane
M,H;M,H-B$_{10}$	5,15-Dimethyl-5,15-dihydro-1,3,7,9,11,13,17,19-octaphenylpentacyclo[11.7.1.1^{3,11}.1^{7,19}.1^{9,17}]-decasiloxane
M$_2$;M$_2$-B$_{10}$	5,5,15,15-Tetramethyl-1,3,7,9,11,13,17,19-octaphenyl-pentacyclo[11.7.1.1^{3,11}.1^{7,19}.1^{9,17}]decasiloxane
M,OE;M,OE-B$_{10}$	5,15-[Di(triethoxysilyl)oxy]-5,15-dimethyl-1,3,7,9,11,13,17,19-octaphenylpentacyclo-[11.7.1.1^{3,11}.1^{7,19}.1^{9,17}]decasiloxane
M,OH;M,OH-B$_{10}$	5,15-Dimethyl-5,15-dihydroxy-1,3,7,9,11,13,17,19-octaphenylpentacyclo[11.7.1.1^{3,11}.1^{7,19}.1^{9,17}]-decasiloxane
MALDI	Matrix assisted laser desorption ionization
M_n	Number-average molecular weight
MS	Mass spectrometry
M_w	Weight-average molecular weight
NMR	Nuclear magnetic resonance spectroscopy
Np	Naphthyl
Ph	Phenyl
POSS	Polyhedral oligomeric silsesquioxane
Q$_8$	Spherooctasilicate
R_f	Relative distance of development of solutes
SEC	Size exclusion chromatography
TBAF	Tetrabutylammonium fluoride
T_{d5}	5% Weight loss by thermogravimetric analysis
TFB	Tris(pentafluorophenyl)borane
T_g	Glass transition temperature
TGA	Thermogravimetric analysis
THF	Tetrahydrofuran
T_m	Melting temperature
TMS	Trimethylsilyl
TMSPh	Trimethylsilylphenyl
T_n	Number of silicon atoms in the POSS frame structure

TOF	Time of flight
Tolyl	4-Methylphenyl
T_s	Softening temperature
Vi	Vinyl
XRD	X-ray diffraction

1 Introduction

Polysilsesquioxanes are a class of compounds having the empirical formula, $[RSi(O)_{3/2}]_n$, in which three oxygen atoms and one alkyl or aryl group are attached to the silicon atom. These compounds are formed under various reaction conditions, and can take various three dimensional structures. They can be random, ladder, or even cage. Cage polyhedral oligomeric silsesquioxane (cage POSS) is usually abbreviated to T_n, in which n indicates the number of silicon atoms in the frame structure.

Scott initially discovered completely condensed methyl-substituted oligomeric silsesquioxane in 1946 [1]. Later, Barry showed the cubic or hexagonal prismatic shape of the completely condensed molecules [2]. The structure of a cage with eight R-substituted silicon atoms(namely R_8T_8) is shown in Fig. 1.

T_8 is often called "cubic." Brown reported the formation of cubic cage-structured octaphenyloctasilsesquioxane (Ph$_8$T$_8$) [3–5]. The other typical cages are hexahedral hexasilsesquioxane, T_6, decahedral decasilsesquioxane, T_{10}, and dodecahedral dodecasilsesquioxane T_{12}.

A cage can be a completely or incompletely condensed structure [3–39]. Over the past decade, incompletely and completely condensed cage POSS, obtained by hydrolysis and condensation of tri-functional alkyl- or arylsilane, have been firstly used to support metallic species [40], and later to enhance the physical properties, such as thermal stability, glass transition temperature, dielectric constant, oxidative resistance, and even the opto-electronic properties of the cage POSS-based hybrid systems [41–51]. A clustering effect, with POSS cages interacting via non-bonded interactions was suggested [52].

Brown also isolated all-*cis* tetraphenyltetrasiloxanetetrol (Ph$_4$T$_4$-tetrol), and mentioned the possibility of the compound as the intermediate to the Ph$_8$T$_8$ cage. Meanwhile, Flory showed that the 8-membered ring, cyclic octamethyltetrasiloxane, and the 10-membered ring, decamethylpentasiloxane, are thermodynamically

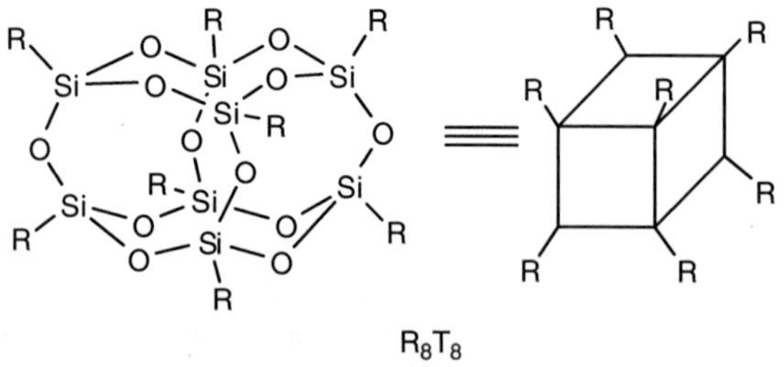

Fig. 1 Framework of R_8T_8 cage with R as the substituent on the silicon atom

Polyhedral Oligomeric Silsesquioxanes with Controlled Structure

the most stable rings and are the major components in the equilibrium mixture of cyclic oligomers and linear polymer of dimethylsiloxane [53–56]. This observation may hold even for the formation of POSS derivatives with controlled structure, i.e., cyclic tetrasiloxane and pentasiloxane frames might be the key to construction of specific structures. It should be commented that all-*cis* cyclic tetrasiloxanetetrol (namely all-*cis* T$_4$-tetrol) or its alkali metal salt is often selectively formed under acidic or basic conditions [57]. All-*cis* cyclic pentasiloxanepentaol (namely all-*cis* T$_5$-pentaol, shown in Fig. 2) or its alkali metal salt might also have been produced in the reaction, although this has not been confirmed.

In the formation of cage or ladder structures of silsesquioxanes, the stereochemical structure around the silicon atom also seems important. When two all-*cis* T$_4$-tetrols condense by forming a siloxane linkage, there are two arrangements, *facing* and *apart* for the two rings, as illustrated in Scheme 1.

One could imagine that the *facing* arrangement might give the T$_8$ cage by further intramolecular condensation. By contrast, the *apart* arrangement might lead

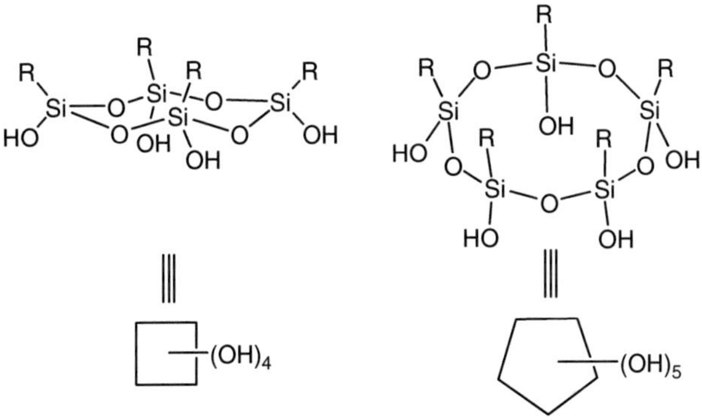

Fig. 2 Cyclic all-*cis* T$_4$-tetrol and T$_5$-pentaol

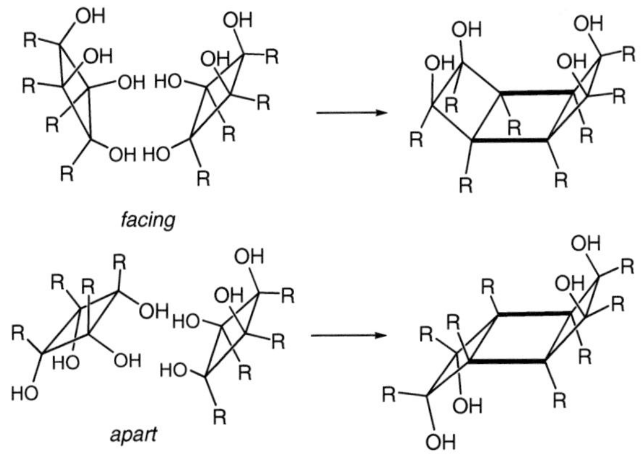

Scheme 1 Two possible *facing* and *apart* arrangements for the two all-*cis* T$_4$-tetrols in forming a siloxane linkage (indicated by *bold line*)

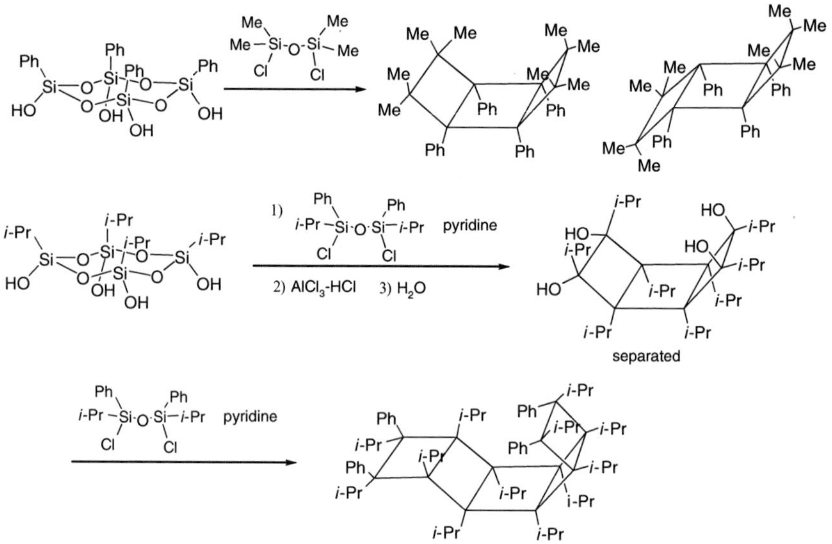

Scheme 2 Step-by-step procedure for synthesis of a ladder structure

to the formation of a ladder structure by successive intermolecular condensation. Unno synthesized a partial ladder structure by step-by-step synthesis, as illustrated in Scheme 2 [58–63].

Cages including hexahedral T_6 and dodecahedral T_{12} could be imagined to form through condensation of all-*cis* T_4-tetrols and T_5-pentaols of *facing* arrangement as one of the key steps in forming a siloxane linkage, as shown in Scheme 3. Kudo reported the molecular orbital calculation of formation energy of such cages, assuming a similar reaction mechanism [64, 65]. Of course, this scheme is an imaginative one without scrambling of cages (Sects. 3.3 and 5.4) and does not express the real reaction mechanism. We found it possible to construct T_{12} from three all-*cis* T_4-tetrols if scrambling of cages was suppressed. Nevertheless, the all-*cis* T_4-tetrol might be a possible key intermediate for the formation of various T_8 structures [3–5, 64–66].

Formation of completely or incompletely condensed POSS is not a simple reaction, but includes many equilibration steps, depending on the reaction conditions [6–10, 14–20, 29, 30, 33–36, 38, 39, 67–69].

In this article, formation of cyclic T_4-tetrol with all-*cis* configuration is described first, followed by description of the formation and functionalization of cages. Finally, applications of incompletely condensed cages are described.

2 Cyclic Tetrasiloxanetetrol (R$_4$T$_4$-tetrol)

We have reported the formation and isolation of all-*cis* Ph$_4$T$_4$-tetrol in *i*-propanol (*i*-PrOH) from phenyltrimethoxysilane in the presence of equimolar amounts of sodium hydroxide [38, 39]. Unno reported the formation of all-*cis* *i*-Pr$_4$T$_4$-tetrol

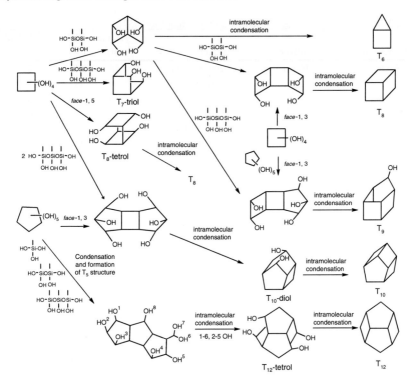

Scheme 3 Imaginary schemes to give completely and incompletely condensed cage structures

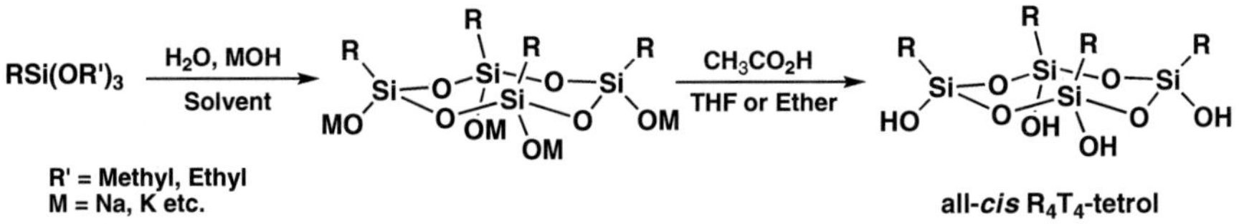

Scheme 4 Formation of all-*cis* R₄T₄-tetrol

[31]. Shchegolikhina [70–72], Klement'ev [73], and Makarova [74] reported the formation [70–72], characterization [70–72, 74], isomerization [73], and derivatization [74] of all-*cis* Ph₄T₄-tetrol, or all-*cis* vinyl(Vi)₄T₄-tetrol. Meanwhile, Matsumoto reported the formation of an isomeric mixture of phenyl (*i*-propyl)tetrasiloxane containing the *cis-cis-trans* isomer as the major component, and commented on the possibility of isomerization of all-*cis* Ph₄T₄-tetrol [75]. A synthetic scheme under basic condition is shown in Scheme 4.

Alkaline hydrolysis of an alkyltrialkoxysilane in the presence of controlled amounts of water gives all-*cis* R₄T₄-tetrol alkaline metal salt as the crystalline major product. Neutralization of the product gives all-*cis* R₄T₄-tetrol. The stability of R₄T₄-tetrol depends on the substituent, and only the alkali metal salt could be isolated in some cases. There are also cases where only small amounts of all *cis* isomer are formed.

Table 1 Formation of all-*cis* R$_4$ T$_4$-tetrols at room temperature [57]

Trialkoxysilane	Solvent	Product	Time (h)	Yield (%)	^{29}Si NMR (ppm)
i-BuSi(OCH$_3$)$_3$	Hexane	[*i*-Bu(OH)SiO]$_4$	48	34	−57.9
		[*i*-Pr(OH)SiO]$_4$		93 [75]	−59.7 [31]
PhSi(OCH$_3$)$_3$	*i*-Propanol	[Ph(OH)SiO]$_4$	20	30–40	−69.7
4-CH$_3$OPhSi(OCH$_3$)$_3$	*n*−Butanol	[4-CH$_3$OPh(OH)SiO]$_4$	48	44	−67.2
4-BrPhSi(OEt)$_3$	EtOH	[4-BrPh(OH)SiO]$_4$	48	Trace	–
		[4-BrPh(ONa)SiO]$_4$		41 [80]	–
ViSi(OEt)$_3$	Hexane-EtOH	[Vi(OK)SiO]$_4$	15	34	–
				74 [72]	–
4-ViPhSi(OEt)$_3$	EtOH	[4-ViPh(ONa)SiO]$_4$		71 [80]	–
NpSi(OCH$_3$)$_3$	*n*−Butanol	[Np(OH)SiO]$_4$	69	30	−68.4
4-CH$_3$ONpSi(OCH$_3$)$_3$	*n*-Butanol	[4-CH$_3$ONp(OH)SiO]$_4$	19	6	−68.8

As a typical example, phenyltrimethoxysilane (9.9 g, 50 mmol) was added to *i*-PrOH (50 mL), water (0.9 mL, 50 mmol), and sodium hydroxide (2.0 g, 50 mmol) at room temperature, and stirred for few hours. Formed crystalline material was collected, dissolved in THF or ether and carefully neutralized with acetic acid. Selected results on the formation of all-*cis* R$_4$T$_4$-tetrol are summarized in Table 1.

Reasonable yields were obtained for all the trialkoxysilanes examined. Vinyl, vinylphenyl, and bromophenyl derivatives gave higher yield, but the products could be isolated only as the sodium or potassium salt, or as trimethylsilyl derivatives. Each aromatic derivative showed only one ^{29}Si signal at −69.7, −67.2, −68.4, and −68.8 ppm. The isobutyl derivative gave only one peak at −57.9 ppm in acetone, which is close to the reported value (−59.7 ppm) for all-*cis* *i*-Pr$_4$T$_4$-tetrol in dimethylsulfoxide [31]. The ^{29}Si signal at −69.7 ppm for Ph$_4$T$_4$-tetrol was thought due to the all-*cis* isomer [17, 71, 74]. Other aromatic derivatives are also considered to be all-*cis*. Methoxyphenyl, naphthyl, and methoxynaphthyl derivatives are good intermediates for further functionalization via electrophilic substitution reactions. By selecting suitable reaction conditions, all-*cis* isomers could be obtained for the substituents examined.

When Ph$_4$T$_4$-tetrol was obtained by the hydrolysis of phenyltrichlorosilane, stereoisomers other than all-*cis* were present in the product, as evidenced by ^{29}Si NMR (Fig. 3a). There are four stereoisomers in the tetramer, which should possibly give six signals. Actually, five signals were observed. Overlapping of the signals might have occurred. Since only limited information was available on the isomerization and separation of the stereoisomers depending on the conditions, isomerization of all-*cis* Ph$_4$T$_4$-tetrol (0.68 g, 1.25 mmol) was carried out in acetone (10 mL) with 1 M hydrochloric acid (4 mL) at room temperature. After 10 min, the products remained as one peak in size exclusion chromatography (SEC), and showed [M+Na]$^+$ = 575.50m/z in matrix-assisted laser desorption ionization time-of-flight mass spectrometry (MALDI-TOF MS), consistent with the T$_4$-tetrol. Four

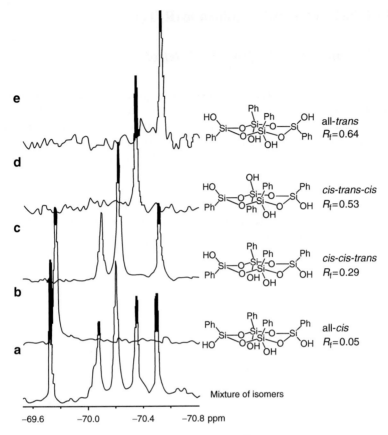

Fig. 3 Identification of stereoisomers of Ph$_4$T$_4$-tetrol by ^{29}Si NMR. Signals of the mixture of isomers (**a**) and of four component isomers (**b–e**) are shown.

new peaks appeared in ^{29}Si at −70.5 (weak), −70.4, −70.2, and −70.1 ppm, with the consumption of the peak of all-*cis* at −69.7 ppm. Configuration of the silicon atom in the ring seems to have been inverted.

A mixture of Ph$_4$T$_4$-tetrol (1.0 g, 1.81 mmol, shown in Fig. 3a) was used for separation. Most of the all-*cis* isomer was removed by precipitation after dissolving in ether (5 mL) and pouring into hexane (20 mL) (∼25%, R_f = 0.05).

After removal of the all-*cis* isomer, three isomers with R_f of 0.29 (∼40%), 0.53 (∼15%), and 0.64 (∼1%) (toluene-ether,1:1) were separated chromatographically. The ^{29}Si scans are also shown in Fig. 3b–d. It was clearly shown that possibly six signals appeared as five overlapped signals (Fig. 3a). The structure of the isomer with R_f = 0.64 (^{29}Si at −70.5 ppm, Fig. 3e) was further confirmed by single-crystal XRD [57]. Matsumoto reported the XRD of *cis-cis-trans* and *cis-trans-cis* T$_4$-tetrol with *i*-Pr substituents. They also reported one ^{29}Si peak at −58.97 ppm for the *cis-trans-cis* isomer and three signals at −59.69, −59.12, and −59.00 ppm for the *cis-cis-trans* isomer [75].

3 Octahedral Octasilsesquioxane (R_8T_8)

3.1 Formation of R_8T_8 from R_4T_4-tetrol

Feher reported the condensation of T_4-tetrol to give T_8 cage under pressurized hydrogen [76]. We and Bassindale reported such condensation in the presence of ammonium salts, where scrambling of the stereochemistry of the tetrol and scrambling of cages occurring [66, 77]. When the all-*cis* R_4T_4-tetrol was treated with benzyltrimethylammonium hydroxide (BzTMAH), a T_8 cage was obtained as shown in Scheme 5.

The typical procedure was as follows: To all-*cis* Ph_4T_4-tetrol (1.384 g, 2.50 mmol) placed in a 30-mL flask under dry atmosphere, benzene (10 mL) and BzTMAH (0.28 mL, 40 wt% methanol solution, 0.60 mmol) were added. After the reaction system was stirred for 2 h at refluxing temperature of the system, the formed solid was filtered and washed successively with benzene and methanol (15 mL each). The product was pure enough for further analysis, confirmed by ^{29}Si NMR and MS, and was determined to be Ph_8T_8. Another effective catalyst was tetrabutylammonium fluoride (TBAF). The results under various reaction conditions are summarized in Table 2.

It is interesting to note that benzene is the best choice as solvent for production of Ph_8T_8 from Ph_4T_4-tetrol, as is the same for the direct synthesis of Ph_8T_8 from

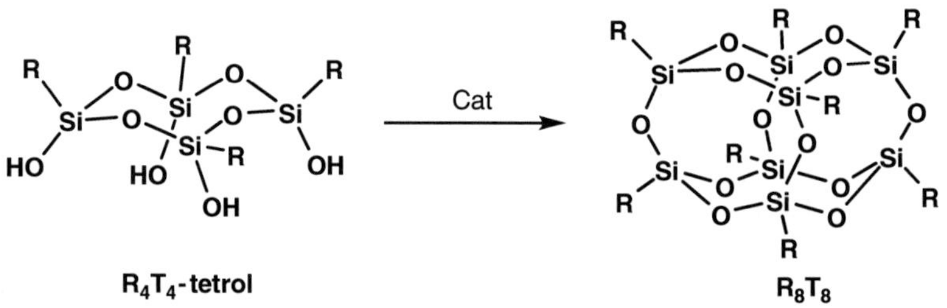

Scheme 5 Formation of R_8T_8 from all-*cis* R_4T_4-tetrol

Table 2 Formation of Ph_8T_8 from all-*cis* Ph_4T_4-tetrol (0.25 M) by ammonium catalysts

Catalyst	[Ph_4T_4-tetrol]/[Catalyst]	Temp.	Time (h)	Solvent	Yield (%)
BzTMAH	4.2:1	Reflux	2	Acetone	1.5
				Methanol	Randomized
				Chloroform	16
				Benzene	95
TBAF	100:1	Reflux	2	Benzene	82
	100:1	r.t.	72	Acetone[a]	85
	1.7:1		24	Acetone[b]	>95 (T_8 and T_{10})
	100:1		72	Methanol[a]	Mix
	100:1		72	Chloroform[a]	61

[a] 0.14 M
[b] 1.7 M, ref [66]
Mix: randomized with identifiable Ph_8T_8

phenyltri(ethoxy)silane in the presence of BzTMAH. TBAF showed higher reactivity than BzTMAH in benzene, as evidenced by the fact that 82% yield was obtained in the presence of 1 mol% of the catalyst compared with 95% yield with 24 mol% BzTMAH [33–36]. Acetone is another choice for selective production of T_8 when TBAF is used as the catalyst. A lower concentration of 0.14 M was selected to obtain reasonable yield under mild reaction conditions. Chloroform can be also used as a solvent. When a higher concentration of TBAF was used, a mixture of cages was formed [29, 66]. Under such conditions, the kinetic rate of the formation and further scrambling and decomposition (Sects. 3.3 and 5.4) seem competitive. The solubility of the products, governing their crystallization out from the solvent, is another important factor in determining the actual products. When the yield is low, the T_4-tetrol was changed into randomized oligomeric unidentified products. The reaction with TBAF was applied to (4-Tolyl-)$_4$, i-Bu$_4$, Np$_4$T$_4$-tetrol derivatives, and the results are shown in Table 3.

(4-Tolyl)$_4$T$_4$-tetrol gave randomized products. In chloroform, T_{10} was formed in rather low yield with hydrochloric acid. In ethanol, T_8 could be produced from 4-TolylSi(OEt)$_3$ under refluxing condition for 5 days with hydrochloric acid. In the case of 4-methoxylphenyl (4-CH$_3$OPh)$_4$T$_4$-tetrol, (4-CH$_3$OPh)$_8$T$_8$ could be identified in the reaction mixture, although the yield was poor.

Table 3 Yield (%) of R$_8$T$_8$ from all-*cis* R$_4$T$_4$-tetrol (0.14 M) in various solvents with 1 mol % TBAF at room temperature for 3 days

Solvent	4-Tolyl	R i-Bu	Np
Acetone	Mix[a]	93[a]	86[b]
	–	44[c,d]	21[e]
Acetonitrile	Mix[a]	91[c]	80[b]
Ethyl acetate	Mix[a]	93[c,f]	63[b]
Chloroform	17[g]	88[c]	80[b]
Hexane	Mix[a]	88[c]	94[b]
Benzene	Trace[g]	74[c]	45[b]
Toluene	Mix[a]	67[c]	10[b]
Ether	–	77[c]	89[b]
Tetrahydrofuran	Mix[a]	91	77[b]
Methanol	Mix[a]	91[c]	83[b]
Ethanol	18[h]	–	–
i-Propanol	1–2	–	–

Mix: randomized with identifiable R$_8$T$_8$
[a] 1 day
[b] With 0.25 M Np$_4$T$_4$-tetrol, 5 days
[c] With 10 mol% TBAF
[d] Reflux, 4 hr.
[e] 0.25 M Np$_4$T$_4$-tetrol, reflux, 4 hr.
[f] 1 day
[g] (4-Tolyl)$_{10}$T$_{10}$
[h] From 4-TolylSi(OEt)$_3$ with 1:1.5 molar ratio of HCl, reflux for 4 days

The isobutyl derivative gave good yield of T_8 in various solvents. Reaction in THF gave 91% yield. A shorter reaction time of one day was sufficient in acetone. Not only acetone or THF, but also acetonitrile and ethyl acetate were good solvents. In addition to these solvents, hydrocarbon solvents, ethers, and even alcohols can be used. In case of Np_4T_4-tetrol, the reaction was slower than for the isobutyl derivative, and higher concentration of the reagents and a longer reaction time were applied. Reasonable yield of T_8 was attained in various solvents. Heating in acetone lowered the yield.

The formation of T_8 from T_4-tetrol was originally intended to synthesize unsymmetrical T_8 from the combination of two different T_4-tetrols. It is interesting that in the formation of T_8 from T_4-tetrol, T_8 could also be attained from the stereoisomeric mixture of Ph_4T_4-tetrol, and that scrambling of the component of T_4-tetrol in produced T_8 had occurred. To study the situation, the mixtures of Ph_4T_4-tetrol and $(Ph-d_5)_4T_4$-tetrol, and of $(4\text{-Tolyl})_4T_4$-tetrol and Ph_4T_4-tetrol were treated in the presence of BzTMAH in benzene. The MALDI-TOF MS of the product (shown in Fig. 4) clearly indicates the random distribution of each component in the produced T_8.

Both Ph and Ph-d_5 units are distributed statistically in the T_8 cage. Decomposition of T_4-tetrol and reassembling to T_8 occur simultaneously in the reaction system. This reaction can be applied to synthesize T_8 with mixed substituents in the cage. When $(4\text{-Tolyl})_4T_4$-tetrol and Ph_4T_4-tetrol were treated with BzTMAH in benzene for 24 h, a mixture of crystalline products was obtained in 30% yield. This product had mass ranging from 1032 m/z [peak of Ph_8T_8 with Na^+] to 1130 m/z [peak of

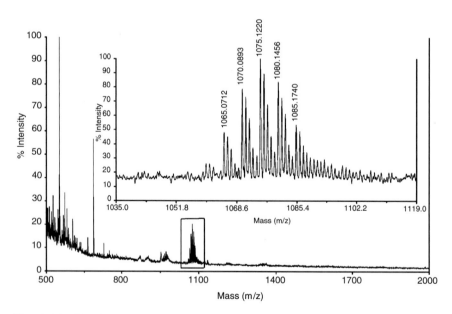

Fig. 4 MALDI-TOF MS of the product Ph_8T_8 from the mixture of Ph_4T_4-tetrol and $(Ph-d_5)_4T_4$-tetrol

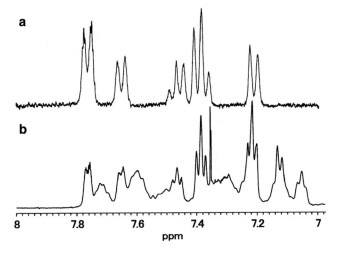

Fig. 5 ^{1}H NMR of the product T_8 from the mixture of Ph_4T_4-tetrol and $(4\text{-Tolyl})_4T_4$-tetrol. Isolated (**a**) and as-produced crystalline product (**b**) are shown

$(4\text{-Tolyl})_7PhT_8$ with Na^+], indicating that the cage product was composed of mixed substituents of 4-Tolyl and Ph. Using proton NMR (Fig. 5a), the ratio of 4-Tolyl to Ph was determined to be 4:1.

Relatively low yield of isolated T_8 cages and absence of $(4\text{-Tolyl})_8T_8$ cage may indicate that complete scrambling of the components had occurred but that at least five to six non-substituted phenyl groups are needed to make the cages crystallize and precipitate from the reaction mixture.

3.2 R_8T_8 from Condensate

To widen the applicability of cage POSS derivatives, it is very important to introduce functional groups in the POSS structure. Chujo used silane-functionalized H_8T_8 cage as the core for the synthesis of a dendritic structure with hydrophilic sphere groups via hydrosilylation [78]. Unno introduced carboxylic acid function to the phenyl group of dimethylphenylsilylated spherooctasilicate (Q_8) core shown in Fig. 6 by chloromethylation and oxidation [79].

The problem of this compound lies in the presence of a rather labile Si-O-Si linkage stemmed from the core. However, it is normally very difficult to directly synthesize functionalized Ph_8T_8 cage, and thus functionalization of the phenyl group directly attached to the core silicon atom has been desired.

There are some cases where the R_4T_4-tetrol cannot be isolated as pure tetrol, or where neutralization of the alkali metal salt with acid gives complex product mixture. A typical example is the 4-bromophenyl derivative [80]. The alkali metal salt of (4-Br substituted phenyl)$_4T_4$-tetrol could be isolated as a solid crystalline material, but neutralization gave complex oligomeric condensed products.

Fig. 6 Framework of Q_8

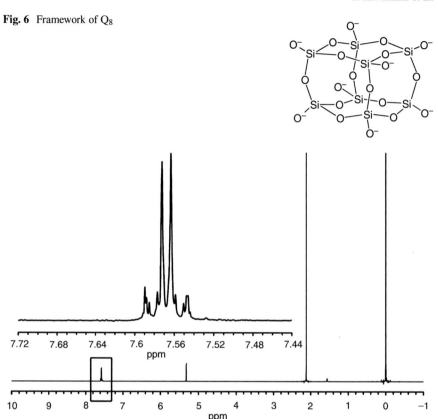

Fig. 7 ^{1}H NMR of (4-BrPh)$_8$T$_8$ obtained by BzTMAH from condensate of (4-BrPh)$_4$T$_4$-tetrol. Boxed area is shown in greater detail above

When the condensed product was treated with BzTMAH, (4-BrPh)$_8$T$_8$ was obtained as the pure crystalline material in reasonable yield (30%); the ^{1}H NMR is shown in Fig. 7.

This is the first report on the direct synthesis of pure (4-BrPh)$_8$T$_8$, although Laine reported the synthesis of 4-BrPh T$_8$ by the bromination of Ph$_8$T$_8$, and obtained a complex mixture of the product [81, 82]. The brominated T$_8$ can be used in the synthesis of new POSS systems.

3.3 Scrambling of the Components in the Formation and Reaction of R_8T_8

The amino group is one of the most versatile functional groups for construction of a new structure by condensation or addition reactions. Introduction of the nitro group is the key step in introducing the amino group directly onto the aromatic group.

Olsson and Laine reported the nitration of Ph_8T_8 by fuming nitric acid [83, 84] and further functionalization and application of the product [84], but the extent and position of nitration was not controlled. This compound has been used without further purification to construct many cross-linked systems. However, the multi- but incomplete functionalization of the phenyl ring often made it unclear how to correlate the properties of the system with the structure.

It is well known that the aryl–silicon bonds are susceptible to cleavage by electrophilic reagents. For an example, we reported synthesis of an optically active bromosilane by cleavage of the naphthyl–silicon bond of optically active silane [85]. Benkeser reported the nitration of methyl(triethylsilyl)-substituted phenyl rings, and showed a stronger directing effect of the methyl group than of the trimethylsilyl group [86, 87]. Eaborn reported that nitration of 1,4-bis(trimethylsilyl)benzene by fuming nitric acid resulted in the substitution of one trimethylsilyl group by a nitro group [88]. It will be interesting to study the possibility of selective cleavage of the phenyl–silicon linkage in phenyl-trimethylsilyl as compared to phenyl-silsesquioxane groups.

$(4\text{-Tolyl})_8T_8$, $(4\text{-}i\text{-PrPh})_8T_8$ and (4-trimethylsilyl-substituted phenyl)$_8T_8$ [$(4\text{-TMSPh})_8T_8$] were obtained by direct acidic hydrolysis of (4-tolyl)triethoxysilane, (4-i-PrPh)triethoxysilane, and (4-TMSPh)triethoxysilane in the presence of hydrochloric acid as shown in Scheme 6.

Simple passage through a silica gel column with hexane as an eluent to remove less-soluble oligomeric products with silanol groups gave pure T_8 cage. Hydrolysis of (4-TMSPh)triethoxysilane in the presence of TBAF gave a mixture of $(4\text{-TMSPh})_nT_8$, T_{10}, and T_{12} cages (n = 8, 10 and 12, respectively for T_8, T_{10}, and T_{12}), with T_{10} as the major fraction (Scheme 7.).

Treatment of the mixture firstly with hexane at $-30°C$ removed the T_8 cage as a crystalline material. Crystallization using ethanol-hexane (1:4) gave the T_{12} cage as crystal. Final crystallization using acetonitrile-hexane (3:5) gave the T_{10} cage as crystal. Each fraction was purified by further recrystallization. The results of the synthesis of 4-TMSPh cages, including the rearrangement of the cage structure, are shown in Table 4.

Formation of T_{10} and T_{12} cages might be the result of the decomposition of the T_8 cage and re-assembling to thermodynamically similarly stable cages, as

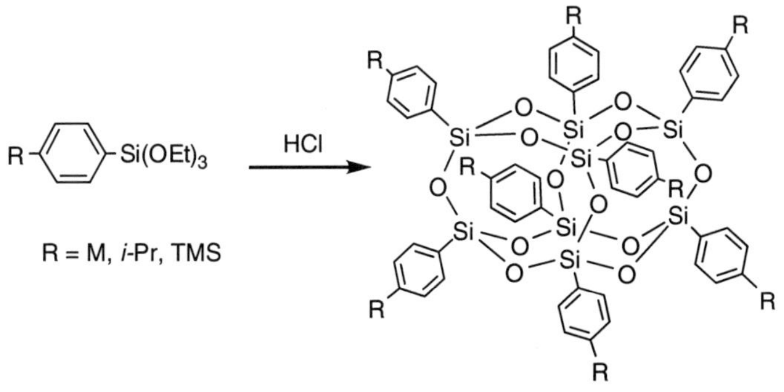

Scheme 6 Formation of (4-substituted phenyl)$_8T_8$ under acidic conditions

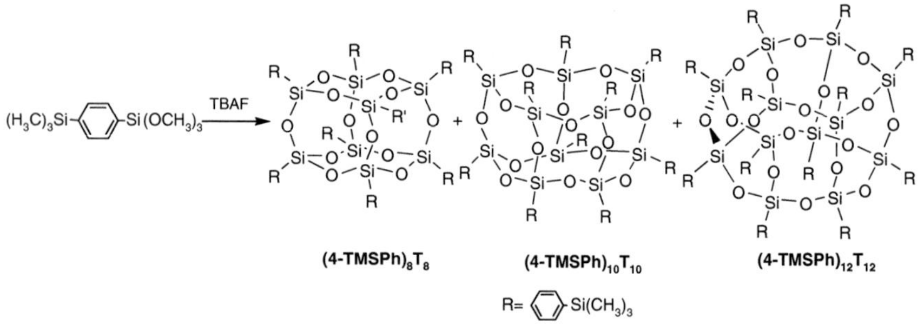

Scheme 7 Synthesis of 4-TMSPh POSS cages in the presence of TBAF

Table 4 Synthesis of 4-substituted Ph-POSS cages

R in R(triethoxy)silane	Catalyst	Solvent	Yield of T_8 (%)	Yield of T_{10} (%)	Yield of T_{12} (%)
4-TMSPh	HCl	EtOH	21	—[a]	—[a]
4-TMSPh	TBAF	CH_2Cl_2	11	30	6.5
4-Tolyl	HCl	EtOH	21	—[a]	—[a]
4-Tolyl	TBAF	CH_2Cl_2	10	40	—[b]
4-i-PrPh	HCl	EtOH	9	—[a]	—[a]

[a] Not formed
[b] Could not be isolated

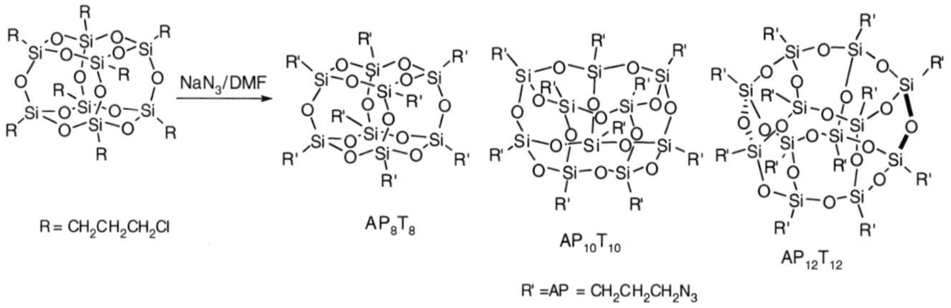

Scheme 8 Cage rearrangement of T_8 to T_{10} and T_{12} in azidation

proposed in Scheme 3. T_8 cage can be constructed only from four eight-membered T_4-tetrol rings, in contrast to T_{10}, which can be constructed under the coexistence of two 8- and 10-membered rings. T_{12} cage can be constructed from four 10-membered rings. They are similarly stable. Such rearrangement and scrambling of POSS cage structures is commonly observed [29, 32, 39, 66]. Yokozawa reported the formation of T_8, T_{10}, and T_{12} in the hydrolysis of cyclic hexasiloxanehexaol or dodecasiloxaneodecaol, phenylsilanetriol, or trialkoxysilanes [32]. Bassindale reported the formation of a mixture of cages in the presence of TBAF [29, 66].

We noticed that such a rearrangement even occurred in the nucleophilic substitution reaction of chloropropyl substituents on T_8 cage by sodium azide [89], as shown in Scheme 8.

Extensive rearrangement of the T_8 cage under azidation of octakis(3-chloropropyl) octasilsesquioxane to produce a thermodynamically stable mixture of octakis(3-azidopropyl)octasilsesquioxane (AP_8T_8), deca(3-azidopropyl)decasilsesquioxane ($AP_{10}T_{10}$), and dodeca(3-azidopropyl)dodecasilsesquioxane ($AP_{12}T_{12}$) is probably because of the necessary use of pure DMF of high polarity as the solvent under severe reaction conditions and a long reaction time to achieve a complete conversion of the eight functional groups.

The first treatment using ethyl acetate-hexane (2:5) as solvent could separate only the T_8 cage ($R_f = 0.60$; 0.25 g) AP_8T_8 as pure form. T_{10} and T_{12} cages gave very similar R_f values (0.55) and could not be obtained as pure form. Second separation using ethyl acetate-hexane (2:9) gave the T_{10} ($R_f = 0.35$; 0.45 g) cage $AP_{10}T_{10}$ and the T_{12} ($R_f = 0.30$; 0.18 g) cage $AP_{12}T_{12}$ as pure form. ^{29}Si NMR peaks of the cage AP_8T_8 and $AP_{10}T_{10}$ appeared at -67.04 and -68.94 ppm, respectively, as singlets. The signal of $AP_{10}T_{10}$ consisting of eight- and ten-membered rings resulted in an up-field shift of about -1.90 ppm from that of AP_8T_8. The ^{29}Si chemical shift of $AP_{12}T_{12}$ appeared at -68.69 and -71.36 ppm, with an integral ratio of 1:2 corresponding to the presence of two different types of silicon atom in the cage of D_{2d} symmetry. Three 10-membered rings were obviously connected and rolled up to each other by sharing two disiloxane linkages (a total of eight Si atoms), leaving one silicon atom in each ring unconnected (total four Si atoms) (see Scheme 3). Once a pair of unconnected silicon atoms was connected via a siloxane linkage, two eight-membered rings would be formed (total four eight-membered rings). Such a cage rearrangement was also reported by Rikowski [90].

4 Nitration of Cage 4-Trimethylsilyl-Substituted Phenyloligosilsesquioxanes

First, nitration of $(4\text{-Tolyl})_8T_8$ was examined. Nitration by fuming nitric acid at room temperature was not clean, and gave various products. Nitration by copper(II) nitrate trihydrate [copper(II) nitrate/octasilsesquioxane = 1.2:0.125 molar ratio at room temperature] seemed clean but slow. When the starting material's peak in ^{1}H NMR had disappeared, three aromatic protons at 8.32, 7.87, and 7.45 ppm, and one CH_3Ph at 2.64 ppm appeared, which strongly supported the formation of 4-methyl-3-nitro-phenyl group. Nitration with fuming nitric acid at $-30°C$ also gave a clean reaction product. ^{29}Si NMR showed only one peak at -79.4 ppm assignable to a T^3 structure in the absence of T^1 (approximately -68 ppm) or T^2 (approximately -70 ppm) structure. Cubic octasilsesquioxane structure seemed to have been maintained during the reaction.

Nitration of $(4\text{-}i\text{-PrPh})_8T_8$ by fuming nitric acid gave unsatisfactory incomplete substitution of the isopropyl group with direct substitution on the ring.

Nitration of $(4\text{-TMSPh})_8T_8$ by fuming nitric acid was carried out. Although the reaction at room temperature gave some decomposed products, the nitration reaction

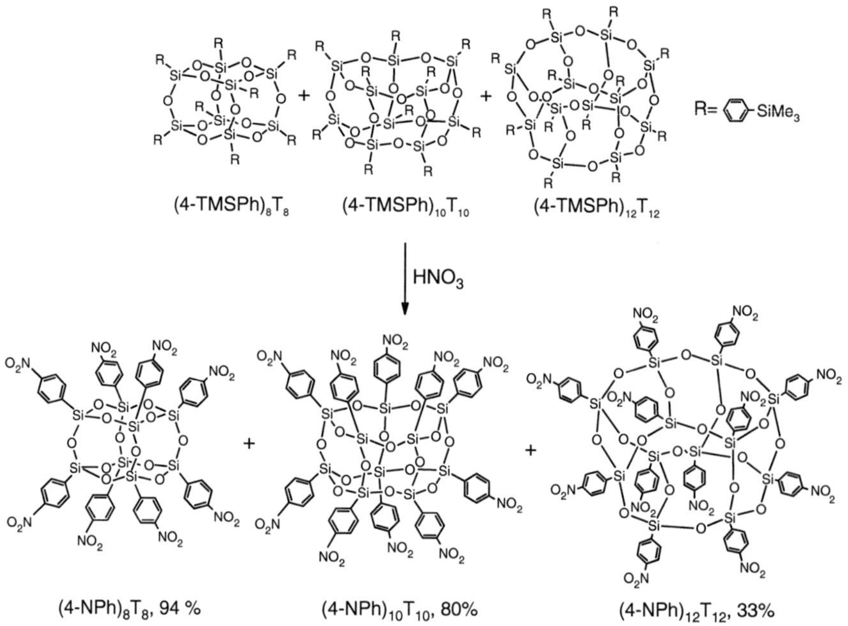

[Scheme 9 diagram showing nitration of (4-TMSPh)₈T₈, (4-TMSPh)₁₀T₁₀, and (4-TMSPh)₁₂T₁₂ POSS cages with HNO₃ to give (4-NPh)₈T₈ (94%), (4-NPh)₁₀T₁₀ (80%), and (4-NPh)₁₂T₁₂ (33%). R = C₆H₄–SiMe₃]

Scheme 9 Clean and complete nitration at 4-position of phenyl groups of 4-TMSPh POSS cages (see also Scheme 7)

itself, by substitution of the trimethylsilyl group attached to phenyl ring, could be completed in 10 h. The reaction at −30°C gave almost pure completely nitrated products. Cleavage of Si–oxygen or Si–phenyl in the silsesquioxane structure was not noticed. The reaction scheme is shown in Scheme 9 [91].

5 Incompletely Condensed Oligomeric Silsesquioxanes

5.1 Heptaphenyltricyclo[7.3.3.1⁵,¹¹]heptasilsesquioxane-3,7, 14-triol (heptaphenylheptasilsesquioxanetriol: Ph₇T₇-triol) and Octaphenyltetracyclo[7.3.3.3³,⁷]octasilsesquioxane-5,11,14,17-tetrol (octaphenyloctasilsesquioxanetetrol: Ph₈T₈-tetrol)

When application to the synthesis of new cage-containing structure is considered, the completely condensed systems are not convenient. Incompletely condensed systems, in which silanol functional groups are included, can be used with more versatility. Feher obtained three incompletely condensed POSS, namely, (cyclo-C₅H₉)₇Si₇O₉(OH)₃, [(cyclo-C₇H₁₃)₇Si₇O₉(OH)₃], and tetrol [(cyclo-C₇H₁₃)₆Si₆O₇(OH)₄], in 29, 26, and 7% yield, respectively [16]. A typical

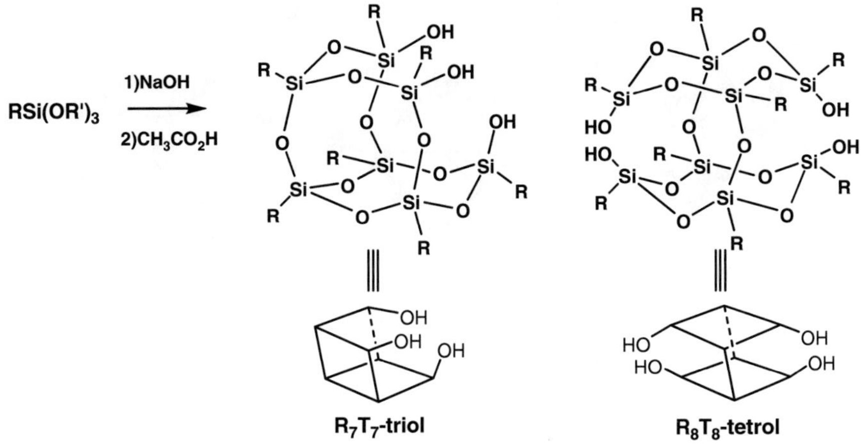

Scheme 10 Formation of incompletely condensed R_7T_7-triol and R_8T_8-tetrol cages

incompletely condensed cage is heptasilsesquioxanetriol (T_7-triol), which is usually used to synthesize a POSS-functionalized monomer like methacrylate [21–28]. As already discussed, it has been well-recognized that the structure of the products – cages, branched or double-chain polymers, completely or incompletely condensed cage – depends on the reaction conditions. Products seem to be generated from multistep processes via many different intermediates [6–10, 14–20, 29, 30, 33–36, 38, 39, 67–69]. T_7-triol can be obtained for various alkyl or aryl substituents. We have paid attention to tetra-functional T_8-tetrol [37–39], possibly produced for aromatic substituents according to Scheme 10.

5.2 Formation of Ph_8T_8-tetrol

As already discussed, when phenyltri(methoxy)silane was reacted with water in the presence of sodium hydroxide (Si:Na:H$_2$O = 1.0:1.0:1.0) in i-PrOH or i-BuOH, all-*cis* Ph_4T_4-tetrol tetra-sodium salt was formed. Neutralization with acetic acid (0.5 M) gave the tetrol. Meanwhile, when phenyltrimethoxysilane (0.24 mol) was treated with sodium hydroxide (Si:sodium = 2:1) at refluxing temperature of i-PrOH (240 mL) for 4 h under nitrogen, a crystalline compound was formed [38, 39].

The compound after treatment with trimethylchlorosilane was determined to be tris(trimethylsilylated) Ph_7T_7-triol using MALDI-TOF MS, ^{29}Si NMR and X-ray single-crystal analysis. After 40 h of stirring, another product with 1379.47 m/z (calculated 1379.24 m/z) and two ^{29}Si signals at −76.12 and −78.94 ppm (O$_3$*Si*Ph) was obtained after trimethylsilylation, which was determined to be tetrakis(trimethylsilylated) Ph_8T_8-tetrol, (68.2% yield). Single-crystal XRD analysis confirmed the double-decker structure of the compound. The crystal of Ph_8T_8-tetrol included two THF molecules. The XRD structure is shown in Fig. 8.

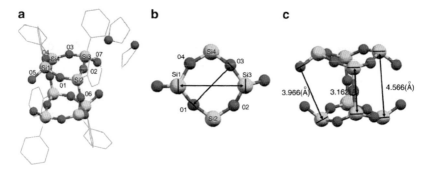

Fig. 8 XRD structure of Ph$_8$T$_8$-tetrol

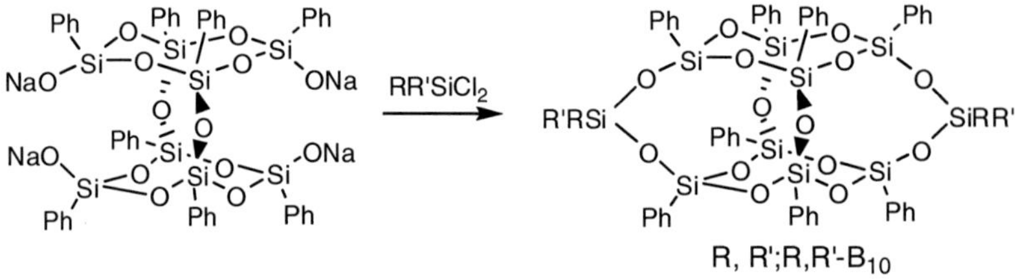

Scheme 11 Bridging of Ph$_8$T$_8$-tetrol by dialkyldichlorosilane

In Fig. 8b, it is shown that the oxygen–oxygen distances in a deck are on average 3.62 and 3.71 Å (O$_1$–O$_3$ and O$_2$–O$_4$), and the silicon–silicon distances are 4.16 and 4.29 Å (Si$_1$–Si$_3$ and Si$_2$–Si$_4$), respectively. The distance between terminal oxygen to facing terminal silicon atoms (3.966 and 4.566 Å) is much longer than the ordinary Si–O bond (1.609 Å). Whereas tetrakis(trimethylsilyl) derivatives completely lost their weight upon heating to 800°C, Ph$_8$T$_8$-tetrol lost less than 2.6 wt% between 240 – 500°C, presumably due to the desorption of adsorbed or condensed water. Decomposition began at 505°C, and the residue yield was 81.5%.

The intramolecular bridging reactivity of T$_8$-tetrol tetra-sodium salt is quite interesting and the general reaction scheme is shown in Scheme 11. Reaction with dimethyldichlorosilane gave completely condensed 5,5,15,15-tetramethyl-1,3,7,9,11,13,17,19-octaphenylpentacyclo[11.7.1.1^{3,11}.1^{7,19}.1^{9,17}] decasiloxane, M$_2$;M$_2$-B$_{10}$ (R, R′ = CH$_3$) in almost quantitative yield. Such high and selective reactivity of *facing* sodium salt of silanol in the Ph$_8$T$_8$-tetrol structure will open the possibility to selectively synthesize unique structures of new polysiloxanes and higher order silsesquioxane structures.

5.3 Reactivity of Four Silanol Groups of Ph$_8$T$_8$-tetrol

The reactivity of each silanol in Ph$_8$T$_8$-tetrol seems quite similar, and selective functionalization was difficult. Treatment of Ph$_8$T$_8$-tetrol with an equimolar amount of

Polyhedral Oligomeric Silsesquioxanes with Controlled Structure 205

dimethyldichlorosilane usually gave bis(dimethylsilyl)-substituted product, $M_2;M_2$-B_{10}, as shown in Scheme 12.

By controlling the reaction conditions, *facing* diol function could be bridged to give capped product at only one side, M_2B_9-diol. Introduction of a polymerizable group to the remaining disilanol gives a new type of monomer. One example is shown in Scheme 13.

This system will give a new silsesquioxane and polycarbosilane hybrid system. Trialkoxysilane functional groups can be also introduced to the Ph_8T_8-tetrol structure, as shown in Scheme 14.

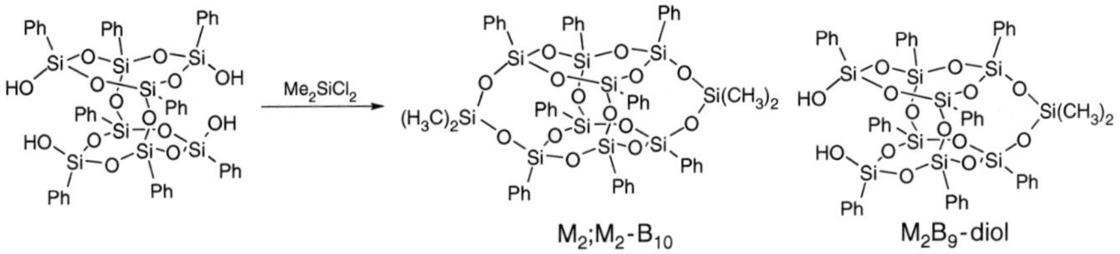

Scheme 12 Selective functionalization of Ph_8T_8-tetrol (See also Scheme 11)

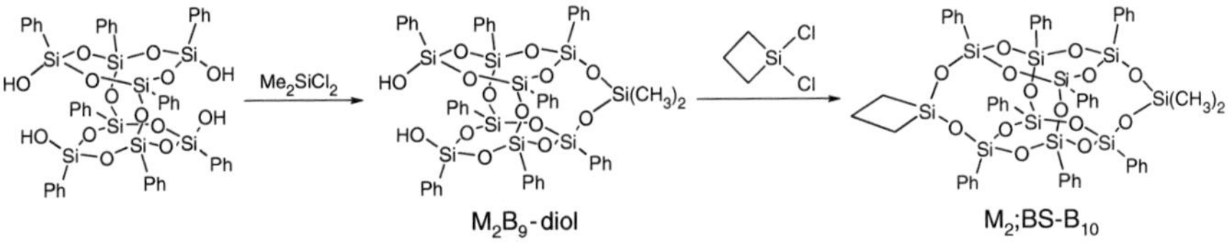

Scheme 13 Introduction of silacyclobutane ring to Ph_8T_8-tetrol structure

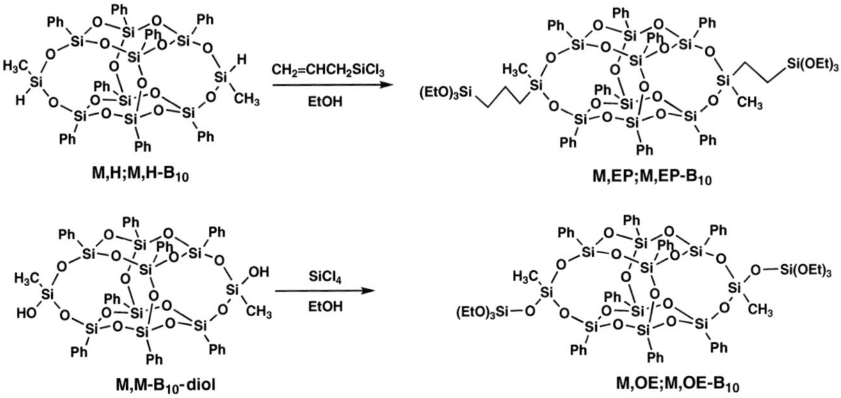

Scheme 14 New materials for construction of nano-periodical silica structures

5.4 Scrambling of the Components in the Formation of Ph$_8$T$_8$-tetrol

The cleavage of Ph$_8$T$_8$ under strongly basic conditions was studied with the intention to obtain Ph$_8$T$_8$-tetrol, which could be a new procedure for forming cyclic tetramer, incompletely condensed Ph$_7$T$_7$-triol, and Ph$_8$T$_8$-tetrol through framework rearrangement of Ph$_8$T$_8$. Apparent reaction paths are shown in Scheme 15 and the synthetic results are summarized in Table 5.

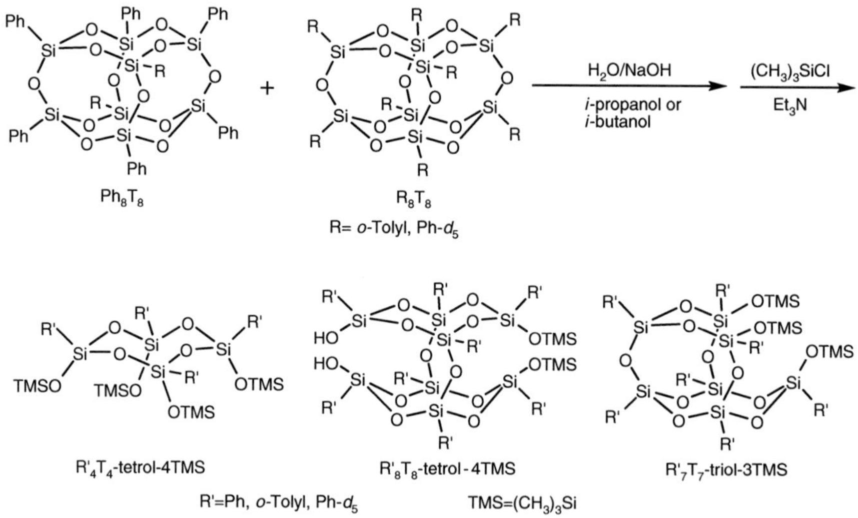

Scheme 15 Formation of incompletely condensed T$_7$-triol and T$_8$-tetrol cages via the hydrolysis of the T$_8$ cage mixture

Table 5 Products in the hydrolysis of the mixture of Ph$_8$T$_8$ and R$_8$T$_8$ followed by TMS-capping

No.	Ph/R	Solvent	Temperature (°C)	Time (hr)	Product
1	Ph/Ph[a]	i-Propanol	r.t.	40	Ph$_4$T$_4$-tetrol-4TMS
2			Reflux	24	Ph$_8$T$_8$-tetrol-4TMS
3		i-Butanol	90	24	Ph$_8$T$_8$-tetrol-4TMS
4	Ph/Ph[b]	i-Propanol	Reflux	24	Ph$_7$T$_7$-triol-3TMS
5	Ph/Ph-d_5[c]		Reflux	24	R'$_8$T$_8$-tetrol-4TMS R'$_7$T$_7$-triol-3TMS
6	Ph/Ph-d_5[d]		r.t.	40	Only R'$_4$T$_4$-tetrol-4TMS
7	Ph/o-Tolyl[e]		Reflux	24	R'$_7$T$_7$-triol-3TMS

[a] Ph$_8$T$_8$:H$_2$O:NaOH = 1:2:4 (molar ratio), 0.17 M
[b] Ph$_8$T$_8$:H$_2$O:NaOH = 1:1:2, 0.1 M
[c] Ph$_8$T$_8$:(Ph-d_5)$_8$T$_8$:H$_2$O:NaOH = 1:1:4:8, 0.04 M
[d] Ph$_8$T$_8$:(Ph-d_5)$_8$T$_8$:H$_2$O:NaOH = 1:1:4:8 (molar ratio), 0.045 M
[e] Ph$_8$T$_8$:(o-Tolyl)$_8$T$_8$:H$_2$O:NaOH = 1:1:4:8), 0.11 M

When Ph$_8$T$_8$ was hydrolyzed with water and sodium hydroxide (Ph$_8$T$_8$:H$_2$O: NaOH, 1:2:4 in molar ratio) in *i*-PrOH at room temperature for 40 h, Ph$_4$T$_4$-tetrol with all-*cis* configuration was produced as the major component. Ph$_8$T$_8$-tetrol was obtained in high yield when the hydrolysis was carried out in refluxing *i*-PrOH or at 90°C in *i*-BuOH for 24 h. When the molar ratio of raw materials (Ph$_8$T$_8$:H$_2$O:NaOH) was changed from 1:2:4 to 1:1:2, Ph$_7$T$_7$-triol became the main product. Consequently, hydrolysis of Ph$_8$T$_8$ by sodium hydroxide in water–alcohol solvents was a selective method for obtaining cyclic tetramer, Ph$_7$T$_7$-triol, and Ph$_8$T$_8$-tetrol frameworks depending on the reaction temperature and the ratio of raw materials. When co-hydrolysis of Ph$_8$T$_8$ and (Ph-d_5)$_8$T$_8$ was carried out, not only T$_8$-tetrol but also T$_7$-triol with mixed substituents was formed, as in the case of hydrolysis of Ph$_8$T$_8$ (Table 5). The molar ratio, estimated from ^{29}Si NMR, was about 1.5:1. The weight ratio after column chromatography was 2.33:1. In ^{29}Si NMR, peaks of silsesquioxane frameworks of both the T$_7$-triol and T$_8$-tetrol were separated into two different chemical shifts. The reason is that phenyl and phenyl-d_5 substituents were randomly distributed in T$_7$ and the T$_8$ frameworks, and that the silicon atoms linked to phenyl and phenyl-d_5 substituents have only slightly different electronic effect. More quantitative data were obtained by MALDI-TOF MS, as shown in Table 6.

Deuterated substituents were distributed randomly in both R$'_7$-T$_7$-triol-3TMS and R$'_8$T$_8$-tetrol-4TMS. The scrambling of substituents occurred even at room temperature for T$_4$-tetrol, although not as extensively as for T$_8$-tetrol and T$_7$-triol at the refluxing temperature of the solvent.

If the hydrolysis of Ph$_8$T$_8$ is a simple decomposition process, the frameworks of Ph$_7$T$_7$-triol-3TMS and Ph$_8$T$_8$-tetrol-4TMS should consist of only one kind of the same substituent (Scheme 16), but this is not the case. Random distribution of substituents in the products in the co-hydrolysis of Ph$_8$T$_8$ with (Ph-d_5)$_8$T$_8$ and

Table 6 MALDI-TOF MS data for R$'_7$T$_7$-3TMS and R$'_8$T$_8$-tetrol-4TMS obtained from Ph$_8$T$_8$ and (Ph-d_5)$_8$T$_8$ co-hydrolysis after capping

	Ph/Ph-d_5	Relative intensity	[M+Na]$^+$, found	[M+Na]$^+$, calc.
R$'_7$T$_7$-triol-3TMS	6/1	25	1174.61	1174.21
	5/2	63	1180.15	1179.25
	4/3	100	1185.22	1184.28
	3/4	99	1189.26	1189.31
	2/5	64	1195.16	1194.34
	1/6	34	1199.09	1199.37
R$'_8$T$_8$-tetrol-4TMS	7/1	11	1384.50	1384.27
	6/2	21	1389.99	1389.30
	5/3	40	1394.82	1394.33
	4/4	45	1399.65	1399.36
	3/5	41	1403.99	1404.39
	2/6	26	1409.81	1409.42
	1/7	11	1415.48	1414.46

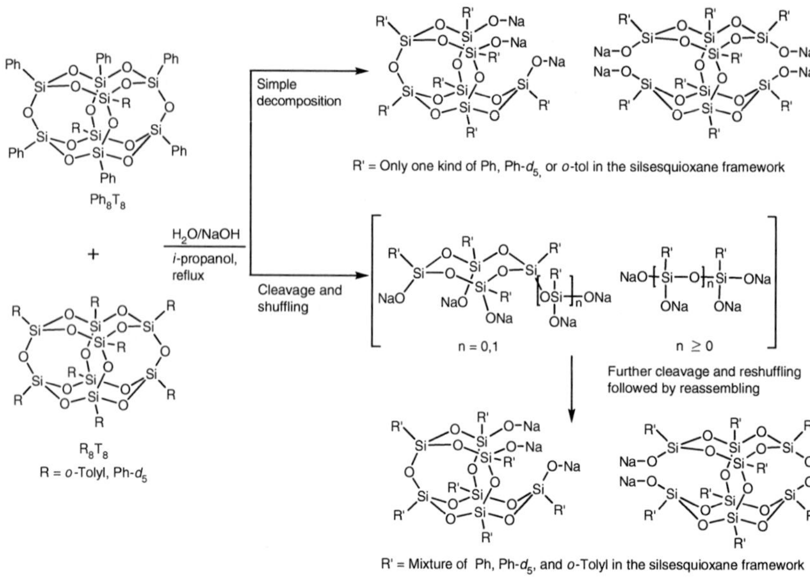

Scheme 16 Decomposition, scrambling, and reassembling of cages

(*o*-Tolyl)$_8$T$_8$ provides strong evidence for the hydrolysis of these POSS as a reshuffling process, by which completely condensed POSS decomposed into smaller fragments, and then reassembled to form T$_7$-triol and T$_8$-tetrol structures.

6 Polymers and Higher Order Structures with Cage Oligomeric Silsesquioxane Core

It is generally accepted that many of the properties of polysiloxanes are a consequence of the static and dynamic nature of the siloxane backbone [92]. Linear poly(dimethylsiloxane)s have been well-known to show particular characteristics such as low glass transition temperature (T_g), thermo-stability, stability against oxidation due to a strong and flexible main chain, and physiological inertness and hydrophobicity based on mainly methyl side chains. However, low T_g and easy cleavage of Si–O–Si bonds under acidic or basic conditions, or by thermal treatment, often limits their applications [93, 94]. Introduction of bulky and rigid moieties like phenylene, naphthylene, anthrylene, phenanthrylene, or adamantylene moieties in the main chain increases T_g and the thermal stability of the polysiloxanes [95–98]. These polymers, especially, poly(arylene-dimethylsiloxane)s are good candidates for high-temperature elastomers because of the flexible Si–O–Si backbone and increased thermal stability given by the presence of an arylene moiety [99–101].

Although the introduction of bulky and rigid organic moieties improves some expected properties, it breaks the continuity of the Si–O–Si backbone in the polymer. If double-decker silsesquioxane consisting of only siloxane bonds was introduced into polysiloxane backbone instead of an organic counterpart, the resulting polymer would be expected to exhibit higher T_g and better thermal stability than found with organic moieties, because of the stronger siloxane bonds. In addition, the cage part might introduce toughness and enhanced gas permeability to the polymer because loose packing leaves some free volume, and also reduce its dielectric constant because of the low polarity of cage silsesquioxane [21, 42–44].

By selecting a suitable bridging group of Ph_8T_8-tetrol, functional groups can be introduced to the cage POSS structure to produce R,R';R,R'-B_{10} with polymerizable groups (See Schemes 11, 13, and 14). A typical functional group is SiH, namely a silane functional group. Usually, hydrosilylation is the choice of reaction for construction of a polymer structure using hydrosilyl-bridged R,H;R,H-B_{10} [102, 103]. We have been interested in constructing higher order structures of polymers or silsesquioxanes by borane-catalyzed siloxane bond formation.

6.1 Siloxane Bond Formation in the Presence of Tris(pentafluorophenyl)borane

To construct the above-mentioned structures, an efficient method of forming the siloxane linkage is essential. Condensation of silane functional group with either silanol, alkoxysilane, or alcohol catalyzed by tris(pentafluorophenyl)borane (TFB) has been proved useful for forming siloxane linkages [104–113]. Transition metal catalysts can be also used [114, 115]. The reaction catalyzed by TFB is considered to proceed by the activation of the silane functional group by TFB (Scheme 17).

Some of the reactions are even stereospecific [116]. However, the steric effect on the reaction is not well studied. Methoxysilane has much higher reactivity than ethoxysilane, and propoxysilane is almost not reactive in the reaction. In some cases, a slightly elevated temperature was required [107].

We noticed that the reactivity of the silane and silanol (alkoxysilane) components in the condensation was very much dependent on the steric environment of the silicon atom bearing the functional group. To construct new polymer structures consisting of only POSS cage, understanding the reactivity of silane and silanol in POSS structures during siloxane formation is essential.

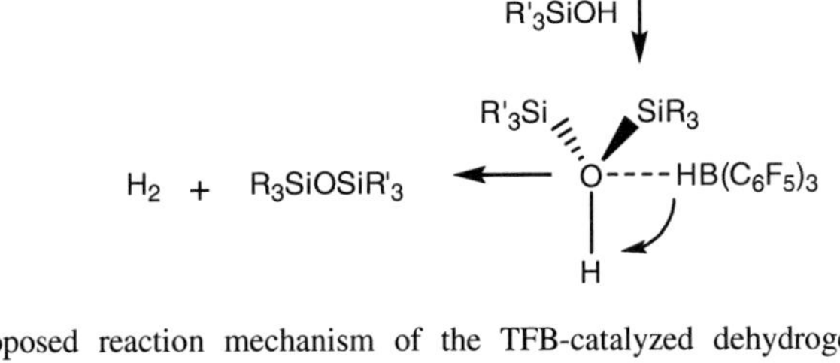

Scheme 17 Proposed reaction mechanism of the TFB-catalyzed dehydrogenative coupling reaction

6.2 Synthesis of Silane-, and Silanol-Functionalized Cage Oligomeric Silsesquioxanes

The structures of the silanes and silanols studied, including dimethylsilyl (DMS) derivative, are shown in Fig. 9.

The synthetic route to *i*-Bu,OH; *i*-Bu,OH-B$_{10}$ is shown in Scheme 18 [113].

The reaction of Ph$_8$T$_8$-tetrol tetra-sodium with *i*-butyltrichlorosilane was performed in THF solution in presence of triethylamine. The capping with *i*-butyltrichlorosilane takes a longer time (12 h) than with methyltrichlorosilane (1 h). Steric hindrance of the bulky *i*-Bu group results in the longer reaction time.

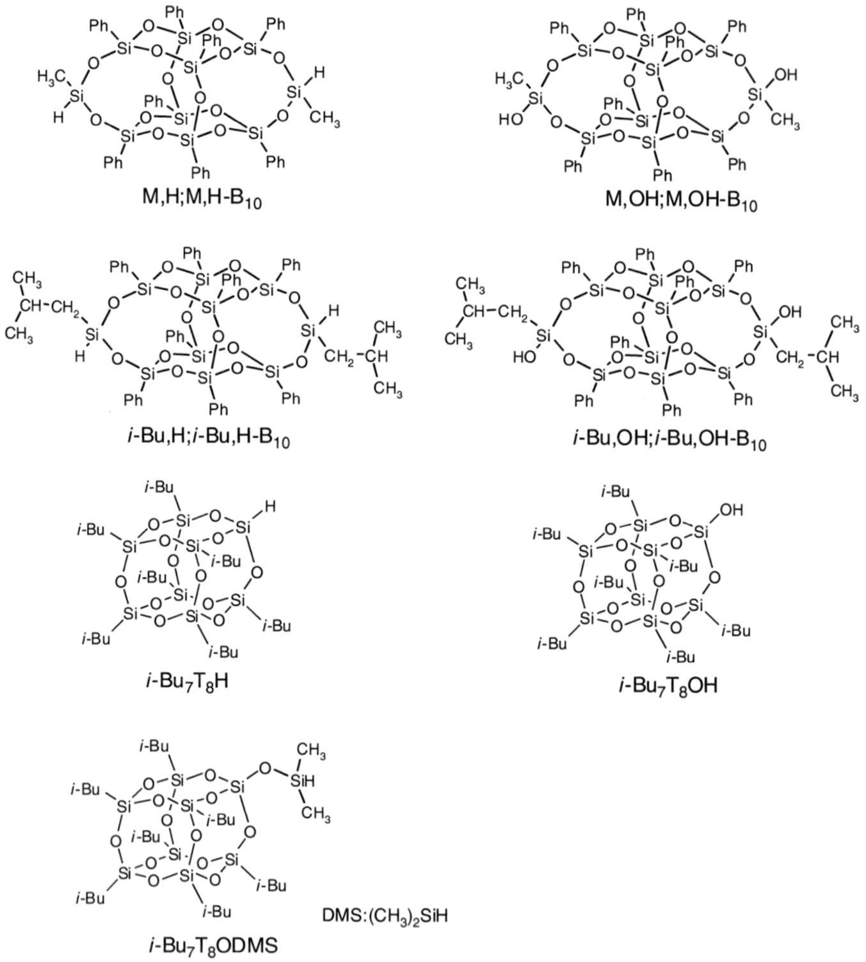

Fig. 9 Structure of silane- and silanol-functionalized cage oligomeric silsesquioxanes

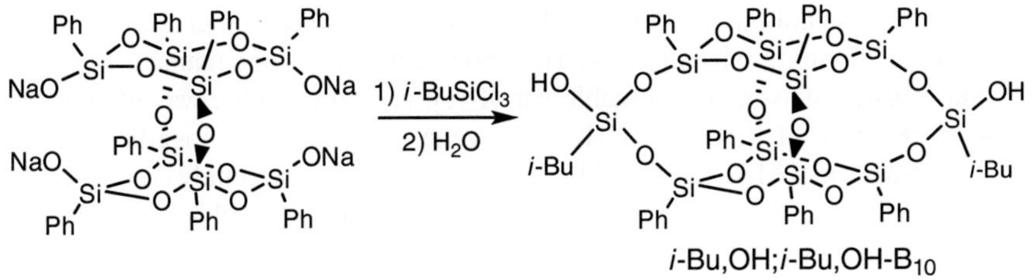

Scheme 18 Capping of Ph$_8$T$_8$-tetrol tetra-sodium salt with i-butyltrichlorosilane

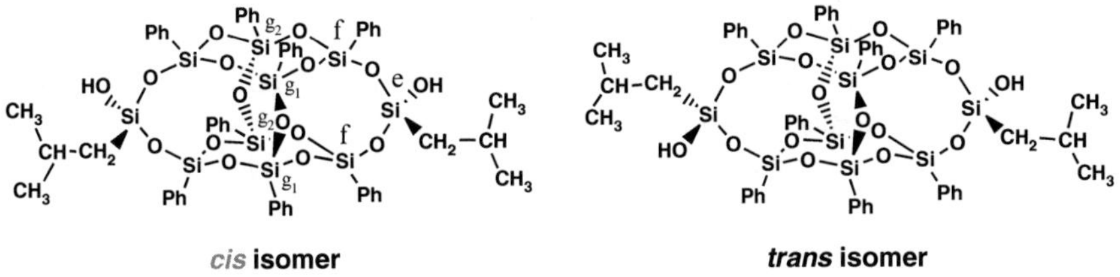

Fig. 10 The *cis* and *trans* configurations of i-Bu,OH;i-Bu,OH-B$_{10}$

Comparatively less polar i-Bu,OH;i-Bu,OH-B$_{10}$ was more soluble in toluene, and formed insoluble products on reducing the volume of toluene. Drying the solvent from the clear filtrate gave i-Bu,OH;i-Bu,OH-B$_{10}$ as colorless solid product.

Three signals were observed in ^{29}Si NMR at −67.27, −78.90, and −79.26 ppm (corresponding to e, f and g, respectively, in Fig. 10) for i-Bu,OH;i-Bu,OH-B$_{10}$. The signal of *trans* isomer should appear as one signal (g), and that of *cis* isomer could appear as two signals (g$_1$ and g$_2$), as can be understood from Fig. 10. Actually, the signal g has split into three peaks, which suggested the existence of isomeric structures for i-Bu,OH;i-Bu,OH-B$_{10}$. Existence of two doublets at 0.81 ppm (CH$_2$) and another two doublets at 0.91 ppm (CH$_3$)$_2$ in ^{1}H also strongly suggests the existence of isomers for i-Bu,OH;i-Bu,OH-B$_{10}$.

Isomers of i-Bu,OH;i-Bu,OH-B$_{10}$ were separated by fractional crystallization and column chromatography. A solution of isomeric mixture of i-Bu,OH;i-Bu, OH-B$_{10}$ (2 g) dissolved in chloroform (10 mL) and hexane (25 mL) was kept in a refrigerator overnight to induce formation of solid. The solid formed was separated by filtration and contained a mixture of isomers with intensity of 8:2 (isomer-1:isomer-2). Hexane (15 mL) was added to the filtrate and the mixture kept at −30°C for 24 h to induce further precipitation, which contained an isomeric ratio of about 2:8 (isomer-1:isomer-2).

The solid fraction [1 g of isomeric mixture of ratio 8:2 (isomer-1:isomer-2)] was developed on a silica gel column using toluene as solvent. The isomers were successfully separated as two fractions [isomer-1, $R_f = 0.29$, 0.72 g (80%) and isomer-2, $R_f = 0.23$, 0.18 g (20%)]. A slight difference in solubility was utilized to separate them. Before final separation of isomers on a silica gel column, fractional precipitation was performed to create a different isomeric ratio in the mixture.

A relatively long column was used for better efficiency. Pure isomer-2 was obtained easily from the mixture having isomeric ratio of 2:8 in the same manner.

Fine crystal of isomer-2, suitable for X-ray crystallography, was obtained by slow evaporation of a 10 wt% CHCl$_3$/benzene (1:2) solution. Together with the ^{1}H, ^{29}Si NMR, and MALDI-TOF MS, single-crystal XRD definitely confirmed that the isomer-2 with $R_f = 0.23$ is the *cis* isomer. The more soluble *i*-Bu,OH;*i*-Bu, OH-B$_{10}$ isomer is *trans* isomer. The *cis* isomer exists as aggregates due to intermolecular hydrogen bonding. The simplified structure is shown in Fig. 11.

Bis-silane, *i*-Bu,H;*i*-Bu,H-B$_{10}$was synthesized similarly to the synthesis of *i*-Bu,OH;*i*-Bu,OH-B$_{10}$ in Scheme 18, as shown in Scheme 19.

Other silane- or silanol-functionalized compounds were also synthesized as shown in Scheme 20 [22, 117].

Trisilanol *i*-Bu$_7$T$_7$-triol was easily converted into *i*-Bu$_7$T$_8$Cl in ca. 70% yield. End-capping of *i*-Bu$_7$T$_7$-trisilanol with trichlorosilane, or with tetrachlorosilane followed by reduction with lithium aluminum hydride gave *i*-Bu$_7$T$_8$H. Hydrolysis of *i*-Bu$_7$T$_8$H readily gave *i*-Bu$_7$T$_8$OH. Treatment of *i*-Bu$_7$T$_8$OH with dimethylchlorosilane gave *i*-Bu$_7$T$_8$ODMS.

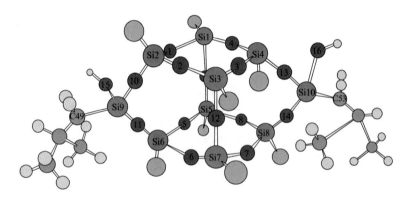

Fig. 11 Simplified configuration of *cis i*-Bu,OH;*i*-Bu,OH-B$_{10}$based on the analysis of single-crystal XRD

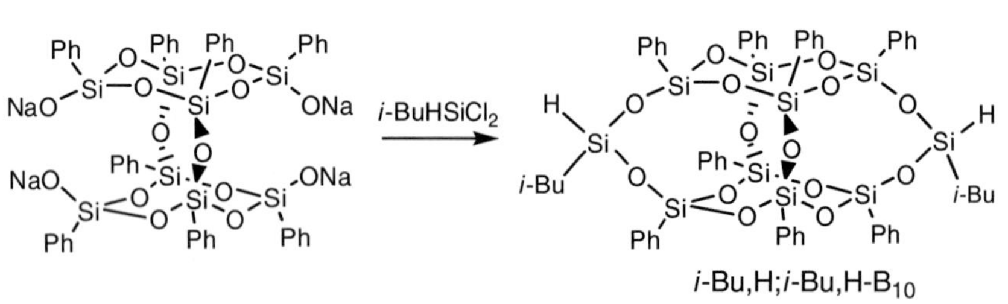

Scheme 19 Synthesis of *i*-Bu,H;*i*-Bu,H-B$_{10}$fromPh$_8$T$_8$-tetrol tetra-sodium salt

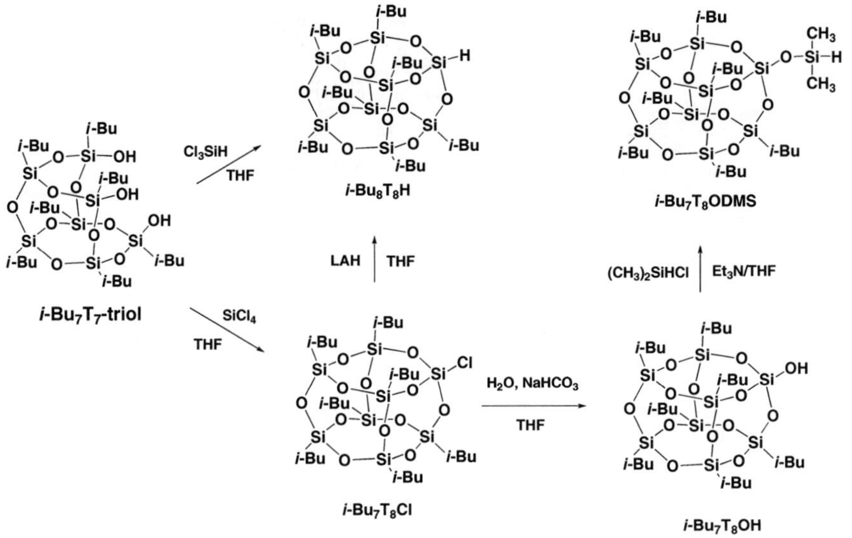

Scheme 20 Synthesis of i-Bu$_7$T$_8$X (X=H, Cl, OH, ODMS) from i-Bu$_7$T$_7$-triol

6.3 Reactivity of Silane and Silanol Functions in Cage Oligomeric Silsesquioxanes

The reaction between triphenylsilane and triphenylsilanol in the presence of TFB proceed without any problem at room temperature. The reactions of triphenylsilane with M,OH;M,OH-B$_{10}$, i-Bu,OH;i-Bu,OH-B$_{10}$, or i-Bu$_7$T$_8$OH, and that of triphenylsilanol with M,H;M,H-B$_{10}$ or i-Bu$_7$T$_8$ODMS proceeded smoothly.

Some effects of the structure were observed for the reactions of cage silsesquioxane derivatives. Reaction of M,H;M,H-B$_{10}$ with M,OH;M,OH-B$_{10}$, or i-Bu,OH;i-Bu,OH-B$_{10}$ also proceeded without major problems. For an example, M,H;M,H-B$_{10}$ (0.97 g, 10.9 × 10^{-4} mol) and an isomeric mixture of i-Bu,OH; i-Bu,OH-B$_{10}$ (0.1 g, 1.8 × 10^{-4} mol) and TFB (0.0074 g, 1.5 × 10^{-5} mol, 2 mol%) were reacted in toluene (5 mL) at room temperature for 24 h. The resulting solution after filtration through a Florisil column was reduced in volume under vacuum and precipitated slowly via slow addition into methanol (40 mL) to obtain a white polymeric solid (0.98 g).

The reaction of triphenylsilanol with i-Bu,H;i-Bu,H-B$_{10}$ or i-Bu$_7$T$_8$H did not proceed well. The reactions of i-Bu,H;i-Bu,H-B$_{10}$ with M,OH;M,OH-B$_{10}$ or with i-Bu,OH;i-Bu,OH-B$_{10}$ did not give any polymers, even at 110°C and a long reaction time (72 h). Steric hindrance around the silane function in i-Bu,H;i-Bu,H-B$_{10}$ seems quite high in the presence of isobutyl substituent, as shown in Scheme 21.

The reaction of i-Bu$_7$T$_8$H with i-Bu,OH;i-Bu,OH-B$_{10}$, or of i-Bu,H;i-Bu,H-B$_{10}$ with i-Bu$_7$T$_8$OH did not occur either. The reaction of i-Bu$_7$T$_8$H with i-Bu$_7$T$_8$OH showed interesting aspects concerning the catalyst activity, as shown in Scheme 22.

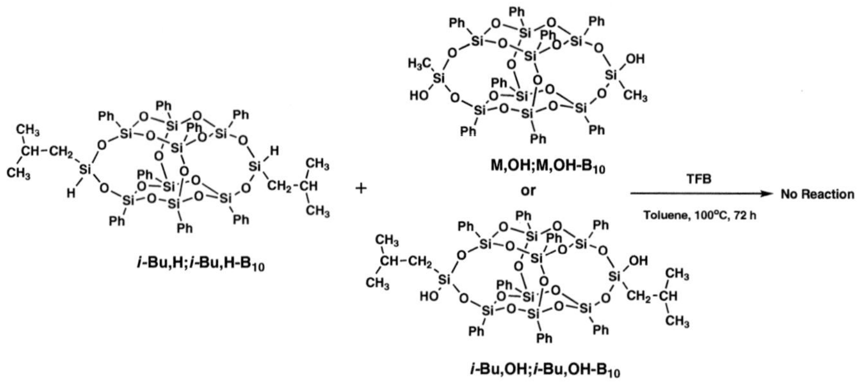

Scheme 21 Attempted polymerization of i-Bu,H;i-Bu,H-B$_{10}$ with i-Bu,OH;i-Bu,OH-B$_{10}$ or M,OH;M,OH-B$_{10}$

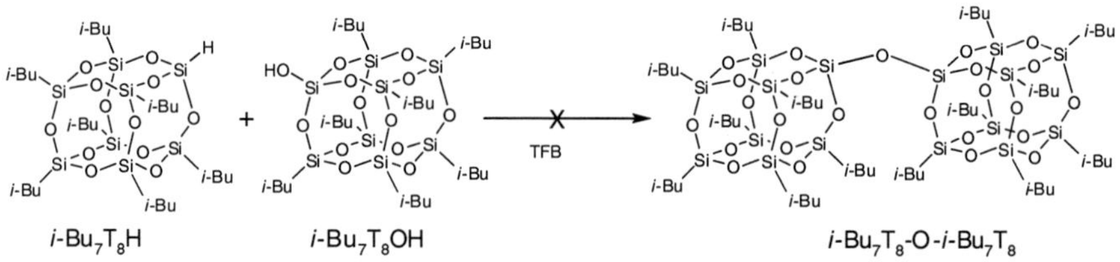

Scheme 22 Reactivity of i-Bu$_7$T$_8$H in the TFB-catalyzed dehydrogenative coupling reaction

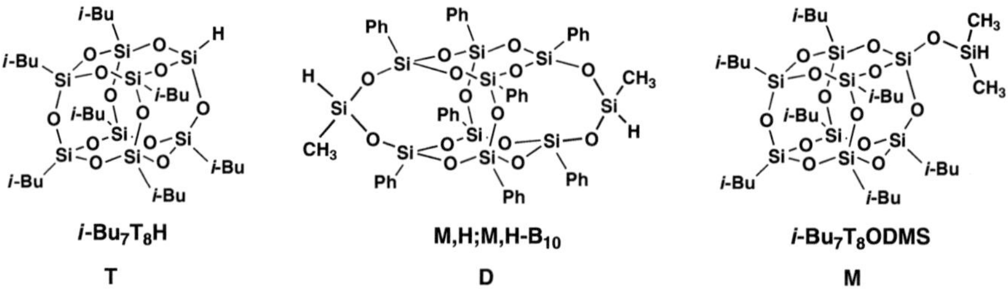

Fig. 12 Structure of the silane function in various cage silsesquioxane structures

The TFB-catalyzed dehydrogenative coupling did not occur, nor any transformation; whereas the reaction in the presence of platinum 1,3-divinyl-1,1,3,3-tetramethyldisiloxane [Pt(dvs)] resulted in formation of the coupling product in very low yield (<2%) as well as unidentified compounds.

The reactivity of the silane is improved in M,H;M,H-B$_{10}$ (D structure). Not surprisingly, if one more silicon atom is covalently introduced onto the T$_8$ framework structure, the reactivity of the produced i-Bu$_7$T$_8$ODMS (M structure) is very much improved compared with that of i-Bu$_7$T$_8$H in which the hydrogen is attracted to the T-structured atom, as shown in Fig. 12.

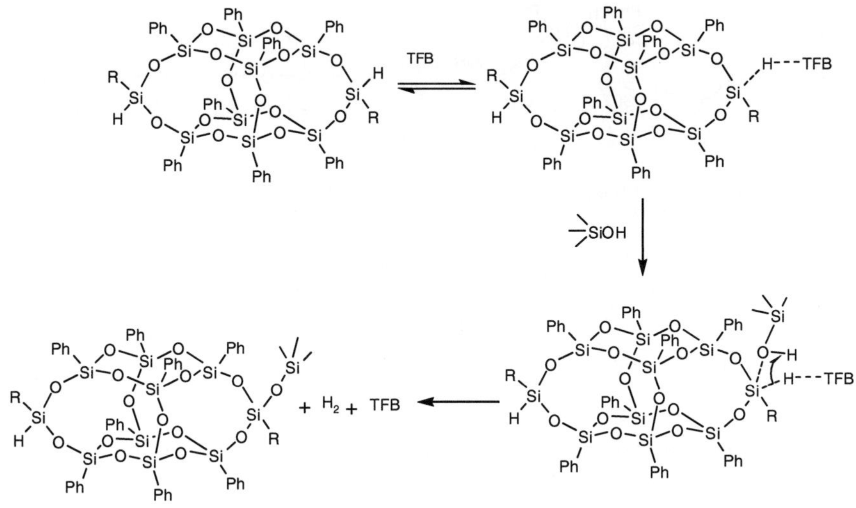

Scheme 23 Tentative reaction mechanism for dehydrocoupling reaction catalyzed by TFB

A tentative reaction mechanism for the TFB-catalyzed dehydrogenative coupling reaction of the POSS silane with POSS silanol is shown in Scheme 23, following the accepted mechanism shown in Scheme 17 [107, 112, 112].

Formation of i-Bu$_7$T$_8$-O-i-Bu$_7$T$_8$ was possibly caused by nucleophilic substitution. Product was isolated in 45% yield, in accordance with the detailed description of synthetic procedure in [117].

6.4 Polymers with Cage Oligomeric Silsesquioxane in the Main Chain

Based on the above-mentioned results, polymers were synthesized using i-Bu,OH; i-Bu,OH-B$_{10}$ and linear oligosiloxane (Scheme 24).

The polymers P$_{cis}$, P$_{tran}$, and P$_{mix}$ were isolated in 92, 87 and 86% yields, respectively, according to the detailed procedure described in [113]. The results of polymerization are summarized in Table 7.

Figure 13 shows the ^{29}Si NMR of polymers. The peak at around −55.95 ppm assigned to the T^2-structured silicon O$_2$Si(i-Bu)(OH) completely disappeared and a new signal at −67.27 ppm was observed, corresponding to the T^3-structured silicon atom, which also confirms the condensation between the hydroxyl group of i-Bu,OH;i-Bu,OH-B$_{10}$ and hydrogen atom (HSi) of tetrasiloxane. Furthermore, two new signals at −21.18 ppm and −20.63 ppm (Fig. 13a–c) are assignable to the silicon chemical shifts of tetrasiloxane. These two signals again support the co-condensation of i-Bu,OH;i-Bu,OH-B$_{10}$ and tetrasiloxane units.

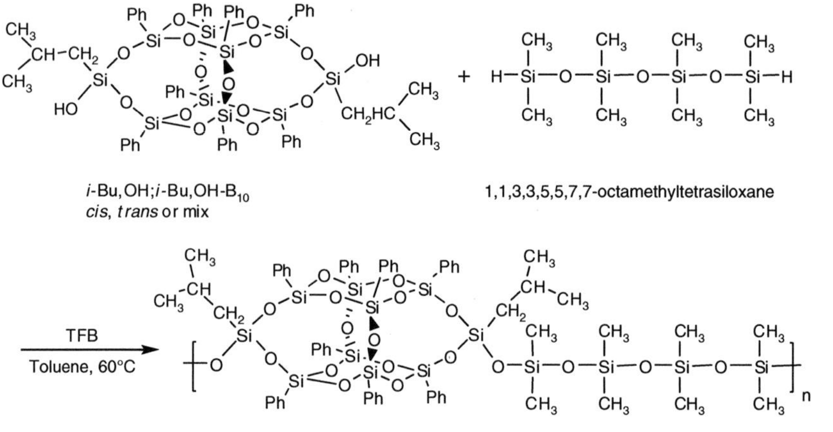

Scheme 24 Cross-dehydrocoupling polycondensation of i-Bu,OH;i-Bu,OH-B$_{10}$ and 1,1,3,3,5,5,7,7-octamethyltetrasiloxane to obtain polymers P$_{cis}$, P$_{trans}$, and P$_{mix}$

Table 7 Polymerization of i-Bu,OH;i-Bu,OH-B$_{10}$ with octamethyltetrasiloxane with feed ratio 1:1 catalyzed by TFB at 60°C and the thermal properties of the formed polymers

i-Bu,OH; i-Bu,OH-B$_{10}$		Polymer properties				$T_{d5}{}^d$(°C)		Residual (wt%)	
Structure	$T_m{}^a$(°C)		$(M_w/M_n)^b \times 10^{-3}$	$T_g{}^a$(°C)	$T_s{}^c$(°C)	N$_2$	air	N$_2$	air
cis	304	P$_{cis}$	50/29	~30	~39	500	470	77	52
trans	357	P$_{trans}$	64/39	~35	~82	500	460	79	53
mix	295	P$_{mix}$	42/27	~34	~45	450	400	71	53

[a] Determined by DSC (30°C/min)
[b] By GPC with polystyrene standard
[c] By TMA (10°C/min)
[d] 5% weight loss by TGA (10°C/min)

Chemical shift of Si atoms of i-Bu,OH;i-Bu,OH-B$_{10}$ (except for capped Si atoms) gave a singlet at −78.90 and a triplet at −79.28 ppm, which after polymerization gave a triplet with an enhanced intensity of the centered signal at −79.63 ppm (Fig. 13c).

The above-described polymerization can be applied to oligosiloxanes with different length and substituents. Longer chain length gave softer polymer having lower T_g. The use of phenyl groups as substitute in linear siloxane instead of methyl groups, or oligosiloxanes shorter than trisiloxane gave insoluble polymer.

The thermal properties of polymers P$_{cis}$, P$_{trans}$, and P$_{mix}$ were studied using DSC, TGA, and TMA, and the results are summarized in Table 7 and Fig. 14.

All the polymers exhibit good thermal stability above 450°C. The high thermal stabilities of these polysiloxane polymers arise due to the presence of a B$_{10}$ unit in the main chain. It can be seen that 5% weight loss temperature (T_{d5}) for the P$_{cis}$ and

Fig. 13 ^{29}Si NMR of polymers (**a**) P$_{cis}$ (**b**) P$_{trans}$, and (**c**) P$_{mix}$

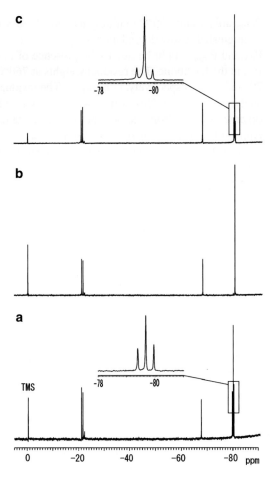

Fig. 14 Thermogravimetric analysis of polymers P$_{cis}$, P$_{trans}$, and P$_{mix}$ in nitrogen and air with a heating rate of 10°C/min

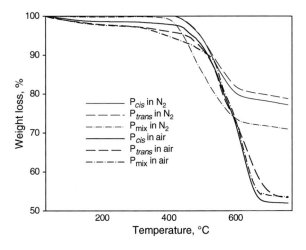

P_{trans} are around 500°C in nitrogen atmosphere, whereas that of P_{mix} is 450°C. The distinguishable low degradation temperature recorded for P_{mix} compared to that of P_{cis} and P_{trans} might be due to the presence of random *cis* and *trans* sequences of B_{10} in the backbone. The residual weights at 760°C for P_{cis}, P_{trans}, and P_{mix} are 77, 79, and 71%, respectively, in nitrogen. The residual weight at the same temperature analyzed in air was 53%, which was consistent with the complete removal of only organic moieties from the polymer structure via oxidation, namely as SiO_2 (53%). Apparently, no cleavage of tetrasiloxane or cage component occurred. Accordingly, some of the organic moieties still remained at 760°C, while maintaining all the siloxane-connected main chain under nitrogen atmosphere.

Although clear T_g values could not be seen for all the polymers at heating or cooling rate slower than 30°C/min, they might be in the range of 30–35°C, even with tetrasiloxane as connecting unit. Since the T_g analysis did not show a clear difference between the polymers, detailed information was obtained by thermo-mechanical analysis on film samples. It is interesting to note that the P_{trans} showed considerably higher softening temperature than P_{cis} or P_{mix}. These T_g or softening temperatures are much higher than those of linear polydimethylsiloxane (-123°C). Poly(arylene-disiloxane)s with disiloxane as connecting unit showed T_g in the range of -20 to 60°C [95, 96]. The present value is comparable with these data. The improved T_g of polymers P_{cis}, P_{trans}, and P_{mix} are due to the presence of the bulky and rigid double-decker silsesquioxane unit in main chain. The double-decker silsesquioxane unit is comparatively bulkier and more rigid than organic moieties. T_g of the polymer synthesized from *i*-Bu,OH-B_{10} and oligosiloxane [HSi(CH$_3$)$_2$O{Si(CH$_3$)$_2$O}$_n$-Si(CH$_3$)$_2$H], average $n = 14$ as determined by ^{1}H NMR] was found to be around -80°C. Although Zhu et al. commented on the toughening effect of short and long polydimethylsiloxane chains in a polysilsesquioxane network [118], detailed discussion on the effects of the incorporation of oligodimethylsiloxane chain in the molecular structure of the polymer backbone could not be made.

Although polymers P_{cis}, P_{trans}, and P_{mix} have a high content of double-decker silsesquioxane (82 wt%), they provide excellent film qualities with high transparency. Films with thickness 50–85 μm were prepared. The optical transparency of a polymer can be seen by comparing photographs of the silsesquioxane symbol taken directly and taken through the polymer, as shown in Fig. 15 for P_{cis}.

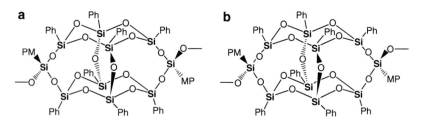

Fig. 15 Photographs of silsesquioxane symbol on white paper, taken (**a**) directly and (**b**) through a thick film (83 μm) of P_{cis}

Polymers were also synthesized from M,H-B$_{10}$ with M,OH-B$_{10}$ or i-Bu,OH-B$_{10}$, based on previously described data. M,H-B$_{10}$ (0.2 g, 0.173 mmol) and M,OH-B$_{10}$ (0.21 g, 0.173 mmol), or M,H-B$_{10}$ (0.18 g, 0.16 mmol) and i-Bu,OH-B$_{10}$ (0.2 g, 0.16 mmol) were reacted in the presence of TBF (0.0035 g, 2 mol %) in toluene (1 mL) at 90°C. After a few minutes of stirring, the monomers went into solution with hydrogen bubbling, which confirmed the progress of reaction. The reacting solution was stirred for a further 12 h, and the mixture was then poured into methanol to give a precipitate, which was dried in vacuum to give 0.28 g and 0.29 g (68%, 72% yield) of polymers P1 and P2, respectively. The crude polymers were purified by repeated re-precipitation from THF into methanol. The results of synthesis and thermal analysis are shown in Table 8 and Fig. 16.

The reaction between M,H;M,H-B$_{10}$ and M,OH;M,OH-B$_{10}$ was also examined in the presence of various transition metal catalysts of Pt, Pd, and Rh. These

Table 8 Polymerization of M,H;M,H-B$_{10}$ with M,OH;M,OH-B$_{10}$ and i-Bu,OH;i-Bu,OH-B$_{10}$

Polymer	M_w/M_n[a]	T_g[b] (°C)	T_{d5}[c] (°C) N$_2$	Air	Residual (wt%) N$_2$	Air
P1	40,000/13,000	245	520	–	76	–
	23,000/9000		484	484	77	35
P2	24,000/9000		522	501	81	51

[a] By GPC with polystyrene standard
[b] Determined by DSC (10°C/min)
[c] 5% weight loss by TGA (10°C/min)

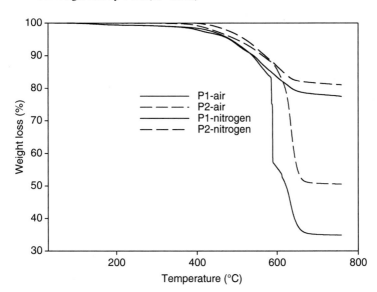

Fig. 16 Thermogravimetric analysis of polymer from M,H;M,H-B$_{10}$ with M,OH;M,OH-B$_{10}$ (P1) or i-Bu,H;i-Bu,H-B$_{10}$(P2) in air and nitrogen with a heating rate of 10°C/min

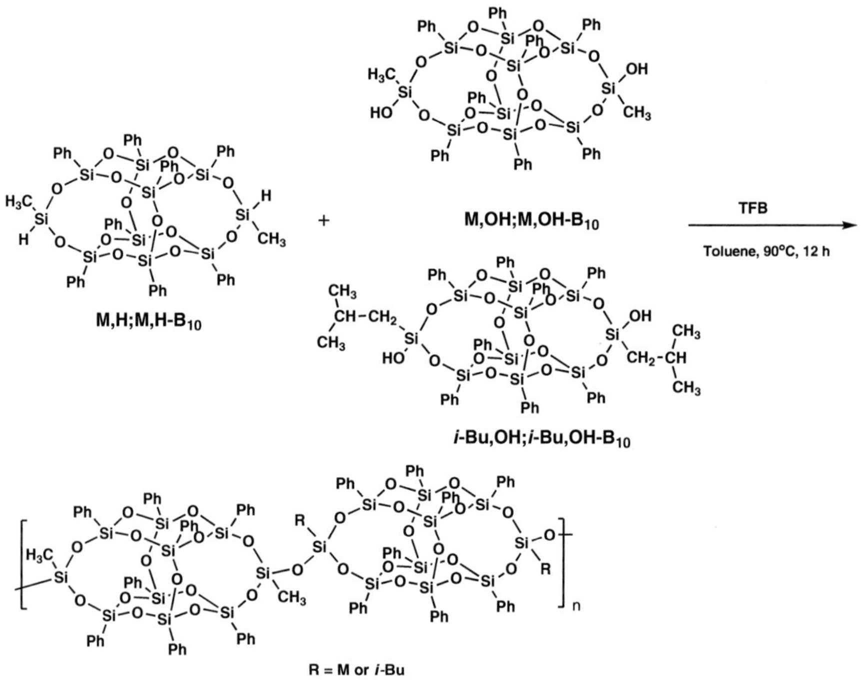

Scheme 25 Polymerization of M,H;M,H-B$_{10}$ with R,OH;R,OH-B$_{10}$

catalysts, especially tris(dibenzylideneacetone)dipalladium [Pd$_2$(dba)$_3$], have been found to be efficient catalysts for the cross-dehydrogenative coupling polymerization of linear hydrosilanes and linear silanols, as described in our previous paper [95], but they were not effective for this reaction. Simple heating of compound M, OH; M, OH-B$_{10}$ did not give polymer, either. Interestingly, the reactions proceeded at a lower temperature (90°C) in 12 h (Scheme 25).

Polymers with molecular weight by gel permeation chromatography (GPC) of $M_w/M_n = 24{,}000/9000$ Da and $M_w/M_n = 23{,}000/9000$ Da were obtained from the combination of M,H-B$_{10}$ and M,OH-B$_{10}$ or i-Bu,OH-B$_{10}$.

7 Higher Order Polysilsesquioxanes with Oligomeric Silsesquioxane as Constitutional Units

Although some interest has been given to the construction of dendritic polysiloxane systems [111, 119], research on the preparation of higher order silsesquioxane systems with cage oligosilsesquioxane as constitutional units has been limited, until now. Wada reported on alkylene-bridged higher order POSS as a model for a silica catalyst surface [120]. This was the first attempt to prepare the giant silsesquioxanes with dendritic structure, in which POSS was incorporated as cores and periphery. However, there have been almost no reports on the construction of higher order silsesquioxane systems composed of POSS with definite structure connected through siloxane linkage, which allows enhanced thermal stability of the materials compared with those made of alkylene linkages.

Until now, some investigations of POSS dimers or trimers covalently linked through siloxane linkage have been reported. The synthesis of POSS dimer was first reported by Fei [121] for bis(silsesquioxanyl) ether [$Cy_7T_8OCy_7T_8$, where Cy is c-C_6H_{11}], prepared by controlled hydrolysis of Cy_7T_8Cl in the presence of triethylamine, Anderson obtained similar compounds by the reaction of R_7T_8Cl and R_7T_8OH (R = Cy or Cp; where Cp is c-C_5H_9) in the presence of n-butyl lithium or triethylamine. The structure of the dimer, $R_7T_8OR_7T_8$, was characterized by XRD and theoretical calculation. A POSS trimer was also synthesized [117]. However, the reactions were extremely slow even when heated to high temperature. We used the knowledge described in Sect. 6 to construct higher order silsesquioxane structures with POSS, starting from either cyclic tetrasiloxanetetrol (R_4T_4-tetrol) or a polyhedral spherosilicate (Q_8) as cores.

7.1 Dehydrogenative Coupling Reactions for Synthesis of the Higher Order Silsesquioxane Structures

Dehydrogenative coupling reaction of cage silsesquioxane silanes with cage silsesquioxane silanols were carried out similarly to the procedures described in Sect. 6 in the presence of TFB. Based on the results that M-type silane showed high reactivity, even toward silanol groups in cage silsesquioxane, i-Bu_7T_8ODMS and Q_8-8DMS were selected as silanes, and R'_4T_4-tetrol and i-Bu_7T_8OH were selected as silanols. The reaction schemes and the structure of the compounds are shown in Schemes 26–28.

The reaction of Ph_4T_4-tetrol with i-Bu_7T_8ODMS was rather slow when equimolar amounts of the reagents were used. POSS dimer ($Ph_8T_8OPh_8T_8$) was formed as a minor product, which should have resulted from a side reaction of the hydrolyzed silanol with i-Bu_7T_8ODMS. Simple passage through a Florisil column

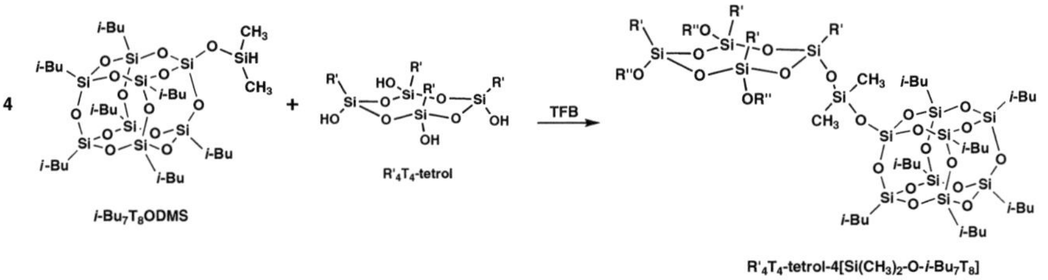

Scheme 26 Dehydrogenative coupling reaction of i-Bu_7T_8ODMS with R_4T_4-tetrol (R=Ph, i-Bu)

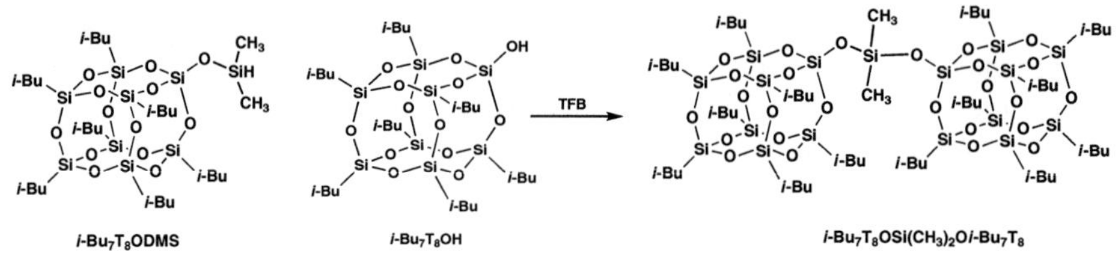

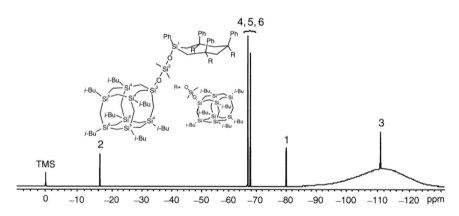

Scheme 27 Dehydrogenative coupling reaction of i-Bu$_7$T$_8$ODMS with i-Bu$_7$T$_8$OH

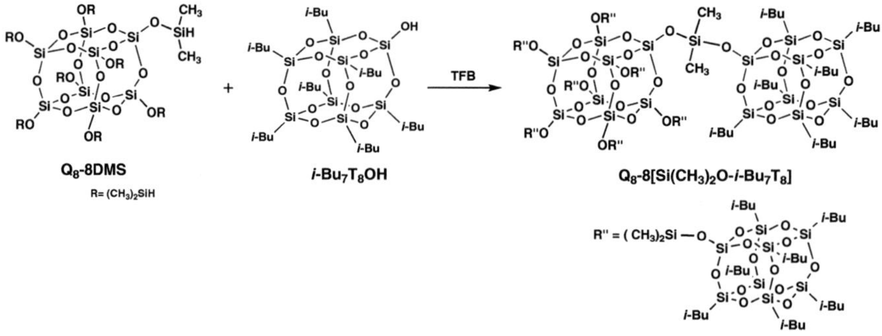

Scheme 28 Dehydrogenative coupling reaction of Q$_8$DMS with i-Bu$_7$T$_8$OH

Fig. 17 ^{29}Si NMR for the product from Ph$_4$T$_4$-tetrol with i-Bu$_7$T$_8$ODMS. The number indicate the position of silicon atom in the structure

was sufficient to obtain the completely substituted product by removing the incompletely reacted compounds with remaining silanol groups, or excess Ph$_4$T$_4$-tetrol. ^{1}H, ^{13}C, and ^{29}Si NMR signals are shown in Fig. 17. MALDI-TOF MS gave the value of 4135 m/z for [M + Na]$^+$ for the calculated molecular weight, based on the most abundant isotope, 4110 m/z. The compound was concluded to be the target compound.

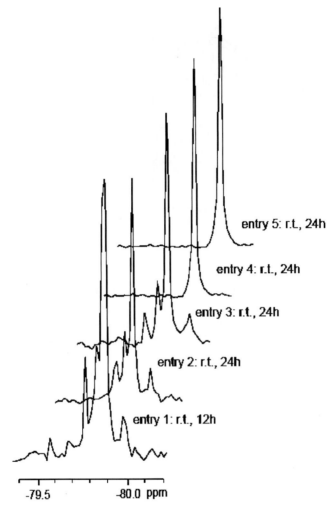

Fig. 18 PhSiO$_3$ ^{29}Si NMR for the product from i-Bu$_7$T$_8$ODMS in 1:4.5 (entry 1–3) and 1:6 (entry 4 and 5) molar ratio at the designated times; *r.t.*, room temperature

The change in the NMR signal during progress of the reaction under various reaction conditions is shown in Fig. 18. The stereochemistry of the starting tetrol, when the mixture of the stereoisomer was used, was maintained in the products in the earlier stage of the reaction at room temperature (entries 1 and 2 in Fig. 18).

Interestingly, when the reaction was carried further at 1:6 molar ratio of Ph$_4$T$_4$-tetrol with i-Bu$_7$T$_8$ODMS for a longer reaction time (24 h, entry 4), the configuration of the product became all-*cis*. This was also true for the reaction of i-Bu$_4$T$_4$-tetrol, as shown in entry 5 in Fig. 18. This is another example of scrambling the constitutional units of the POSS cage.

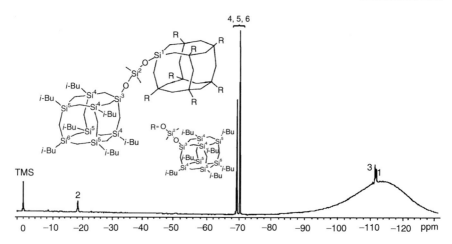

Fig. 19 ^{29}Si NMR for the product from Q_8–8ODMS with i-Bu$_7$T$_8$OH. The numbers indicate the position of silicon atom in the structure

The reaction of all-*cis* Ph$_4$T$_4$-tetrol with i-Bu$_7$T$_8$ODMS gave a slightly lower yield (ca. 70%, entry 5) than the reaction of the mixture of stereo-isomers under the same reaction conditions (ca. 80%, entry 4). All-*cis* i-Bu$_4$T$_4$-tetrol gave a similar yield of ca. 70%. The MALDI-TOF MS also gave a reasonable value (4051 m/z for 4025 m/z).

Various transition metal catalysts of Pt, Pd, and Rh were not effective in this dehydrogenative cross-coupling condensation reaction, even though some of them had been reported as efficient catalysts for linear hydrosilanes and silanols in our previous paper [95, 115].

To further investigate, the reaction of Q_8-8ODMS with i-Bu$_7$T$_8$OH was performed at room temperature or at 60°C. When 14 equivalents of i-Bu$_7$T$_8$OH were used, a product was obtained after passing through a long Florisil column to remove excess i-Bu$_7$T$_8$OH. The NMR signals of this product are shown in Fig. 19.

Hydrosilane resonance (δ 4.7 in ^{1}H) completely disappeared, and a new siloxane methyl signal appeared at 0.2 ppm. ^{29}Si NMR indicates the presence of Q- (−118 ppm), T- (−70 ppm), and D-type (−18 ppm) silicon atoms. The presence of D-type Si signal also clearly indicates the formation of siloxane linkage between Q and T$_8$ cages. The Q-type signal was shifted to a higher field compared with its original position (−108 ppm), and the T-type signal was also shifted from −100 ppm. Thus, the NMR signals qualitatively support the formation of the desired product.

However, MALDI-TOF MS gave the considerably lower ion mass peak at 6867 m/z ([M+Na]$^+$) than expected (7658 m/z as [M]$^+$). It is not clear whether such inconsistency was caused by the incomplete structure of the compound, or by the decomposition of the compound in MS analysis. Because quantitative analysis of the ratio of A and T cages was not possible due to the absence of proton signal of the Q cage, and inconsistent values in MS analysis were obtained, we admit that there is still doubt about the accurate structure of the product. Nevertheless, we can conclude that a higher order POSS structure was constructed from T$_4$ and Q$_8$ cores.

8 Conclusion

In the first half of this article, the features of the formation of cage oligomeric silsesquioxanes, including several new functional cages, were described and a possible reaction mechanism was proposed. Synthesis of selectively functionalized phenyl POSS at the 4-position was reported, based on the new synthetic procedures. In the second half of the article, the formation and utilization of incompletely condensed POSS to construct new polymer structures or higher order structures was described.

Possible future applications of completely condensed cages will be found in reinforcing of nanohybrid materials. Introduction of functional groups in a controlled manner is important. The functionalized cage will find application in the synthesis of new higher order silsesquioxane structures. Incompletely condensed cages will find applications in functionalization of silsesquioxane derivatives as core material for the construction of opto- and electroactive materials. Such compounds can be used for improvement of the performance of holography gratings and high resolution next-generation resist materials. These materials will also find application as transparent insulating materials or optoelectronics materials. Application in hard coating materials is also interesting.

References

1. Scott DW (1946) J Am Chem Soc 68:356
2. Barry AJ, Daudt WH, Domicone JJ, Gilkey JW (1955) J Am Chem Soc 77:4248
3. Brown JF Jr, Vogt LH Jr, Katchman A, Eustance JW, Kiser KM, Krantz KW (1960) J Am Chem Soc 82:6194
4. Brown JF Jr, Vogt LH Jr, Prescott PI (1964) J Am Chem Soc 86:1120
5. Brown JF Jr (1965) J Am Chem Soc 87:4317
6. Baney RH, Itoh M, Sakakibara A, Suzuki T (1995) Chem Rev 95:1409
7. Harrison PG (1997) J Organomet Chem 542:141
8. Clarson SJ, Fitzgerald JJ, Owen MJ, Smith SD, ed., (2000) Silicones and silicone-modified materials. ACS symposium series 729. American Chemical Society, Washington DC
9. Chandrasekhar V, Boomishankar R, Nagendran S (2004) Chem Rev 104:5847
10. Hanssen RWJM, Van Santen RA, Abbenhuis HCL (2004) Eur J Inorg Chem 675
11. Li G, Wang L, Ni H, Pittman CU Jr (2002) J Inorg Organomet Polym 11:123
12. Pielichowski K, Njuguna J, Janowski B, Pielichowski J (2006) Adv Polym Sci 201:225
13. Feher FJ, Newman DA, Walzer JF (1989) J Am Chem Soc 111:1741
14. Feher FJ, Budzichowski TA, Weller KJ (1989) J Am Chem Soc 111:7288
15. Feher FJ, Newman DA (1990) J Am Chem Soc 112:1931
16. Feher FJ, Budzichowski TA, Blanski RL, Weller KJ, Ziller JW (1991) Organometallics 10:2526
17. Feher FJ, Schwab JJ, Soulivong D, Ziller JW (1997) Main Group Chem 2:123
18. Feher FJ, Soulivong D, Lewis GT (1997) J Am Chem Soc 119:11323
19. Feher FJ, Terroba R, Ziller JW (1999) Chem Commun 2153
20. Feher FJ, Terroba R, Ziller JW (1999) Chem Commun 2309
21. Haddad TS, Lichtenhan JD (1996) Macromolecules 29:7302
22. Shockey EG, Bolf AG, Jones PF, Schwab JJ, Chaffee KP, Haddad TS, Lichtenhan JD (1999) Appl Organomet Chem 13:311

23. Lichtenhan JD, Gilman JW, Feher FJ (1996) US Patent 5,484,867 A, 16 Jan 1996
24. Lichtenhan JD, Schwab JJ, Reinerth W, Carr, MJ, An Y-Z, Feher FJ, Terroba R (2001) WO 01/10871 A1, 15 Feb 2001
25. Lichtenhan JD, Schwab JJ, An Y-Z, Reinerth W, Feher FJ (2004) US Patent 2004/0068075 A1, 8 Apr 2004
26. Lichtenhan JD, Schwab JJ, An Y-Z, Liu Q, Haddad TS (2005) US Patent 6,927,270 B2, 9 Aug 2005
27. Lichtenhan JD, Schwab JJ, An Y-Z, Reinerth W, Carr MJ, Feher FJ, Terroba R, Liu Q (2005) US Patent 6,972,312 B1, 6 Dec 2005
28. Lichtenhan JD, Hait SB, Schwab JJ, Carr MJ (2006) US Patent 2006/0263318 A1, 23 Nov 2006
29. Bassindale AR, Liu Z, MacKinnon IA, Taylor PG, Yang Y, Light ME, Horton PN, Hursthouse MB (2003) Dalton Trans 2945
30. Bassindale AR, Chen H, Liu Z, MacKinnon IA, Parker DJ, Taylor PG, Yang Y, Light ME, Horton PN, Hursthouse MB (2004) J Organomet Chem 689:3287
31. Unno M, Takada K, Matsumoto M (1998) Chem Lett 27:489
32. Kawakami Y, Yamaguchi K, Yokozawa T, Serizawa T, Hasegawa M, Kabe Y (2007) Chem Lett 36:792
33. Kawakami Y (2007) React Funct Polym 67:1137
34. Pakjamsai C, Kawakami Y (2004) Polym J 36:455
35. Pakjamsai C, Kobayashi N, Koyano M, Sasaki S, Kawakami Y (2004) J Polym Sci Part A Polym Chem 42:4587
36. Pakjamsai C, Kawakami Y (2005) Design Monom Polym 8:423
37. Yoshida K, Morimoto Y, Watanabe K, Ootake N, Inagaki J, Ohguma K (2003) WO 03/24870 A1, 27 Mar 2003
38. Lee DW, Kawakami Y (2007) Polym J 39:230
39. Li Z, Kawakami Y (2008) Chem Lett 37:804
40. Feher FJ (1986) J Am Chem Soc 108:3850
41. Cheng W-D, Xiang K-H, Pandey R, Pernisz UC (2000) J Phys Chem B 104:6737
42. Xu H, Kuo S-W, Lee J-S, Chang F-C (2002) Macromolecules 35:8788
43. Huang J-C, He C-B, Xiao Y, Mya KY, Dai J, Siow YP (2003) Polymer 44:4491
44. Pellice SA, Fasce DP, Williams RJJ (2003) J Polym Sci Part B Polym Phys 41:1451
45. Imae I, Kawakami Y (2005) J Mater Chem 15:4581
46. Lo MY, Zhen C, Lauters M, Jabbour GE, Sellinger A (2007) J Am Chem Soc 129:5808
47. Iacono ST, Vij A, Grabow W, Smith DW Jr, Mabry JM (2007) Chem Commun 4992
48. Létant SE, Herberg J, Dinh LN, Maxwell RS, Simpson RL, Saab AP (2007) Catal Commun 8:2137
49. Froehlich JD, Young R, Nakamura T, Ohmori Y, Li S, Mochizuki A (2007) Chem Mater 19:4991
50. Laine RM, Roll M, Asuncion M, Sulaiman S, Popova V, Bartz D, Krug DJ, Mutin PH (2008) J Sol-Gel Sci Technol 46:335
51. Wahab MA, Mya KY, He C (2008) J Polym Sci Part A Polym Chem 46:5887
52. Anderson SE, Baker ES, Mitchell C, Haddad TS, Bowers MT (2005) Chem Mater 17:2537
53. Carmichael JB, Kinsinger J (1964) Can J Chem 42:1996
54. Brown JF Jr, Slusarczuk GMJ (1965) J Am Chem Soc 87:931
55. Flory PJ, Semlyen JA (1966) J Am Chem Soc 88:3209
56. Suter UW, Mutter M, Flory PJ (1976) J Am Chem Soc 98:5740
57. Ito R, Kakihana Y, Kawakami Y (2009) Chem Lett 38:364
58. Kyushin S, Tanaka R, Arai K, Sakamoto A, Matsumoto H (1999) Chem Lett 28:1297
59. Unno M, Suto A, Takada K, Matsumoto H (2000) Bull Chem Soc Jpn 73:215
60. Unno M, Suto A, Matsumoto H (2002) J Am Chem Soc 124:1574
61. Unno M, Matsumoto T, Matsumoto H (2007) J Organomet Chem 692:307
62. Shklover VE, Chekhlov AN, Struchkov YT, Makarova NN, Andrianov KA (1978) Zh Strukt Khim 19:1091
63. Shklover VE, Klement'ev IY, Struchkov YT (1981) Dokl Akad Nauk SSSR 259:131

64. Kudo T, Machida K, Gordon MS (2005) J Phys Chem A 109:5424
65. Itoh M (ed) (2007) Shirusesukiokisan Zairyo no Kagaku to Oyo Tenkai. CMC, pp 22–36
66. Zhi-hua L, Bassindale AR, Taylor PG (2004) Chem Res Chin Univ 20:433
67. Mantz RA, Jones PF, Chaffee KP, Lichtenhan JD, Gilman JW, Ismail IMK, Burmeister MJ (1996) Chem Mater 8:1250
68. Hasegawa I, Sakka S, Sugahara Y, Kuroda K, Kato C (1989) J Chem Soc Chem Commun 208
69. Pescarmona PP, Van der Waal JC, Maschmeyer T (2004) Eur J Inorg Chem 978
70. Shchegolikhina O, Pozdniakova Y, Antipin M (2000) Organometallics 19:1077
71. Shchegolikhina OI, Pozdnyakova YA, Molodtsova YA, Korkin SD, Bukalov SS, Leites LA, Lyssenko KA, Peregudov AS, Auner N, Katsoulis DE (2002) Inorg Chem 41:6892
72. Shchegolikhina OI, Pozdnyakova YA, Chetverikov AA, Peregudov AS, Buzin MI, Matukhina EV (2007) Russ Chem Bull Int Ed 56:83
73. Klement'ev IY, Shklover VE, Kulish MA, Tikhonov VS, Volkova EV (1981) Dokl Akad Nauk SSSR 259:1371
74. Makarova NN, Petrova IM, Petrovskii PV, Kaznacheev, AV, Volkova LM, Shcherbina MA, Bessonova NP, Chvalun SN, Godovskii YK (2004) Russ Chem Bull Int Ed 53:1983
75. Unno M, Kawaguchi Y, Kishimoto Y, Matsumoto H (2005) J Am Chem Soc 127:2256
76. Feher FJ, Budzichowski (1989) J Organomet Chem 373:153
77. Tateyama S, Kakihana Y, Kawakami Y (2010) J Organomet Chem 695:898
78. Naka K, Fujita M, Tanaka K, Chujo Y (2007) Langmuir 23:9057
79. Liu H, Kondo S, Takeda N, Unno M (2008) J Am Chem Soc 130:10074
80. Ronchi M, Pizzotti M, Orbelli Biroli A, Macchi P, Lucenti E, Zucchi C (2007) J Organomet Chem 692:1788
81. Laine RM, Tamaki R, Choi J (2002) WO 02/100867 A1, 19 Dec 2002
82. Brick CM, Tamaki R, Kim S-G, Asuncion MZ, Roll M, Nemoto T, Ouchi Y, Chujo Y, Laine RM (2005) Macromolecules 38:4655
83. Olsson K, Grönwall C (1961) Ark Kemi 17:529
84. Tamaki R, Tanaka Y, Asuncion MZ, Choi J, Laine RM (2001) J Am Chem Soc 123:12416
85. Suzuki K, Kawakami Y, Velmurugan D, Yamane T (2004) J Org Chem 69:5383
86. Benkeser RA, Brumfield PE (1951) J Am Chem Soc 73:4770
87. Benkeser RA, Landesman H (1954) J Am Chem Soc 76:904
88. Deans FB, Eaborn C (1957) J Chem Soc 498
89. Ervithayasuporn V, Wang X, Kawakami Y (2009) Chem Commun 5130
90. Rikowski E, Marsmann HC (1997) Polyhedron 16:3357
91. Miyazato A, Pakjamsai C, Kawakami Y (2010) Dalton Trans 39:3239
92. Clarson SJ, Semlyen JA, ed., (1993) Siloxane polymers. Prentice Hall, Englewood Cliffs, pp 309
93. Ziegler JM, Gordon FW, ed., (1990) Silicon-based polymer science, Advances in chemistry series 224. American Chemical Society, Washington DC, pp 47–90
94. Brook MA (2000) Silicon in organic, organometallic, and polymer chemistry. Wiley, New York, chapter 9
95. Li Y, Kawakami Y (1999) Macromolecules 32:8768
96. Otomo Y, Nagase Y, Nemoto N (2005) Polymer 46:9714
97. Sato I, Takeda S, Arai Y, Miwa H, Nagase Y, Nemoto N (2007) Polym Bull 59:607
98. Ito H, Akiyama E, Nagase Y, Yamamoto A, Fukui S (2006) Polym J 38:109
99. Lewin M, Atlas SM, Pearce EM, ed., (1975) Flame retardant polymeric materials. Plenum Press, New York
100. Dvorinc PR, Lenz RW (1990) High temperature siloxane elastomers. Huthig & Wepf Verlag, Basel
101. Zeigler JM, Fearon GFW, ed., (1990) Silicon-based polymer science: a comprehensive resource. American Chemical Society, Washington DC
102. Aminuzzaman M, Watanabe A, Miyashita T (2008) J Photopolym Sci Technol 21:537
103. Aminuzzaman M, Watanabe A, Miyashita T (2008) J Mater Chem 18:5092
104. Blackwell JM, Foster KL, Beck VH, Piers WE (1999) J Org Chem 64:4887
105. Parks DJ, Piers WE (1996) J Am Chem Soc 118:9440

106. Harrison DJ, MacDonald R, Rosenberg L (2005) Organometallics 24:1398
107. Chojnowski J, Rubinsztajn S, Cella JA, Fortuniak W, Cypryk M, Kurjata J, Kaźmierski K (2005) Organometallics 24:6077
108. Rubinsztajn S, Cella JA (2005) Macromolecules 38:1061
109. Chen X, Cui Y, Yin G, Liao L (2007) J Appl Polym Sci 106:1007
110. Longuet C, Joly-Duhamel C, Ganachaud F (2007) Macromol Chem Phys 208:1883
111. Thompson DB, Brook MA (2008) J Am Chem Soc 130:32
112. Zhou D, Kawakami Y (2005) Macromolecules 38:6902
113. Hoque MA, Kakihana Y, Shinke S, Kawakami Y (2009) Macromolecules 42:3309
114. Li Y, Kawakami Y (1999) Macromolecules 32:3540
115. Wu S, Hayakawa T, Kikuchi R, Grunzinger SJ, Kakimoto M, Oikawa H (2007) Macromolecules 40:5698
116. Shinke S. Tsuchimoto T. Kawakami, Y (2007) Silicon Chem 3:243
117. Anderson SE, Mitchell C, Haddad TS, Vij A, Schwab JJ, Bowers MT (2006) Chem Mater 18:1490
118. Zhu B, Katsoulis DE, Keryk JR, McGarry FJ (2004) Macromolecules 37:1455
119. Uchida H, Kabe Y, Yoshino K, Kawamata A, Tsumuraya T, Masamune S (1990) J Am Chem Soc 112:7077
120. Wada K, Watanabe N, Yamada K, Kondo T, Mitsudo T (2005) Chem Commun 95
121. Fei Z, Fischer A, Edelmann FT (2003) Silicon Chem 2:73

Index

A
Addition polymerization 111
Alkane-*co*-silicone polymers 178
Alkoxysilane 163
Anthracene, silylsubstituted 43
Anthradithiophenes 33, 48
2-(Arylsilyl)aryl triflates 56

B
$B(C_6F_5)_3$ 161
Benzo-1-silacyclobutenes 126
Benzo-diphenyl-1-silacyclobut-2-ene 127
Benzophenone 24
Benzosiloles 56
Bis(bromoaryl)disilanes 86
Bis(dimethyl-*tert*-butylsilyl) oligothiophenes 39
Bis(1-methyl-2,3,4,5-tetraphenyl silacyclopentadienyl)ethane (2PSP) 51
4,4'-Bis[*N*-1-naphthyl-*N*-phenylamino]-biphenyl (NPD) 47
Bis(silsesquioxanyl) ether 221
Bis(*o*-silyl)diphenylacetylenes 57
Bis-silylphenyl derivatives 175
Bis(styryl)benzenes, tetrakis-silicon-bridged 57
Bis(triisopropylsilylethynyl) pentacene (TIPS-5AC) 44
2,3-Bis(trimethylsilyl)norbornadiene-2 145
Bithiophene 38
Bithiophenesilane monodendrons 66
Boranes 164
2-Bromo-5-(trimethoxysilyl)thiophene (TMOS-T-Br) 70

C
Cage oligomeric silsesquioxane 185
Cage scrambling 185
Chlorosilanes 4, 162
Controlled 3D silicone structures 161
Cycloolefins, ROMP 133

D
Decamethylpentasiloxane 188
Decasilsesquioxane 188
Dehalogenative coupling 4
Dehydrocarbonative condensation 161
Dehydrocoupling 1, 12
Dendrimers 33, 65
Dialkyloligothiophenes 39
Diaryldithienosilole 86
Dibenzosilole (BS) 37, 53
Di-(3-biphenyl)-1,1-dimethyl-3,4-diphenylsilacyclopentadiene (2PS2P) 51
Dibromodithienosilole 86
Di-*iso*-butyldimethyldisilene 10
5-(Dicarbazolylmethylsilyl)norbornene 137
Dichlorodihexylsilane (DCDHS) 5
Dichlorodimethylsilane 55
Dichlorodisilane 6
Dichlorohexylmethylsilane (DCHMS) 5
Dichloromethyloctylsilane (DCMOS) 5
Dichloromethylphenylsilane (DCMPS) 5
Dichloromethylpropylsilane (DCMPrS) 5
4-Dicyanomethylene-2-methyl-6-(*p*-dimethylaminostyryl)-4*H*-pyrane (DCM) 24
Diethyl-5,11-bis(triethylsilylethynyl)anthradithiophene (TES-ADT-ET) 48
Diethynyl-2,3,4,5-tetraphenylsilole (DE4PS) 71

Index

Dihalobiaryls 56
Dihydrosiloxane-dimethoxysilane reactions 178
Dimethyl-2,3-benzo-1-silacyclobutene 126
Dimethyl-3,4-diphenyl-2,5-bis(2'-thienyl)silole (TST) 94
Dimethyl[1]ferrocenophane 128
Dimethyl-silacyclobutane (MSCB) 113
Dimethyl[1]silaferrocenophane 131
Dimethylsilaindan 128
Dimethylsilole 50
1-(Dimethylsilyl)-4-(trivinylsilyl)-benzene (DMS-1,4-Ph-TVS) 71
Dioctylfluorene (FO) 94
Diphenyldimethoxysilane 170
2,5-Di(2-pyridyl)-1,1-dimethyl-3,4-diphenylsilacyclopentadiene (PySPy) 51
Dipyrrolylsilole 51
1,3-Disilacyclobutanes, ROP 123
Disilanolates 120
Disilenes, masked 6
Dithienosilole (TS) 37, 53

E
Electroluminescence 33
Epoxide ring opening 180

F
Ferrocenophane 116, 131
Field effect transistor (FET) 27
Fullerenes 26
Functionalization 185

H
Heptasilsesquioxanetriol 203
Hexaethyldisilane 4
Hexaphenylsilole (HPS) 53
Hole-transporting layer (HTL) 51
Hydrolysis 185
Hydroquinone derivatives 175
Hydrosilanes 163
Hydrosilylation 179
Hyperbranched conjugated organosilicon polymers 70

L
Ladder oligo(*p*-phenylenevinylene)s (LOPVs) 57
Lewis acidity 164
Limonene 5

M
Magnesia silicide 3
Masked disilenes 1
Metallocenes 13
Metathesis 165
Methacryl-oxypropyltriethoxysilane 22 3-
1-Methyl-1,2,3,4,5-pentaphenylsilole (PSP5) 52
1-Methyl-1-norbornenylmethyl-1-silacyclobutane 136
Methyldiethynylsilane (MDES) 71
Methylsilicones 170
Monodendrons 66

N
Norbornadienes 134, 137
Norbornenes 134
Norbornenyltriethoxysilane 142

O
Octaethyl-s-hydrindacen-4-yl 58
Octakis(3-chloropropyl) octasilsesquioxane 201
Octamethyltetrasiloxane 216
Octasilsesquioxane 201
    octahedral 194
OFEDs 1
OLEDs 1, 23
    pn bulk heterojunction 26
    polysilanes 23
Oligo(silole)s 50
Oligo(*p*-phenylenevinylenes) 37, 57
    disilene analogs 58
Oligoacenes 43
Oligosilanes 3
Oligothiophene 33
    silylated 38
Oligothiophenesilane, monodendrons 66
    nanosized star 61
Organic field-effect transistor (OFET) 27, 33
Organic light-emitting diode (OLED) 33
Organic photovoltaics (OPV) 23
Organic thin film transistors (OTFTs) 36
Organosilicon dendrimers 65
Organosilicon oligomers, linear conjugated 37

P
PEDOT:PSS 24
Pentacene 33, 44
    anthradithiophenes (ADT) 47
Pentacenequinone 44

Index

Pentamethyldisilanyl-oligothiophenes 38
Pentaphenylsilolyl-containing polyacetylenes 75
Phenyl-1-methyl-1-silacyclobutanes 113
Phenyl nonamethyl cyclopentasilane 11
Phenyl(*i*-propyl)tetrasiloxane 191
Phenylene-bis(benzosilole) PBBS 56
Phenylenevinylene–carbosilane dendrimers 65
Phenyloligosilsesquioxanes, 4-trimethylsilyl-substituted 201
Phenylsilane 171
Phenylsiloles 52
Phenyltrichlorosilane 192
Phenyltrimethoxysilane 190
Photoluminescence 33
Photovoltaics 26
Piers-Rubinsztajn reaction 161, 163
  metathesis 165
Platinum 1,3-divinyl-tetramethyldisiloxane 214
Poly(9,9′-alkyl-3,6-silafluorenes 91
Poly(arylene-dimethylsiloxane)s 208
Poly(arylene-disiloxane)s 218
Poly(bis(4-butylphenyl)silane 25
Polybis(trimethylsilyl)tricyclononene 147
Poly(carbosilarylene)s 71
Poly(9,9-dihexyl-2,7-dibenzosilole) 90
Poly(4,4-di-*n*-hexyldithienosilole) (TS6) 88
Poly(dimethylsiloxane) (PDMS) 177
Poly(9,9-dioctyl-3,6-dibenzosilole) (PSFC8) 92
Poly(di-n-hexylfluorene-*co*-4,4-diphenyldithienosilole) (PF-TS) 90
Poly(dithienosilole-2,6-diyl)s 89
Poly(ethylene dioxythiophene) 86
  poly(styrene sulfonate) 24
Poly(ferrocenylsilanes) 132
Poly(methylpentyne) (PMP) 147
Poly(1,4-phenylene vinylene) 33
Poly(penylenesilolene)s, hyperbranched 71
Poly(propylene oxide) 175
Poly(3,6-silafluorene)s 91
Poly(silylene)oligothiophenes 83
Poly(silylenevinylene), hyperbranched 71
Poly(2,5-silylthiophene)s 70
Poly[(1,2-tetraethyldisilanylene)-9,10-diethynylanthracene] (DSDEA) 83
Poly[(tetraethyldisilanylene)oligo(2,5-thienylene)] 86
Poly(1-trimethylsilyl-1-propene) 74
Poly(1-trimethylsilyl-1-propyne) 146
Poly(vinyl alcohol)-silicone 180

Polyacetylenes 74
Polycarbosilanes, heterochain 112
Polydihydrosiltrimethylene 117
Polyhydrosilanes 3
Polymethylhydrosiltrimethylene 117
Polymethylphenylsilane 5, 21, 25
Polynorbornadienes 138
Polynorbornenes 135
Polyphenylhydrosiltrimethylene 117
Polyphenylsilanes 12
Polysilane–anthracene 26
Polysilane–fullerene 26
Polysilane–polyolefin 23
Polysilane–titania 22
Polysilanes 1
  synthesis 1
Polysilastyrene 9
Polysiloxane resins 170
Polysilsesquioxanes 188, 220
Polysiltrimethylenes 119, 136
Polythiophenevinylene, silyl-substituted 75
Protection–deprotection 137

Q
Quaterthiophenes 40
Quinquethiophenesilane 39

R
ROMP 111
ROP 111
Ru-carbene Grubbs 136

S
SAMFETs 41
Semiconductors 27
Sexithienylsilanes 39
Silaarenophanes 131
Silacarbocycles 111, 126
Silacycles, ROP 11
Silacyclobutanes 114
Silacyclopentadiene (silole) 37, 50
Silacyclopentenes 128
Silacyclopentyl-tetraphenylsilole (CPSP4) 52
Silaferrocenophanes 115, 131
Silafluorene (SiF) 33, 53, 96
Silaindans 128
Silametallocenophanes 131
Silane, trifunctional 185
Silanol 213
Silanylene 83
Silicon carbide 22

Silicon-bridged biaryls (SBArs) 56
Silicon-carbon heterocycles (SCHs) 112
Silicones, controlled 3D structures 172
  copolymers 174
  synthesis 161, 170, 178
Silole (silacyclopentadiene) 33, 37, 50, 88
Siloxanes, double-decker 175
Silsesquioxane, oligomeric 185, 220
    cage 185
    phenyl 185
Siltrimethylene, amphiphilic 118
Silylcycloolefins 111
Silylnorbornenes 111
Silylzirconocenes 17
Solar cells, organic 26, 33
Spherooctasilicate, dimethylphenylsilylated 197
Spiro-bis-septithiophene 60
Spiro-9-silabifluorenes (SSF) 59
Stannacenophane 132
Stannylbenzosilole 56
Stilbenes, bis-silicon-bridged 58
Strain 111
Surface-initiated ring-opening metathesis polymerization (SIROMP) 142

T
Tetrakis(2-thienyl)silane 66
Tetramethyl-1-(4′-(trivinylsilyl)-phenylene-1)-disiloxane (TMDS-1,4-Ph-TVS) 71
Tetramethyldisilacyclobutane (DSCB) 113, 135
  ROP 125

Tetramethyldisiloxane 171
Tetramethyldivinyldisiloxane 121
Tetraphenyltetrasiloxanetetrol 188
Tetrasiladistyrylbenzene 59
Tetrasiloxanetetrol, cyclic 190
Thiophene oligomers, silicon-containing 38
Thiophenesilane dendrimers 66
(4-Tolyl)$_4$T$_4$-tetrol 195
Triethoxysilylnorbornene 143
Triethylsilylnorbornene 139
Trimethylsilylnorbornene 135
Trimethylsilyloligothiophenes 38
Trimethylsilyloxymethylnorbornene 137
Triphenylsilanol 213
Triplet harvesting 25
Tris(pentafluorophenyl)borane 209
Tris(8-quinolinolato) aluminum(III) 47
Tris(2-thienyl)methoxysilane 66

U
UV emitters 24

V
Vinylsilanes 168

W
Wurtz-coupling 1

Z
Zr-hyride 17